CHOICE CUTS:

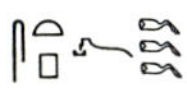

MEAT PRODUCTION IN ANCIENT EGYPT

for

Ama and Aba

ORIENTALIA LOVANIENSIA
ANALECTA
——— 69 ———

CHOICE CUTS:

MEAT PRODUCTION IN ANCIENT EGYPT

by

SALIMA IKRAM

UITGEVERIJ PEETERS en DEPARTEMENT OOSTERSE STUDIES
LEUVEN
1995

D. 1995/0602/93
ISBN 90-6831-745-8 (Peeters Leuven)
ISBN 2-87723-271-9 (Peeters France)

CONTENTS

ACKNOWLEDGEMENTS

I owe debts of gratitude to many people who have helped me in this work. Firstly I wish to thank my supervisor Barry Kemp for suggesting the topic of 'Food', and giving me freedom, bones, and encouragement to pursue my research in several unconventional ways.

I am indebted to all the museums and their curators who have provided me with information, photographs, and research facilities: the Metropolitan Museum of Art, the Boston Museum of Fine Arts, the Lowie Museum of Anthropology (now the Phoebe Hearst Museum), the California Academy of Science, San Francisco, the American Museum of Natural History, New York, the British Museum, the Petrie Museum, University College London, the Ashmolean Museum, Oxford, the Fitzwilliam Museum, Cambridge, the Egyptian Museum, Turin, the Louvre Museum, the Griffith Institute, Oxford, the Egyptian Museum, Cairo, the Dokki Agricultural Museum, and the Luxor Museum, and all the people who answered my queries, and willingly, though bemusedly, searched their collections for meats and meat boxes. I am especially grateful to C. Roehrig and M. Hill of the MMA, P. Podzorski of the LMA, J. Malek of the Griffith Institute, and P. Lacovara, P. Der Manuelian, and T. Kendall of the BMFA for all of their help prior to, during, and after my visits to their respective institutions, and to May Trad and Drs. Khattab and el-Akkad in Cairo and Dokki. D. Ryan of the Pacific Lutheran University generously gave me access to all the joints from his excavation in KV60, and the AMNH kindly gave me a meat sample for examination under the SEM.

Several specialists have answered questions, guided me through, and taught me about their subjects. For my faunal training (from giving the chance to dismember beasts to the final computer processing of the material) I am deeply indebted to M. Levine who gave many hours of patience, reading lists, parties and practicals; E. Yannouli, who suffered many fragments and long hours at the computer centre; J. Rippengal for cut-and-boil specials; L. Chaix and S. Goodman for sharing their knowledge of Egyptian fauna and butchery; and T. Legge who not only gave me access to what seem to be the only set of gazelle bones in England, but also discussed the Malkata material with me. For actual butchery, S. Fabish, Andrew's butchers, Nimr, the butchers of the covered market in Oxford, J. Page, the Harrods butchers, and the butchers of Sohag, Minya, and the Khan in Cairo have all provided me with invaluable assistance from showing me how they jointed meat to speculating about the joints shown in Egyptian representations. Several nutritionists have given me guidance as to how to and when not to speculate on amounts consumed, and what tests can be run on 3000 year old meat, especially A. Black, J. Loughbridge, Drs. Heath, Parker and Southon, Mrs. Clark and L. Green. D. Samuel and C. Putnis have very kindly worked on SEM with me. C. Cecil, S.Lange, and Dr. Stephens and Burk all generously and enthusiastically provided me with x-rays and CAT-scans of meat joints.

Several other people have patiently answered questions, provided information on a number of topics, and/or objects and references. I would like to thank R. Miller for references to meat preservation all over the world, an introduction to JSPR, snails, and for reading portions of this work; L. Leahy for Late Period references; S. Fairgrieve for information on human bone and diet; C. Duhig for the same and moral support; M. Marciniak for his drawings from the tomb of Ramesses III; M. Davis for information on the Plains Indians; M. Serpico for discussions on redistribution and rancidity; J. Ogden for metals; G. Killen for furniture and for making a very beautiful and sharp Egyptian knife for experimental purposes; C. Keller; I. Peacock; J. Bourriau and P. Rose for pottery; J. Ray; J. and E. Romer; L. Kaye; N. Reeves; H. and N. Strudwick; T. Wilkinson; T. Breckwoldt; H. Whitehouse; M. Verner; Dr. Valloggia; Y. Harpur; A. Warner for spatch-cocking; S., I., and S. Koraiem for information and a home in Cairo; P. and J. Spencer; T. Cadbury and D. Phillipson for providing me with flints for experimental work.

Much of the ethnographic work could not have been done without grants from Christ's College, the Thomas Mulvey Fund and the Seven Pillars of Wisdom trust.

I also wish to thank the people with whom I have stayed during the course of my resarches and those who have cheered me on throughout the writing process: S. Hussain, A. Ghosh, D. Cockburn, N. Llewellyn, T. Moore, J. Knudsen, A. Everall, W. Monkhouse, K. and E. Brush, A. Qadir, A. Ikram, L. Marlow, S. and B. Ahmed, T. Afzal, A. Scalley, S. Quie and N. O'Connor.

I cannot sufficiently express my gratitude to Geraldine Pinch who provided references, encouragement, proverbs, chocolate, and read my work; to Aidan Dodson who has been, as always, helpful and encouraging in countless ways, from reading my work, helping trace references, the use of his library, multiple 'phone-calls during dinner and during "sleep-time", and an introduction to Genesis; to Nicholas Warner for drawings, reading, willingly being dragged to markets and butcher's shops around the world, and being endlessly helpful and patient — and marrying me in spite of it all.

Lastly, I would like to thank my parents, Khalid and Shirin Ikram, to whom this work is dedicated, for giving me a chance to be "in my thesis", all their support and encouragement, and the opportunities and the desire to do comparative ethnography.

Cairo
13 July 1993

LIST OF ABBREVIATIONS

AE	*Ancient Egypt and the East.* London.
AMNH or A	American Museum of Natural History, New York
Antiquity	*Antiquity.* Cambridge.
ASAE	*Annales du Service des Antiquités de l'Égypte.* Cairo.
Ash.	Ashmolean Museum, Oxford
BIE	*Bulletin de l'Institut d'Égypte.* Third Series. Cairo.
BIFAO	*Bulletin de l'Institut Français d'Archéologie Orientale.* Cairo.
BM	British Museum, London
BMFA or B	Museum of Fine Art, Boston
C	Egyptian Museum, Cairo
CAS or CA	California Academy of Sciences, San Francisco
CAT	Computerized Axial Tomography
CdE	*Chronique d'Égypte.* Brussels.
CG	Catalogue Générale, Cairo Museum and Dokki Agricultural Musuem
D	Dokki Agricultural Museum, Cairo
Fitz.	Fitzwilliam Museum, Cambridge
GM	*Göttinger Miszellen.* Göttingen.
JAOS	*Journal of the American Oriental Society.* Chicago.
JARCE	*Journal of the American Research Center in Egypt.* New York.
JAS	*Journal of Archaeological Science.* London.
JE	Journal d'Entrée, Cairo Museum and Dokki Agricultural Museum
JEA	*Journal of Egyptian Archaeology.* London.
JHE	*Journal of Human Evolution.* London.
JNES	*Journal of Near Eastern Studies.* Chicago.
JSPR	*Journal of Stored Products Research.* Slough.
JSSEA	*Journal of the Society for the Study of Egyptian Antiquities.* Toronto.
Kemi	*Kemi: Revue de Philologie et d'Archéologie Égyptiennes et Coptes.* Paris.
KMT	*K.M.T: A Modern Journal of Egyptology.* San Francisco.
Kush	*Kush: Journal of the Sudan Antiquities Service.* Khartoum.
KV	Valley of the Kings.
LÄ	Helck, W. and E. Otto (1975-92) *Lexikon der Ägyptologie,* 1-7. Wiesbaden.
LMA	Lowie Museum of Anthropology, Berkeley
Louvre	Louvre Museum, Paris
MDAIK	*Mitteilungen des Deutschen Archäologischen Instituts Abteilung Kairo.* Mainz, etc.
MMA or M	Metropolitan Museum of Art, New York
N	Catalogue numbers, Appendix I.
NARCE	*Newsletter of the American Research Center in Egypt.* New York.
O	Oriental Institute, Chicago.

OIC	*Oriental Institution Communications.* Chicago.
PM	Porter, B. and R. Moss (1934-81) *Topographical Bibliography of Ancient Egyptian Hieroglyphic Texts, Reliefs, and Paintings*, 1-7. Oxford.
QV	Valley of the Queens.
R	Catalogue numbers, Appendix II.ii.
RdE	*Revue d'Egyptologie.* Paris, etc.
RdT	*Recueil de Travaux Relatifs à la Philologie et à l'Archéologie Égyptiennes et Assyriennes.* Paris.
S	Catalogue numbers, Appendix II.i.
SAK	*Studien zur Altägyptischen Kultur.* Hamburg.
SEM	Scanning Electron Microscopy
Sphinx	*Sphinx.* Uppsala.
T	Egyptian Museum, Turin.
TT	Theban tomb.
UC	Petrie Museum, University College London
Urk. I	Sethe, K. (1903) *Urkunden des Alten Reichs.* Leipzig.
Urk. IV	Sethe, K. and W. Helck (1905-58) *Urkunden des 18.Dynastie.* Leipzig and Berlin.
WB	Erman, A. and H. Grapow (1925-35) *Wörterbuch der Aegyptischen Sprache,* 1-5. Leipzig.
ZÄS	*Zeitschrift für Ägyptische Sprache und Altertumskunde.* Berlin.

LIST OF ILLUSTRATIONS

Unless otherwise stated, all photographs and drawings are by the author, with acknowledgements to the responsible authorities.

INTRODUCTION

"The pleasures of the table belongs to all ages, to all conditions, to all countries and to all eras; it mingles with all other pleasures, and remains at last to console us for their departure".

Brillat-Savarin

That food was important in ancient Egyptian culture is obvious from its repeated appearance and mention of food on offering-tables, offering-lists, in prayers for the dead, as part of temples' foundation deposits, and as funerary offerings. In fact, Davies thought that the "carnal preoccupation … and concern for food … monopolized the forethought of the Egyptians" (1923: 15). Thus, it is strange, that while reams have been written about temples, tombs, Pharaohs and hieroglyphs, the extant literature on Egyptian diet is slight. It has been long established that the Egyptians ate cattle, birds and fish as part of their daily fare. However, insufficient attention has been paid as to which animals were eaten, the persons responsible for slaughtering them, or the way in which any of these highly perishable comestibles were preserved and stored. The questions of who ate meat, and the amount they consumed have also not been addressed. This study attempts to answer these questions by going beyond the traditional Egyptological sources, utilising as well the results of scientific analysis of preserved meat specimens and involving, where appropriate, ethnographic observations.

The traditional sources are primarily artistic. Tomb and temple decorations are an extremely rich source of information for meat production, whilst three-dimensional representations of butchery are helpful in determining the sequence of butchery and jointing. The pictorial record is, however, not necessarily an illustrated instruction booklet, but one symbolic of the acts depicted. Indeed, certain scholars believe that all scenes in tombs and temples are allegories awaiting interpretation rather than straightforward scenes of the daily activities of the Egyptians or those directly and practically connected with their burial. Though it is very likely that Egyptian tomb decoration could be interpreted on many different levels, the interpretations are not mutually exclusive, and one must use whatever evidence is available. The artistic record may be particularly misleading, since the pictures might not represent reality, but rather an ideal state or desired truth. Furthermore, as already noted, representations might have symbolic significance above and beyond the secular acts they depict.

Robins writes, "as many layers of meaning and symbolism can exist side by side in Egyptian art, the two interpretations (symbolic [several] and real) are not mutually exclusive" (Robins 1991: 105). There is also the question of distortions of detail and sequence introduced into the work by the artist's ignorance or regard for representational conventions.

One other factor to keep in mind is that the artistic evidence is by no means even (Tables I-IV). From the Old Kingdom to the early Nineteenth Dynasty, private tomb chapels are a key source of information. By comparison, the paucity of Middle Kingdom burials adds to the imbalance. Furthermore, during the Ramesside period changes in the decorative scheme greatly reduce the non-religious content of scenes — during the Third Intermediate Period there are hardly any representations of food processing at all. Thus, changes in practice after the New Kingdom may be concealed by this factor. During the Late Period, representations reappear, but it should be noted that many of these representations in tomb chapels are copies of, or are heavily influenced by, earlier depictions, thereby reducing their value as first-hand evidence. Accordingly they have not been much employed in this work. Neither has the method of butchery and preservation during the Late Period been discussed. The reason for this is, in addition to restrictions of time and space, that, during the Late Period and subsequent centuries, Egypt was no longer entirely Egyptian. It was conquered and ruled in turn by Nubians, Libyans, Assyrians, Persians, and Greeks, all of whom had undoubted influenced and altered the life-style and habits of the indigenous Egyptians, with each invading people contributing to the diet of the native population, either by the introduction of new kinds of food, or by new methods of preparing/processing it (e.g. the increased use of honey for preservation that was especially popular in the Graeco-Roman period).

The written record contains only limited amounts of information: there are no known ancient Egyptian manuals for butchering, fish gutting, or bird strangling. Neither have any recipe books been recovered although it is probable that they existed. Certainly medical papyri contain cures that read like, and indeed are recipes; perhaps similar papyri dealing strictly with food will come to light at a later date. In Graeco-Roman times Alexandria was famous for its food and had its own cuisine, as is shown by the various Alexandrian recipes of Apicus (Flower and Rosenbaum 1958: 77, 95, 221). Of course, these were probably a combination of Greek, Egyptian, and other cuisines, rather than Pharaonic Egyptian food. Thus the literary contribution to the present work is essentially limited to the information derived from jar labels and some economic and administrative texts.

From the foregoing, the importance of the archaeological record becomes plain. However, even this has its limitations. Many of these are affected by the philosophy held by individual excavators. In the early days of Egyptian archaeology, beautiful and/or inscribed objects were collected, while other items of a more humble or unattractive nature were, for the most part, ignored and discarded. Such items included animal bone and foodstuffs found in tombs and more occasionally in settlements. Needless to say, those enlightened excavators who *did* save such finds have provided us with a great fund of information. Published records of food remains suffer from the fact that assumptions were made frequently about the contents of wrapped food-offerings without actually properly examining them. Or, even worse, we are left with such tantalizing statements as that of de Morgan regarding one of the princesses' tombs at Dahshur: that it "contained offerings of food" (1903: 24). What were they, and what was done with them? Such practice is only now being rectified by modern scientific work. Other archaeological evidence including architecture, ceramics, tools and furniture used for meat production are all better preserved and thus better represented than the actual faunal elements, and are used in the interpretation of the process of meat production.

Archaeological methodology and retrieval systems, as well as taphonomic forces, also impose limits upon the information that can be confidently derived from faunal remains, as does the awareness that findings in funerary contexts might represent a desired reality rather than actual truth. Certain bones are hardier than others, and this may contribute to a skewing of the faunal record. As is only now being fully realised, a sophisticated faunal recovery regime, involving sieving and flotation, is required if one is to have any hope of obtaining a meaningful picture of the animals present at a site.

The majority of the faunal evidence used in this thesis comes from B. J. Kemp and D. O'Connor's excavation of the New Kingdom site of Malkata. Mr. Kemp has kindly given me permission to work on these bones which I have examined and identified the species present, their number in each area, and used this information toward determining who consumed what (Chapter X). I have also examined additional faunal material from a funerary context, that of the Predynastic cemetery N7000 at Nag-ed-Deir. This material, made available by the Director and Registrar of the Phoebe Hearst (formerly Lowie) Museum of Anthropology and Archaeology, has also been analysed in the same way as the Malkata material. In addition, it has proved possible to examine New Kingdom victual mummies in museum collections, sometimes with the aid of scientific procedures, namely Scanning Electron Microscopy (SEM), x-rays and scanning by Computerized Axial Tomography (CAT).

Where appropriate I have applied ethnographic evidence. An ever-present danger is that of making direct correlations between ancient and modern practice. The avoidance of such pitfalls has always been uppermost in my mind when employing ethnography in this study. It is my hope that I have been successful in limiting my inferences to those that are strictly justified. I have done most of my ethnographic work in Cairo and Middle Egypt, with only a cursory examination of the work of butchers in Upper Egypt; thus it is possible that additional work in Upper Egypt and in Nubian areas will augment current information on butchery practices.

In addition to observing butchers at work in Egypt and other countries (USA, UK, Pakistan, Greece, France and Morocco), I have done some experimental work with butchering animals (an entire horse and a wolf, as well as joints of meat) in various situations and using different types of tools, ranging from flint to copper-bronze alloys, and stainless steel. This has provided some practical information that is helpful in interpreting reliefs as well as in evaluating types of implements used and their relative strengths.

Through combining these various threads this thesis will successively discuss the various animals available to all classes of Egyptians to eat, and how they were slaughtered, processed, and preserved. The "how?", "where?", and "by whom?" questions will feature strongly in these discussions. To these fundamental matters will be appended studies on the transportation and storage of processed meat, and the socio-economic implications of meat consumption.

The importance of food to man and society was well summed up by Jean Anthelme Brillat-Savarin when he first wrote in 1825: "Dis-moi ce que tu manges, je te dirai ce que tu es" (1982: 19).

I

EDIBLE ANIMALS

Prior to animal domestication the only access the Egyptians had to meat was through hunting. In early Egyptian history this was relatively easy as "the moister climate prevailing in late Predynastic and Early Dynastic time supported a considerable and diversified fauna in and along the margins of the Nile Valley, as well as in the Red Sea Hills" (Butzer 1976: 13). All Egypt once supported many forms of wildlife, of which most are now locally extinct; "the Dakhleh area was, from 10,000 to c. 2,000 BC, a treed savannah with a typically African suite of grazing and browsing ungulates and birds" (Churcher 1980: 391). Lions, perhaps leopards, cheetahs, many different types of antelope, hippopotami, crocodiles, pigs, ostrich, small game, hyenas, wild asses, sheep, goats, and wild cattle all supplied the Egyptians with food and other essentials. Hunting was not limited to mammals; the Nile was rich in many kinds of fish and waterfowl which were captured and consumed.

From these wild animals the Egyptians derived their main domesticates: sheep and goats, cattle, donkeys, and poultry. In most cases domestication occurred before the dynastic period. Some of the domesticated species of herd animals (sheep, goat, pig, possibly cattle)[1] might have come to

[1] There is some discussion as to the origins of most of the Egyptian domesticates. It is believed by some scholars that wild cattle (long horned) came to Egypt via Asia between six and four thousand BC, and were domesticated thereafter (Ghoneim 1977: 17; Epstein 1971a: 554). Epstein believes that the short horned cattle came to Egypt from Asia towards the middle of the third millenium BC (Epstein 1971a: 288) and that the first archaeological evidence for them comes from Upper Egypt (e.g. Badari [Brunton & Caton-Thompson 1928: 38, 44]) and the Fayum (Epstein 1971a: 213; Caton-Thompson & Gardner 1934: 34). However, the evidence collected by Darby and his colleagues (1977: 90), supported by a study of fossils by both Boessneck alone, and together with von den Driesch (1988: 13; 1982) dates the appearance of cattle in Egypt from the Pleistocene. This is confirmed by the findings of Gautier who also has archaeological evidence of domestic cattle in Egypt dating from the late Paleolithic onward (1980: 323; 1987: 433; Wendorf et al 1987: 447). Thus, the date of introduction of the earliest bovids in Egypt is disputed, although a very early date (maybe sometime in the seventh millennium BC) seems feasible. It is probable that cattle were domesticated in Egypt independent of similar, and perhaps earlier, activities in Asia. Scholars generally agree that ovicaprids were imported from Asia, via Syria, in the Neolithic, and then domesticated (Gautier 1980: 336). It is also very difficult to establish when animals become truly domesticated (i.e. humans controlled their breeding, selecting for particular qualities, tamed them, and

Egypt via Asia, and were bred with local stock to become the regular agricultural staples (Butzer 1976: 8). Initially, however, the Egyptians did not limit their domesticates to these animals. In the Old Kingdom, tomb decorations show that the Egyptians attempted to domesticate the hyena (Bissing 1905: 11; Smith 1969). There are several reliefs showing these animals in a farm context where they are being held or tied up and force-fed (Fig. 1). Whether they were to be eaten — and if this were the case, the Egyptians would have been the only people to eat hyenas in the ancient world — or used for hunting (a doubtful possibility given that hyenas are primarily scavengers) is unclear[2]. As hyenas are never depicted thus again, are never mentioned in food/offering-lists, and rarely appear in the faunal record, it seems that this was an unsuccessful attempt at domestication. Antelopes, notably the oryx (e.g. Beni Hasan tomb of Khnumhotep [#3]) were domesticated more successfully and were used for food or kept as pets.

Fig. 1

Once the Egyptians had domesticated their "staple" animals (sheep, goats, cattle and pigs [Fig. 2]), and established their herds, they had a fairly dependable source of protein at hand and were no longer entirely

trained the animals to serve humans) rather than wild animals being used for work as there are very few obvious indications of this change. The only visible changes are those wrought upon bone structure and teeth; unless bones from Egyptian sites are extensively studied nothing can be determined about when animals were domesticated.

[2] However dogs and cats, the latter being a relative of the hyena, are eaten in parts of the Far East, e.g. China, and Africa (Nigeria), while North American Indians eat prairie wolves, and skunk and foxes are consumed in Italy (Simmonds 1859: 51). It is also reported that hyena is consumed by some Arabs as well as Africans (Simmonds 1859: 51).

reliant on hunting. People did, however, continue to hunt for sport[3] (mainly the nobility and royalty), as well as to augment their meat supply without depleting their herds. Hunting decreased by the Middle Kingdom as wild animals became scarcer in Egypt due both to climatic changes, which forced them down into the then more lush Sudan and East Africa, and due to "over-hunting and competition from herd animals" (Butzer 1976: 27) which started to occur in the Old Kingdom. The poorer classes probably hunted small game for food as they would have owned fewer domesticates (if any) and would have preserved them for their by-products or as a measure of personal wealth, in preference to slaughtering them for their meat. It is likely that herd meat was enjoyed only on feast occasions (see Chapter X).

Meat (mammalian, piscian, and avian) is mainly the muscle tissues of the animal. Muscle consists of about 20% protein, 20% fat, 60% water, and contains a small percentage of vitamins and minerals, all of which are important for the nourishment and growth of the human body. The constituents of meat differ depending on the animal, the type of feed it has eaten, the specific joint of meat, the age of the animal when it was slaughtered, and the leanness of the piece. Muscles are made of myofibrils which form the individual muscle fibre — They are composed of the proteins: myosin, tropomyosin, the so called X protein, and actin. Between the separate muscle fibres is a juice called the sarcoplasm (Pyke 1970: 26). The most important component of meat is protein. Protein is "a polymeric compound made up of various amino acids as the monomeric unit...proteins form the primary organic basis of structures such as hair, tendons, muscle, skin, and cartilage" (Lapedes 1977: 527). It also supplies the necessary amino acids for the formation of new and depleted body protein, which builds, maintains and repairs tissue, as well as providing 11% of human energy intake. Although protein is available from other sources, e.g. legumes, animal protein is recommended for humans as it is more efficiently absorbed by the human body than protein of vegetable origin. Protein is the most important element of a diet as any deficiency is responsible for: poor muscle tone and posture, lowered resistance to disease, premature aging, anaemia, stunted growth in children, tissue degeneration, edema, and slow recovery from illness or surgery. Thus meat (mammalian, avian, or piscian), would be an important part of the Egyptian diet.

[3] As shown in tomb reliefs of the Middle (Beni Hasan) and New Kingdoms, and commemorated on scarabs, e.g. Amenhotep III's cattle hunt scarabs.

It was fortunate, therefore, that the Ancient Egyptians had access to a wide variety of food animals. The following discussion of the species involved covers both domesticates and wild animals[4]: some creatures, particularly pigs and certain fish, are alleged to have been the subject of taboos, and this matter will also be dealt with below.

I.1 Bovines

The most important animal of the ancient Near East was the bovid. Cows and bulls were the animals most prized and worshipped in the ancient world, not surprisingly as they fulfil most of the nutritional and material needs of man (Fig.2). Beside meat and bone marrow, cows provide milk and thus other dairy products, such as cream, butter and cheese. Furthermore, horns, bones, and hide provide the raw materials for many artefacts, from weapons to clothes. Fat was burnt as fuel for lamps, as well as rubbed in to protect one's skin from the elements. The hooves were reduced to make gelatine and glue (Lucas 1962: 3), and even the manure of these animals can be used as fuel, fertilizer, and building material (Clutton-Brock 1987: 62). Indeed, it is not surprising that in many cultures people prize their herds above all other material possessions and measure their wealth by the number of cattle they possess, In Egypt this is illustrated by the importance of the cattle census in dating (e.g. the Palermo Stone [Breasted 1906]). As indications of status, cattle were offered as gifts to the gods and to the deceased. Cows represented deities (Hathor, Mehetweret), and there were several cow cults, especially in Lower Egypt (e.g. Sekhat-Hor, patroness of herds and supplier of cow products [Moens and Wetterstrom 1988: 171]). In fact, sacred *zentet* cattle, some of whose remains have been recovered, were kept at the Hathor sanctuaries of Dendera and Deir el-Baheri (Petrie 1900: 47-8; Naville 1907: 46; Pinch 1992). In Egyptian art, bulls could represent the king and figured amongst his epithets, *kȝ-nḫt* being an integral part of the royal Horus name from the Eighteenth to Twenty-Second Dynasty. In writing there are at least five different types of cattle hieroglyph — the most of any single animal other than humans. Cattle appear in the names of four delta nomes (Moens and Wetterstrom 1988: 159). Of all animals represented in Egyptian tombs and temples oxen feature most

[4] All scientific names and divisions used to refer to the animals come from Whitfield 1984 and Nowak & Paradiso 1983.

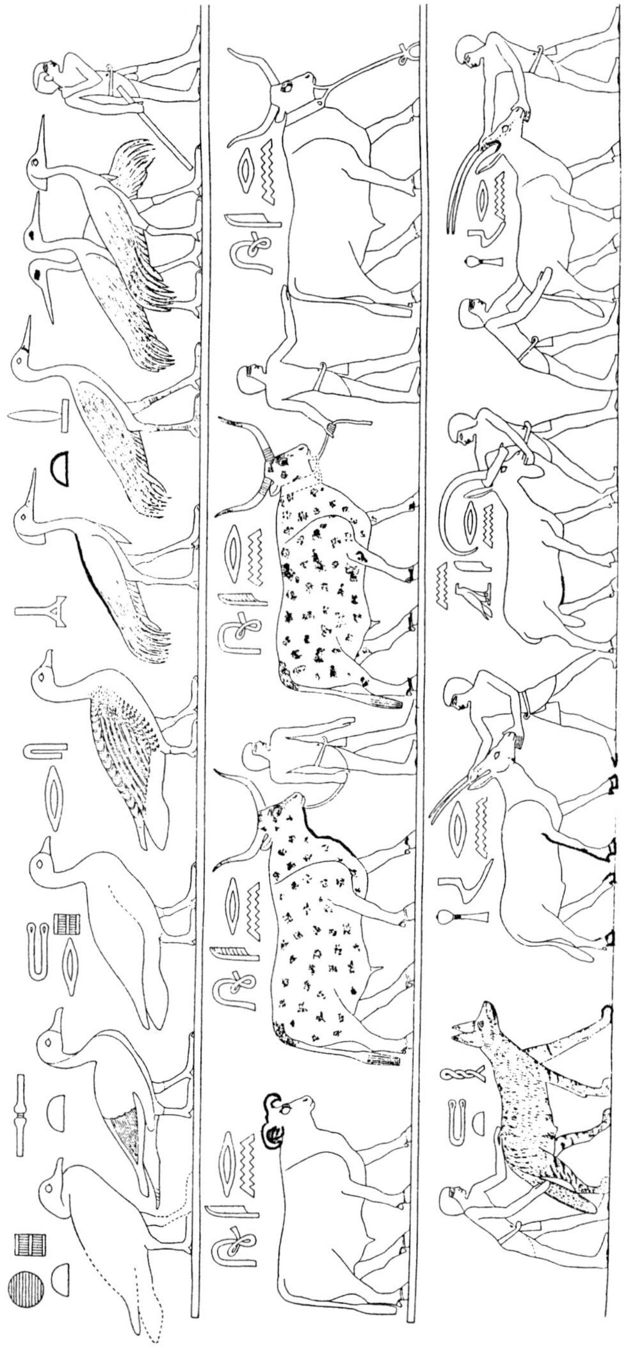

Fig. 2

frequently, whether in the field, slaughterhouse, or on the offering-/feast-table.

As tomb offerings, offering lists on temple walls, and temple records attest, most cattle ultimately belonged to the pharaoh either in his capacity as king or as high priest (Table I; P. Harris I. 8). He possessed species from all over Egypt as well as from neighbouring or "vassal" states, such as Kush (Pendlebury 1951: 174). Nobles also owned cattle, as is shown in countless tomb paintings and models, as well as by the legends bearing the name of private estates on labels from jars containing meat (Hayes 1951: 92). Whether owning cattle was a prerogative granted by the king is unknown. No doubt cattle could be given as gifts by the king to his nobles, just as he could give them to temples. It is known that temples rented out their cattle to private individuals, sometimes in perpetuity (Kruchten 1981: 94).

According to pictorial and textual evidence, the beef that the Egyptians ate was either from oxen — castrated (*sʿb*) male animals — or bulls (P. Anastasi IV: 15,5 [Caminos 1954: 200]). The animals were generally labeled *iwꜣ, ngꜣw,* or *tpiw* (Faulkner 1976: 12, 141 and 298), words commonly translated as oxen. The animals are frequently depicted with their scrotum and penis intact. The fact that the scrotum was intact might suggest that the animals depicted were bulls. However, it is possible and practicable to castrate bovines without removing the scrotum: an operation performed in cattle-producing areas today (Fabish 1989; Koraiem 1990; Boulos 1991). "The scrotum is cut, and the testicles drawn out and severed. There is little loss of blood and the animal soon recovers....The operation (is performed preferably) in the dry season for there is less chance of inflammation than in the rains"; the animals were probably castrated between the ages of eighteen months and two years (Evans-Pritchard 1940: 33). In modern villages the Egyptians use this method of castration and claim to have used it from ancient times. This operation is likely to have been known to the Egyptians and was relatively simple to perform. An alternative method of castration, involving damaging or destroying the testicles without removing them, was to bind the scrotum with a cord so that the testicles atrophied (Cranstone 1969: 255). Given that many of the animals being led to the slaughter that are shown in Egyptian reliefs still have their scrota, it is doubtful whether this last method was used. Castration rendered the males biddable, and "is an essential aid in the human control of animals...It turns an unmanageable, aggressive bull into a placid and submissive ox" (Clutton-Brock 1987: 25).

Occasionally, cows were slaughtered, as can be seen in donation-lists (P. Harris I: pl. IV, line 8), although, on the whole, oxen would be

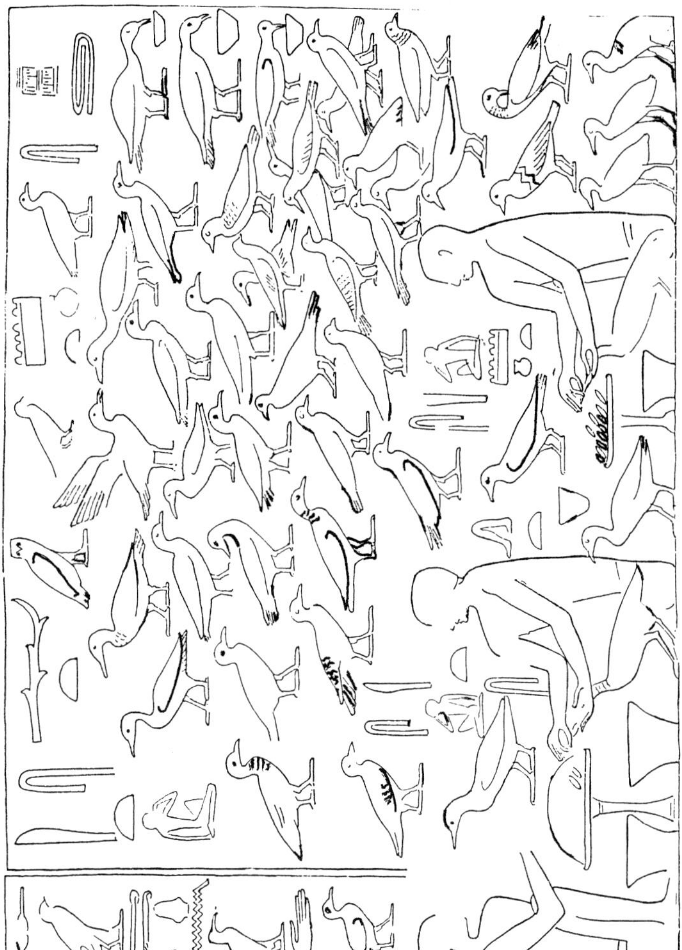

Fig. 3

favoured as meat animals. "After a young male animal has had its testes removed changes occur in its growth pattern so that its bones continue to grow in length rather than in girth, and more fat is laid down in the body tissues. In a castrated ox...the body will be fatter than in a bull (or cow) fed on the same diet but the limbs and the horns will be longer and more slender" (Clutton-Brock 1987: 25). Oxen had more muscle (i.e. meat) than cows, and that, too, was more tender than that of a comparably aged cow, with the added advantage of costing the same in amount of feed. In general, cows were considered too precious to eat as they provided milk (and from it butter, cheese and, yogurt), and were vital for breeding purposes, thereby leaving oxen as the only viable bovine food source. Porphyry writes, "With the Egyptians ... any one would sooner taste human flesh than the flesh of a cow. The cause ... is that this animal (is) useful ... Hence, though they eat bulls, and offer them in sacrifice as first-fruits, yet they spare cows for the sake of their progeny" (Taylor 1965: 69, II:11). Whilst artistic and circumstantial evidence thus points to oxen as the principal source of ancient Egyptian beef, the evidence of faunal remains is, as yet, inconclusive regarding the type of the slaughtered cattle found on archaeological sites.

There were several types of cattle in Egypt, though all were members of two genera, *Bos taurus primigenius* and *B. t. brachyceros*. This division of the taurine group is based on "differences in cranial conformation" and horns (Epstein 1971a: 189). The most salient features about the *B. t. primigenius* are its long horns with several variations in their shape (lyre, small, double) within the breed. *B. t. brachyceros* is distinguished from its cousin by short horns (with variations in direction) along with some cranial features (Epstein 1971a: 192). The Egyptians were very fond of the horns of their cattle, and with the longhorns they teased them into improbable shapes, as was done by the Nuer and Longarims (Evans-Pritchard 1949; Kronenberg 1961), which were apparently aesthetically appealing (as in Luxor temple and the tomb of Ti at Sakkara [Huard 1964]). There is another variety of Egyptian cattle which is actually part of the *B. t. primigenius* breed, but phenotypically very different: a hornless type which was seems to have been bred by the Egyptians from the longhorn, an example of which is seen in the Fifth Dynasty tomb of Akhethotep (Epstein 1971a: 221) and is mentioned in the tomb of Khafra-ankh (Andersson 1912: 148). In addition, there is a polled variety, derived from *B. t. brachyceros* and distinct from the hornless breed (such as the modern Aberdeen Angus [Foaden and Fletcher 1910: 767]), as the horns were removed manually (Epstein 1971a: 221). Generally, in most species, both sexes were horned, although the females had smaller horns than the males. All the animals were of varied colouring (Epstein 1971a: 221).

There is some speculation that the zebu, *Bos indicus*, was known in ancient Egypt. They would be especially useful as they are strong draught animals, produce rich milk, and are a good source of fat. They are larger than most breeds, and if consumed would provide more meat than the smaller (but tenderer) breeds. According to Epstein, it is possible that the "cervico-thoracic-humped zebu cattle were introduced into Ancient Egypt from the Somali coast, less likely from Syria, from the XIIth Dynasty on… [however] only one thoracic-humped zebu is known from ancient Egypt in an Dynasty 18 tomb" (Epstein 1971a: 556; TT101; TT78; TT86). In the second half of the Eighteenth Dynasty and thereafter several tomb reliefs (e.g. TT162 and TT367, among others) show what might be a type of humped cattle. There is still much debate as to whether this represented a zebu (or zebu-derived breed), or whether it was merely a variation of the shorthorn. Lortet and Gaillard also support the presence of zebu in Egypt (1903: 50; 58-9), as does Osborn (1991), although they do not have solid evidence. Unfortunately, the only way of differentiating a zebu from other cattle is by an examination of its dorsal vertebrae, the superior spines of which, "from the sixth vertebra caudally are bifid, the tip being divided medially and also compressed anteroposteriorly" (Epstein 1971a: 388). Another faunal clue as to the presence or absence of zebu is that the orbital rims are flat rather than sharply angled (Marshall 1989: 235). The superior spines as well as the orbitals are difficult and fragile anatomical elements to recover from an archaeological context, consequently, even with more faunal analysis it is doubtful if one can definitely resolve the question of whether zebu were present in ancient Egypt.

One of the earliest types of cattle found in Egypt was the wild bull, *Bos primigenius opisthonomus*, the forerunner of the domesticated breeds. They seem to have died out in Egypt at some time after the reign of Amenophis III (Epstein 1971a: 233), perhaps due to excessive hunting, much like the deer and antelope that were hunted to extinction by King Faruq in the Twentieth century. Thus, later scenes of wild bull hunting, e.g. that at Medinet Habu, are probably archaisms rather than events that actually occurred, unless they took place somewhere other than Egypt.

Even today the exact scientific names of the Egyptian breeds of cattle are not certain and now, as in ancient Egypt, the cattle are identified by their place of preponderance/origin, i.e. upper or lower Egyptian animals,

Baladi bulls, etc (Epstein 1971a). The ancient Egyptians also had many names for cattle. Unhappily, one cannot always match a modern zoological (taxonomical) name to an animal or an Egyptian word, although some correlations may be made. Some words describe the use of an animal rather than its species, although frequently only a specific species was used for a given activity. Some of the words associated with the different types of animals are listed below[5].

ỉwꜣ (*WB* I: 49): associated with stabled cattle that were fattened for slaughter, especially with large-stomached animals. Some scholars identify these animals with the *B. t. africanus* (Montet 1954 p. 43), while others favour *Bos taurus macrocerus* (Paton 1925: 2). In the Eighteenth Dynasty, the *ỉwꜣ* bull seems to be the most valuable type of cattle (Peet 1934: 194).

ng (*WB* II: 348): identified as *B. primigenius opisthonomus,* the wild bulls that were considered ferocious and roamed free (Montet 1954: 46).

ngꜣ (*WB* II: 349): a variation of *ng*. These animals were probably from a long-horned stock derived from the wild population. It has been suggested that the word indicates three year old or mature cattle[6] (Griffith 1898: 104). They were thinner than other breeds and used in agricultural labour rather than as food. They were probably consumed, but the fatter *ỉwꜣ* were preferred for sacrificial offerings (Paton 1925: 4).

šmꜣ (*WB* IV: 478): appears to be related to the *ng* in that it is perhaps best translated as a wild bull (Paton 1925: 5).

kꜣ (*WB* I: 154): the word for bull. These were generally not consumed until the end of their days as a stud or work animal. The meat of a bull would be very tough and much less pleasant to eat due to its age (and sometimes, activity) than that of a cow or an ox.

iḥ(t) (*WB* I: 119): the generic term for cattle, cow, ox, etc (Paton 1925: 2).

idr(w) (*WB* I: 152) identified as general stock (Griffith 1898: 104; Paton 1925: 5).

ḥr ʿnḏ and *ḏbʿ* (*WB* IV: 567): identified as hornless cattle (Paton 1925: 3) or possibly veal (Montet 1925: 155).

[5] In the case of all Egyptian names for animals the *WB* has been used, as well as other sources. As the focus of this thesis is not lexicographical only some of the words have been listed.

[6] Mature cattle, i.e. cattle with all its epiphyses fused would be over four years old (Silver 1969).

Fig. 4

wn-ḏw (*WB* I:326): supposed to be polled animals (Paton 1925: 4), although the WB suggests cow as well.
d-t: generally accepted as meaning a yearling (Paton 1925: 5).
ḥr-sꜣ/st (*WB* III: 135): identified as a two year old (Paton 1925: 5).
smsw or *sr* (*WB* III: 462): believed to be an adult animal (Paton 1925: 5).
wḏww (*WB* I: 398): the Egyptian equivalent of free-range cattle (Paton 1925: 5).
ḥm-t, *iwꜣt* (*WB* I: 49): the words for cow (Paton 1925: 7).
bḥs (*WB* I: 469): is the word for calf, although these would not generally be consumed until their maturity (Paton 1925: 7).
gn (*WB* V: 172): another word for cattle (Andersson 1913: 76).
rnn (*WB* II: 435): another word for cattle.
pḏ.t (*WB* I: 569): another word for cattle.
tpi (*WB* V: 296): perhaps a name used for long-horned cattle.

There is no distinct word for ox (save when castration is specified — generally only when bulls are also included in a list [Caminos 1954: 200]), although it is almost certain (see above) that most of the food-animals were oxen.

Fig. 5

I.2 Ovicaprids

Sheep and goats were domesticated and herded in Egypt from at least the Neolithic period[7]. These animals are grouped together due to their

[7] For a more complete discussion of domestication of ovicaprids, see D. Brewer et al., 1995, Houlihan 1995.

close biological/zoological relationship, and the similarity in their bones which makes it very difficult to distinguish between the two in faunal assemblages[8], although they are very easy to differentiate when they show their hides, beards, and horns. They are prominent in the archaeological record. Sheep and goat bones have been found in the Neolithic levels from Merimde Beni Salamah and Maadi; wool was found in el Omari (Epstein 1971b: 56). In the Fayum, evidence of these animals is found in Fayum A levels, and they have been found in areas of Nagada I date in Upper Egypt (Epstein 1971b: 56; Peet 1914; Petrie and Quibell 1896). They occur in the faunal remains at most Egyptian sites from these early periods onward.

Several predynastic and early dynastic pots, knife handles, and slate palettes are decorated with ovicaprids. In the dynastic period these animals also figure in wall decoration: they are shown being brought as tribute, frolicking, being herded, being slaughtered, and working in the fields, as well as figuring in offering-lists for both the deceased and deities. Sheep were associated with a number of gods, the Ram of Mendes *Bꜣ-nb-ḏdt* (*Ovis platyra*), Khnum (*O. longipes palaeoaegypticus*), and the curly horned ram associated with Amun (*O. ammon*).

These animals were second in importance only to cattle, providing the Egyptians with meat, milk, dairy products, wool, gut, hair and skins. They were also of use in the fields where they trampled in the newly sown grain (as depicted, for example, in the tomb of Meresankh in Giza [PM III.1: 197]).

There were several varieties of sheep and goat in Egypt, but it is hard to identify them all. In fact, as mentioned above, it is very difficult to differentiate between the two, in the archaeological record, hence the frequent joint classification (Wenke 1988: 19). A few of the species of sheep and goat are listed below.

Sheep: As with cattle, there is some debate as to the types of *Ovis* found in Egypt, with different scholars offering their own classifications and ideas concerning these. According to Epstein, both thin- and fat-tailed sheep were found in ancient Egypt, *Ovis longipes palaeoaegypticus* being the most common. One can trace the route of their domestication

[8] As far as bones and teeth are concerned it is quite difficult to distinguish between the two types of animals. S. Payne's work (1973) provides some clues on how to distinguish between them; otherwise, the most reliable differentiators are horn cores, some of the sutures on the skull, and the presence in sheep of a well developed lacrimal pit (Epstein 1971b: 57-59).

in Egyptian art; "in early depictions the ears are shown lateral and backward, signifying wild animals, while in later depictions the ears are pendulous indicative of domesticated stock" (Epstein 1971b: 59). Both genders are horned, though the horns of the rams are larger and stronger than those of the ewes (Epstein 1971b: 59). These are the sheep used to represent the god Khnum, with "horns growing in a horizontal direction to a length of 20 to 25cm. and forming a spiral of two complete turns" (Epstein 1971b: 59). Some of the animals are polled (Epstein 1971b: 118). In the Middle Kingdom, *O. platyra* may have been known (Smith 1969: 308). The Amun ram, *O. ammon*, which is not thought to be indigenous to Egypt, must have been present as it is so frequently represented in the art. *O. aries,* the most common type of domesticated sheep, have been found in the Fayum (von den Dreisch 1986: 2-3) as well as sites all over Egypt, and were probably present in Egypt from a very early date (see note 7). All these sheep are a very good source of wool. Their meat is very edible, though fairly fatty; a quality which in modern Egypt is considered commendable. In the ancient diet fat of any kind was much sought after and is nutritionally very important (see Chapter VIII).

The Egyptian words for sheep are: *b3* (*WB* I: 4414)*; sr* (*WB* III: 462)*; w3p.t* (*WB* I: 251), and *ẖnmw*. Ewes, when differentiated from rams, were called *sr(i)t* (*WB* III: 462).

Goats: As with sheep there is much discussion of the origin of goats in Egypt. Epstein believes that the Egyptian goats were derived from *Capra strepsiceros*, the screw-horned goat (1971b: 222-3; see note 7). Several Egyptian goats had screw horns, for example, *C. hircus girgentana*. To add variety to this group, foreign screw-horns were also being introduced into the country even in dynastic times (Libyan screw-horned goats and sheep are brought to Sahure as tribute in his Valley Temple). There are some "early pictorial records of a small scimitar-horned goat (for example, on Predynastic slate palettes, notably the one from Hierakonpolis)"[9] (Epstein 1971b: 223) too. This might be a distant relative of the type of goat that is pictured in the hieroglyph *sʿḥ*, a bearded goat with backward curving horns, carrying a seal about its neck.

Other goats found in Egypt were the dwarf goat, *Hircus reversus*, pictured on the slate palette from Hierakonpolis, a mummy of which has been discussed by Lortet and Gaillard (Epstein 1971b: 222), the Syrian or Mamber goat, *C. hircus mambricus*, and the Theban goat, *C. hircus*

[9] Ashmolean Museum E3924.

thebaicus, though there is still some debate about the identification (and nomenclature) of the latter (Epstein 1971b: 299).

The common domesticated goat, *C. hircus*, has been found in several archaeological contexts from the predynastic period onward. There is also some evidence that it was found in the neolithic period.

The Egyptian word for goat is *wnḏw* (*WB* I: 326), or *ꜣnḫ* (*WB* I: 205) and is associated with *C. hircus mambrinus*, the Syrian or Mamber goat with undulating horns and a beard.

I.3 Hunted mammals

Prior to domestication the initial source of meat for the Egyptians was the hunted mammal. The fauna of ancient Egypt was so extensive and so varied that it is impossible to give an exhaustive list of the animals that the Egyptian hunted — especially as some of the animals killed might have been hunted for their hides or because they posed a threat, rather than for food. I shall only discuss the animals that appear frequently in pictorial representations, texts, and the archaeological record.

Initially these animals would have been available for everyone to hunt, but with the stratification of society and the advent of more complex advanced weapons and modes of transportation, hunting of the larger animals would have been restricted to the wealthier classes — ironically those that had less need of the meat — much as it did in Europe in historic times. Smaller animals as well as birds and fish are likely to have been commonly available to all classes.

Antelopes

The largest class of hunted animal is the antelope, a member of the bovid family (Fig. 2). There were several varieties that formerly inhabited Egypt, many of which are pictured in hunting and offering scenes, frequently in the act of being slaughtered. Some of the differences between the various species and sub-species of antelope are hard to determine from the pictorial evidence, since the Egyptian artist was not always sufficiently detailed in his rendering to allow a precise identification of the fauna he depicted. It is also difficult, though not impossible, to identify these various antelopes from the archaeological record, provided that sufficient portions of diagnostic elements are preserved. Luckily this has been the case often enough to confirm the existence of many of the antelopes featured in Egyptian art. Sometimes butchery marks on bones prove that these animals were slaughtered and consumed. Antelope would have been fairly easy to hunt, were plentiful, and the meat of an antelope (depending on the species) is equivalent to the meat of three to twelve sheep (Pyke 1970: 31). Crop-reservation has been another reason cited for the killing of antelopes (Ruffer 1919: 23). Several varieties of antelopes were offered regularly in New Kingdom temples as is shown by wall reliefs as well as offering-lists (*Urk.* IV: 754, 1255, 1272).

Fig. 6

Oryx: Oryx are frequently featured in Egyptian tombs. Two kinds of oryx were found in Egypt in the Pharaonic Period, *Oryx leucoryx*, the Arabian oryx, and the Scimitar oryx, *Oryx dammah*. The meat of the oryx, especially *O. leucoryx*, is highly prized even today (Whitfield 1984: 148). These animals are shown in domesticated as well as hunted contexts from the Old Kingdom onward, and their Egyptian names were *mꜣhḏ* (*WB* II: 11) or *rn ḥḏ*. At one time the Oryx was worshipped as Satis, a female deity of the first cataract (Smith 1969: 311).

Addax: The Addax, *Addax nasomaculatus*, ideally suited for desert dwelling, was hunted in Egypt. Although its body is smaller than that of an oryx (c. 1.3m., rather than c. 1.6m.), it too would have provided meat suitable for consumption (Whitfield 1984: 148); its bones have been found on several sites. Its Egyptian name is *nw-ḏw* (*WB* II: 226; Paton 1925: 14). Another possible name for it was *ḏbnw* (*WB* V: 568), although this name has occasionally been ascribed to the Bubalis (Paton 1925: 14).

Bubalis: The word more commonly used to label Bubalis (or Hartebeest), *Alcelaphus buselaphus,* is *šsꜣw* (*WB* IV: 543), although *ḏbnw* (*WB* V: 320) is also used. These animals with their distinctly inward-curving horns were frequently shown in hunting scenes, and their bones have been recovered in settlement sites throughout Egypt and in the oases (Boessneck and von den Driesch 1988: 117-18; Churcher 1980: 388).

Gazelle: The Egyptian *gḥs* (*WB* V: 191), generally translated as 'gazelle', is used to identify several species of gazelles, the most common of these being *Gazella dorcas*, the Dorcas gazelle. It also identifies other types, such as the mountain gazelle, *G. rufrifrous* or *G. albonotala* (Paton 1925: 11), the Yellow Mountain gazelle, *G. gazella*, the Addra gazelle, *G. addra*, or the Sand gazelle, *G. leptoceros* (Seton-Williams and Stocks 1988: 129). Various scholars also suggest that representations of *G. Soemneringi* (Paton 1925: 14) are also denoted by *gḥs*.

It is clear that *gḥs* is a generic term used for smaller antelope rather than any specific species. These various gazelles are shown being hunted and offered, and feature in offering-lists, and are also shown as pets (TT320), and in other domestic contexts. The dorcas gazelle is thought to have originally represented Anukis, a deity of the First Cataract (Smith 1969: 311).

Ibex: The ibex was also very common in Egypt and its bones have frequently been found in settlement sites. The types of ibex found in Egypt were *Ibex nubiana* or *I. beden*. Their Egyptian title is *niꜣ*, *niꜣw*, or *nrꜣw* (*WB* II: 202; *WB* II: 28; Paton 1925: 11). They would provide milk and dairy products as well as meat.

Deer: *ḥnn* (*WB* II: 495) has been identified as a Fallow deer, either *Cervus dama* or *Dama mesopotamica* (Paton 1925: 13). These animals are more common in Egyptian representations of the Old and early Middle Kingdoms, although there is one known example from a Theban tomb of the mid-Eighteenth Dynasty, TT20. It is possible that this is a representation of an imported animal such as the one brought from Mesopotamia as booty by Thutmosis III (Paton 1925: 13). It is generally agreed that due to the changing environment and excessive hunting these animals had become extinct in Egypt by the New Kingdom.

The foregoing list of antelope and deer is doubtless incomplete, but includes all the species definitely represented in Egyptian art, and their associated names which appear frequently in offering-lists. Other words associated with antelopes in a general way are *hbn* (*WB* II: 487) and *gsꜣ* (*WB* V: 206). Furthermore, bones of these animals have been recovered at various Egyptian sites in domestic contexts, some with butchery marks, supporting the view that these were meat animals.

Hippopotami: Amongst other animals hunted, in predynastic and dynastic times, is the hippopotamus. The species of hippopotamus found in Egypt would have been *Hippopotamus amphibius*. These animals figure prominently in aquatic hunting scenes, especially in the Old Kingdom, where they have the added symbolic meaning of the deceased vanquishing the Typhonic forces of evil and chaos (Säve Söderbergh 1953: 5-24). They also figure in several literary and religious texts such as the Coffin and Pyramid Texts (Säve-Söderbergh 1953: 26-40). They are not, however, mentioned in offering-lists, probably due to their negative association with Seth, but if successfully hunted they were no doubt consumed. Hippopotamus flesh is quite edible and due to their size plentiful in quantity, making them worthwhile animals to kill: the meat of one hippopotamus is equivalent to the meat of sixty sheep (Pyke 1970: 31). Historically, they have been eaten in many parts of Africa. Several neolithic (and slightly later) sites in Egypt (Merimde, Omari, Abu Ghalib) contain hippopotami bones. These were used as steps and at thresholds to buildings, and show evidence of butchery (von den Driesch 1986: 2-3). Von den Driesch found hippopotamus bones at what is believed to be a butchering site (Fayum). Other delta sites have also yielded hippopotamus bones (Boessenek and von den Driesch 1988: 117-18; van den Brink 1989). Obviously these animals were captured and consumed primarily in the marshy areas where they abounded for they are not (so far) found in the faunal assemblages of Upper Egypt. Of course they would have been more commonly found in Egypt in earlier periods, their number slowly dwindling as they were killed off and the climate became drier and more inhospitable. Also, with the increase in domesticated animals the Egyptians would have been less dependent on hunted mammals for meat. The Egyptian names for hippopotamus are *db-t, db* (*WB* V: 423), *nhs* (*WB* II: 287), *ḫ3b* (*WB* III: 229), and *rrt* (*WB* II: 438).

The Egyptians also hunted the small game that was abundant in the desert and along the Nile Valley. This would have included the desert hare, *Lepus aegyptiaca*, called *wn* and sometimes *sẖʿt* (*WB* IV: 268), by the Egyptians, as well as other small rodents, desert foxes, and possibly even the common hedgehog (*Erinaceus aethiopicus* or *E. paraechinus aethiopicus*) the meat of which is quite edible when prepared by coating in mud and baking.

Hyenae: Although the hyena (*Hyaena hyaena, H. striata*), is rarely shown being hunted (a notable example is in the tomb of Ukh-hotep's son Senbi in Meir), it is frequently depicted in tombs being force-fed [Sakkara tombs of Mereruka and Kagemni; Fig. 1], or on a leash (Ptahhotep, D64 Sakkara), thereby leading one to believe that the species was being domesticated (see above). It is being included in this section on the basis of its being a wild animal that was never successfully domesticated, and which might have been periodically hunted. Hyenas were definitely kept in domestic contexts throughout the Old Kingdom, and maybe into the Middle Kingdom. Their meat would probably have tasted quite edible as their diet consisted of fresh meat (see the tombs of Kagemni and Mereruka in Sakkara) rather than of carrion. There is one mention of hyena as a meat-offering, so far a unique example, in a Fourth dynasty offering-list (Montet 1925: 114). It is quite possible that these animals were being kept to help with the hunting as they are shown on leashes along with hunting dogs. Hyenas are not only scavengers — "they are...predators in their own right..., and by hunting in cooperative packs can bring down animals as large as zebras" (Whitfield 1984: 98), so they were potentially useful as hunting animals. However, this attempt at domestication must have proved unsatisfactory as the domestic scenes featuring hyenas die out by the Middle Kingdom. It is also possible that this attempt at domestication might have been limited to the Memphite area. However, as the only evidence is wall reliefs, this remains a tenuous hypothesis. No concentration of hyena bones has been found on any archaeological site. The Egyptian name for hyena was *htt* (*WB* II: 203) or *ḥḏr* (*WB* III: 214).

I.4 Avifauna

The avifauna of Egypt was and is very rich and varied due not only to the Nile, an ideal habitat for a vast number of aquatic birds augmenting the usual stock of birds in the area, but also due to the fact that Egypt is a stop-over point for several species of migratory birds on their journey between Europe and Africa, and Africa and the Middle East (Fig. 7). Egyptian hieroglyphs attest to this wealth of avifauna — there are nearly thirty different types of birds listed in Gardiner's sign list [G]. As with the other animals, it is sometimes difficult to identify the precise species of Egyptian bird, especially in the case of ducks and geese, as, in the absence of colour, and therefore distinctive markings, size is frequently the only indicator of difference between them.

Birds not only figure as hieroglyphs, but are also subjects of paintings, reliefs, and sculpture. From predynastic times, various kinds of birds are featured on Egyptian artifacts and monuments. They are shown as part of the wild environment being hunted with throwsticks and clapnets, in domestic circumstances in gardens, or as pets, in poultry yards and markets, in the acts of being slaughtered, preserved and cooked, and finally, brought as offerings or on offering-tables. Various birds — particularly geese, ducks, and pigeons — are regularly mentioned in offering-lists. Unlike most other animals (save cattle) there are several three-dimensional representations of birds (e.g. the limestone Egyptian geese from Deir el-Medina [Houlihan and Goodman 1986: fig. 89] and the wooden swans from the tomb of Tuthmosis III (KV 34) and the tomb of Horemheb (KV 57) [Houlihan and Goodman 1986: figs. 69-70]). Birds were important in Egyptian religion. Thoth and Horus/Re were both represented by birds. Other deities such as Amun and Geb had birds associated with them. Several birds, such as ibises and falcons, were consequently worshipped and in the Late Period thousands of them were mummified. Manifestations of the human soul (*bꜣ*) are shown as half man and half bird.

The Egyptians consumed a great deal of poultry. Fowl were plentiful and would have been available to people of all classes. Anyone could go to the banks of the Nile and hunt water fowl with throw-sticks, nets, small catapults or even by stalking them[10]. On wealthy estates men are shown trapping several birds in large nets. In Graeco-Roman times the Egyptians were famous for their netting and preserving of quails and other small birds. Birds were raised in poultry yards, which are commonly depicted on tombs walls from the Old Kingdom onward. Sometimes the birds are shown being force fed, clearly for the table. Birds are also regularly offered to the deceased and the deities. Several mummified and wrapped (sometimes cooked and/or pickled) birds have been found in Egyptian tombs. These have been examined by the author and appear in the catelogue in Appendix II.1.

Birds not only provided the Egyptians with meat, but also with eggs. Eggs are very rich in protein and thus an ideal source of nutrition; an egg can provide an adult human with all necessary daily nutrients with the exception of vitamin C (Pyke 1970: 62). Duck and goose eggs, the most common in Egypt, are the richest in nutrients. The Egyptians possibly had the technology to hatch eggs artificially, much like their supposed descendants, the Copts. Until quite recently the Copts (and other Egyptians) have ovens topped by rooms with vaulted roofs and apertures for light. The ovens are scattered with bran and the eggs are placed there. After a few days the eggs are moved into the rooms above the

[10] Not all parts of the river were accessible to hunters and fishermen; the Nauri decree records that fishing and fowling rights of certain areas were under temple jurisdiction (l. 35-42, 55-65). This leads one to speculate that perhaps wild-fowl and river fish were not as accessible to people as one would think, thereby limiting available food sources.

ovens where they remain until they hatch. The entire process takes from somewhere between eighteen to twenty-one days (Wilkinson 1854: 171). Although there is no direct evidence that this was done in ancient times, several Copts claim that this has been handed down from antiquity (Boulus 1992). Beside figuring as offerings on tomb walls (e.g. TT78), eggs have been found in Egyptian tombs. In the tombs of Gebel Sedment, "in small niches on the right sides of the entrance to the chambers, little baskets containing the fruit of the dom-palm, poppy-heads, and pieces of bread. Pigeon eggs were deposited in small holes in the walls". The date of these deposits is uncertain as the tombs were reused from the Eighteenth dynasty on (Naville 1894: 12). Some mummified eggs (eggs wrapped in linen bandages, some covered with traces of salt, and others covered with salt and what might be pitch) have also been found, presumably in tombs (e.g. Louvre E3329). An especially important type of egg sometimes available to the Egyptians was the ostrich egg, the body of which is equivalent to approximately twenty-three hen's eggs.

Birds also provided feathers to be used for arrows as well as for fans and were a constant inspiration for decorative motifs. It is quite possible that birds such as cormorants and pelicans were occasionally used by fishermen as aids in fishing. Other birds such as bitterns and some ducks were used as decoy birds by fowlers to attract waterfowl into nets (Houlihan and Goodman 1986: 21).

It is difficult to provide a definitive list of birds consumed by the Egyptians, since a number of species known to exist are not included in the corpus of faunal remains and representational and literary evidence. A further problem derives from our different ethnocultural standpoint that might exclude species that an Egyptian would have found quite palatable. Below are listed the types of birds most likely to have been consumed by the Egyptians although the avifauna they ate is not restricted to this list.

Herons: There were approximately seven types of herons available for possible consumption in Egypt: the Grey heron (*Ardea cinerea)*, the Goliath heron (*A. goliath)*, the Purple heron (*A. purpurea)*, the Western Reef heron (*Egretta gularis*), the Squacco heron (*Ardeola ralloides*), the Striated heron (*Ardeola striata*), and the Night heron (*A. Nycticorax*) (Houlihan and Goodman 1986: 146). Some words for herons are: *ʿhʿw* (*WB* I: 222), *šn.tỉ* (*WB* IV: 519), *gꜣ* (*WB* V: 149), *nwr* (*WB* II: 233), *wḏ* (*WB* I: 407), *ḏꜣ.t* (*WB* V: 516), and *ʿw* (*WB* I: 170). Most of these words could also have been applied to cranes and storks.

Cranes: Two types of cranes were found in Egypt, the Common crane (*Grus grus*), and the Demoiselle crane (*Anthropoides vigo*). Cranes are among the animals that are shown being force-fed in wall reliefs, suggesting that they were for the table (Épron 1939: pl.8). In addition to the words used for herons that might also apply to cranes, the word *snsn* (Griffith 1898: 2-3, pl. VIII l. 30-62) seems to apply to cranes.

Storks: Four types of storks were commonly found in ancient Egypt. The White stork (*Ciconia ciconia*), Black stork (*C. nigra*), Saddlebill Stork (*Ephippiorhynchus senegalensis*), and possibly the Whale-headed Stork (*Balaeniceps rex*). (Houlihan and Goodman 1986: 147).

Ibis: It is debatable whether these birds were eaten. Generally the Sacred Ibis was associated with Thoth rather than with food. However, there is a New Kingdom representation that shows ibises being kept with domestic birds in a poultry yard which might indicate that they were consumed (Houlihan and Goodman 1986: 30; Klebs 1934: 70, fig. 51). The two types shown in the yard are either the Sacred ibis, (*Threskiornis aethiopicus*) or the Glossy ibis (*Plegadis falcinellus*) (Houlihan and Goodman 1986: 26-32).

Chicken: There is very little evidence of the Red Junglefowl (*Gallus gallus*) the ancestor of the modern domestic chicken in Egypt. There is a New Kingdom (Nineteenth Dynasty?) ostrakon found in the Valley of the Kings which shows a Red Junglefowl (Houlihan and Goodman 1986: 79; BMEA68539), a figure of one embossed on a silver bowl, and a mention of what might be such a bird in the Annals of Thutmosis III (Houlihan and Goodman 1986: 80). So far no bones of the fowl have been found in any faunal assemblage. It definitely had become part of the Egyptian diet by the Ptolemaic period, if not by the Thirtieth dynasty. There is virtually no evidence of chickens anywhere else in the Near East until the Hellenistic period, except in Sweyhat (2400 BC), Syria (Macdonald 1992: 304), although, recently the excavations at the Late Bronze Age (1550-1200BC) site of Tel Miqne-Ekron have turned up chicken bones and eggs in what might be a ritual or palace structure (Rosen 1991: 101).

Partridges: There are two types of partridges known from Egypt. The Barbary partridge (*Alectoris barbara*) and the Sand partridge (*Ammoperdix heyi*). Both of these were definitely eaten (Houlihan and Goodman 1986: 152).

Quail: The Common quail (*Coturnix coturnix*) was eaten in Egypt. Preserved quails (and other small birds such as partridges) were salted and/or pickled (Herodotus II: 77; Godley 1946: 365). The letter *w* is supposed to represent a quail chick, and is indeed its name (*WB* I: 243).

Swans: Three types of swans are found in Egypt. It is probable that they were consumed. The swans found in ancient Egypt were the Mute swan (*Cygnus olor)*, the Whooper swan (*C. cygnus)*, and Bewick's swan (*C. bewickii)* (Houlihan and Goodman 1986: 52).

Geese: There were at least nine types of geese in ancient Egypt; it is certain that many of them were domesticated and that all were eaten. They are the Greylag goose (*Anser anser)*, The White-fronted goose (*A. albifrons)*, the lesser White-fronted goose (*A. erythropus)*, the Bean goose (*A. fabalis)*, the Barnacle goose (*Branta leucopsis)*, the Red-breasted goose (*B. ruficollis)*, the Brent goose (*B. bernicla)*, the Egyptian goose (*Alopochen aegyptiaca)*, and the Spur-winged goose (*Plectopterus gambensis)*. Unfortunately as most geese are only 39% meat, they are not an ideal source of nutrition. However, their fat is very useful for cooking, meat preservation (VIII.4.2), as well as in medicine. In the Ebers Papyrus goose-fat appears in several recipies for colds, soothing of joints and orifices, scurvy, eye-diseases, impotence, and digestive disorders, among other things (Ebers XXXI, XXXII, XLIII, LI, LXI, LXXXIII, LXXXV, and LCI [Ebbell 1937]). There are several names for geese, some of which are shared by ducks. Some of the names for geese are: *bsbs* (*WB* I: 477), *bḏꜣ* (*WB* I: 488), *pḫ.t* (*WB* I: 542), *msꜣt* (*WB* II: 136), *r/rꜣ* (*WB* II: 393), *ḥḏ* (*WB* III: 210), *ḫꜣr* (*WB* III: 232), *ḫt-ꜥꜣ* (*WB* III: 342), *st* (*WB* III: 407), *ḏnw* (*WB* V: 575), *ṯrp* (*WB* V: 367), *gbb* (*WB* V: 164), *stp* (*WB* IV: 337), *s* (*WB* IV: 1), *smn* (*WB* IV: 136), *sr* (*WB* IV: 191), and *srt* (*WB* IV: 192).

Ducks: These birds are most commonly depicted in Egyptian art, perhaps because several species inhabited ancient Egypt. They are shown being reared in fowl-yards and are often shown being hunted not only by poulterers, but also by noblemen in the marshes with the aid of what might possibly be a 'decoy' duck (elsewhere referred to as an "erotic" duck) perched in the prow of his skiff. Ducks are very similar to geese — zoologically their differences seem to be determined by their size, feeding habits, and feeding environments (Peterson 1974: 121). The species of duck found in Egypt were the Ruddy Shelduck (*Anas ferruginea)*, Common Shelduck (*A. tadorna)*, Mallard (*A. platyrhynchos)*, Green-winged Teal (*A. crecca)*, Gadwall (*A. strepera)*, European Wigeon (*A. penelope)*, Pintail (*A. acuta)*, Marbled duck (*A. angustirostris)*, Garganey (*A. querquedula)*, Shoveler (*A. clypeata)*, Red-crested Pochard

(A. rufina), Common Pochard *(A. ferina)*, White-eyed Pochard *(A. nyroca)*, Tufted duck *(A. Fuligula)*, Velvet Scoter *(A. fusca)*, Smew *(Mergul albellus)*, Red-breasted Merganser *(Mergus serrator)*, and White-headed duck *(A. leucocephala)* (Houlihan and Goodman 1986: 65-74). Nutritionally ducks are a very good source of protein and fat, and are an excellent source of major ammino acids. A duck contains most vitamins and elements, with the notable exception of D, E, and C vitamins. (Paul and Southgate 1978: 106-9). In addition to the words they share with geese, ducks are identified by separate words, too. They are: *wꜣḏ-ḥꜣ.t* (*WB* I: 270), *wḥʿ.t* (*WB* I: 350), *mꜣʿ* (*WB* II: 24), *ḥp* (*WB* III: 69), *s.t* (*WB* III: 407), *sw* (*WB* IV: 59), *šs.t* (*WB* IV: 59).

Other aquatic birds: The following is a list of other aquatic birds that were consumed: the Purple Gallinule *(Porphyrio porphrio)* was hunted and consumed (Houlihan and Goodman 1986: 89). The European Coot *(Fulica atra)* were popular fowl (Houlihan and Goodman 1986: 91). The Spotted, Little, and Baillon's Crakes *(Porzana porzana, P. parva, P. pusilla*, respectively) (Houlihan and Goodman 1986: 153) were all available for consumption. The Avocet *(Recurvirostra avosetta)* (Houlihan and Goodman 1986: 99) also figures amongst the birds trapped by fowlers. It is eaten in Egypt today "even though its flesh is said to have a pungent fishy flavor" (Houlihan and Goodman 1986: 100). The Greater flamingo (*Phoenicopterus ruber*) (Houlihan and Goodman 1986: 147) and the Dalmatian pelican *(Pelecanus crispus)* (Houlihan and Goodman 1986: 13) were also eaten. The flamingo was a popular dish in Roman times, too (Flower and Rosenbaum 1958: 149).

Pigeons and Doves: Pigeons were and still are a very popular food in Egypt. They are now generally roasted and stuffed with rice or some other grain such as barley or emmer; a possible method of preparation in ancient times. There are seven types of these birds that might have been available to the Egyptians, though not all of these have been identified in Egyptian art. The various types of pigeons and doves available in Egypt are the Stock pigeon *(Columba oenas)*, the Rock pigeon *(C. livia)*, the Pink-headed Turtle Dove *(Streptopelisa roseogrisea)*, the Collared Turtle dove *(S. decaocto)*, the Turtle dove *(S. turtur)*, the Laughing dove *(S. senegalensis)*, and the Namaqua dove *(Oena capensis)* (Houlihan and Goodman 1986: 100). Egyptian names for these words are: *mnw.t/mny* (*WB* II: 79), and *ʿšꜣ* (*WB* I: 229).

Larks: Larks were probably eaten in ancient Egypt — they certainly were a popular dish in historic times in Europe as well as in Egypt (Drummond and Wilbraham 1958: 35, 53-61; Ashtor 1970; Rodinson 1949). The species of larks available in Egypt are the Black-crowned Finch Lark (*Eremopterix nigriceps)*, the Desert lark (*Ammomanes deserti)*, the Bar-tailed Desert lark (*A. cincturus*), the Hoopoe lark (*Alaemon alaudipes*), Dupont's lark (*Chersophilus duponti)*, the Short-toed lark (*Calandrella cinerea* and *C. rufescens)*, the Calandra larks (*Melanocorypha calandra; M. bimaculata*), Thick-billed lark (*Rhamphocorys clotbey)*, Temminck's Horned Lark (*Eremophila bilopha)*, the Crested lark (*Galerida cristata)*, Thekla lark (*G. theklae*), Wood Lark (*Lullula arborea)*, and the Skylark (*Alauda arvensis)* (Houlihan and Goodman 1986: 164).

Other birds: There are several other species extant in Egypt that the Egyptians could have eaten. There are thirteen types of plovers (*Charadrius* of various types), six types of Sandgrouse (*Pterocles)*; several Pipits (*Anthus)* and Wagtails (*Motacilla)*; twelve Wheateaters (*Oenanthe)* which would most probably have been netted in the fields as were the quails; and at least four types of Thrush (*Monticola)* that would have been edible (Houlihan and Goodman 1986).

The Egyptians have several names for birds but these tend to be more generic than specific. They are names for sets of birds, e.g. pigeon, ibis, etc. but there is definitely no differentiation between species. This is the case in contemporary Egypt where most animals have very general names. It is possible that some of the birds have more specific names — some of the hieroglyphic signs that are birds might be the name of the bird as well as the phonogram (e.g. the *st* duck) — but a study of bird names is beyond the scope of this thesis.

The general word for bird is *ꜣpdw*, *ꜣpdw-n-mw* is possibly a term used to identify water fowl, be it duck, goose or Avocet. Other generic terms used for birds are: *wnwn* (*WB* I: 317), *wgs* (*WB* I: 377), *ꜥšꜣ* (*WB* I: 229), *ꜥnnbw* (*WB* I: 192), *ꜥbw* (*WB* I: 175), *it-ḥꜥꜣ* (*WB* I: 142), *ꜣmꜥ* (*WB* I: 10), *im* (*WB* I: 78), *imtr* (*WB* I: 88), and *irḥnn.t* (*WB* I: 116).

I.5 Swine

A great many misconceptions surround the pig, *Sus scrofa*, in ancient Egypt[11]. Until very recently pigs were considered taboo animals that

[11] Wild pigs were found in Egypt in the Fayum and the Wadi Natrun upto 1892 when they became extinct due to hunting (Keimer 1937: 150-1).

were abhorrent to the Egyptians and which they neither kept nor ate. Herodotus is largely responsible for this view. He reports that the Egyptians never touched pigs. If by accident an Egyptian touched a pig he would plunge into the river to cleanse himself of the impurity thus acquired (Herodotus II: 47; Godley 1946: 333-6). Given these strong feelings about brushing against a pig, it is natural, he says, that they never consumed these animals although they were available to them. Egyptologists hypothesize that this is due to the pig's association with Seth, god of chaos and evil. The only time that the Egyptians ate pigs was on certain ritual full-moon days (Herodotus II: 47). Although they were Egyptians, swineherds were looked down upon and had to live apart, were forbidden entry to temples, and forced to socialise exclusively with one another. The only time Herodotus mentions that pigs were put to use by Egyptians is when they were turned into the field to tread in the seed, much as sheep and goats were in the Dynastic period. He also says that pigs were used for trampling in seed when it was sown (II: 14; Godley 1946: 291). Most early Egyptologists relied heavily on the work of the Greeks, especially Herodotus, who wrote about Egypt; consequently their ideas about Egyptian diet and taboos, and therefore pigs, were formed by Herodotus. The fact that in Muslim Egypt the pig is considered taboo (*haram*), as well as by the Jews, previous residents of the Nile valley, has only served to reinforce Herodotus' statements concerning the ban on pig consumption, although the evidence does not support his views.

The idea that pigs were not eaten in Egypt has been further emphasized by a paucity of pictorial representations of the animal. Pigs are depicted in Kagemni's tomb in Sakkara, in Khety's tomb (No.17) at Beni Hasan, in the tombs of Ineni (TT81) and Nebamun (TT24) and several others at Thebes, and in the tombs of Renni and Paheri at el-Kab (Table IV). Other representations appear throughout Egyptian history, albeit less frequently in the Old Kingdom. Clay models of pigs have been found at Hierakonpolis and Abydos from the predynastic period onward (Quibell 1900: pl. 22.8; Petrie 1903: 25, pl. 6.66). Tomb representations show pigs in a positive light as participating in daily activities that helped rather than hindered the tomb owner (Table IV).

The only time that the pig is actively criticised is in some of the Egyptian religious texts such as the Book of the Dead (Chapter 112; 125) and the contendings of Horus and Seth (e.g. in Edfu where Seth is sometimes shown as a pig, although this might be a later Graeco-Roman development) where the pig's association with Seth is underlined. This association with Seth does not, however, exclude pigs from offering-lists.

Pigs, Egyptian *š3ỉ* (*WB* IV: 405) or *rri/rr.t* (*WB* II: 438), are mentioned in list 45 in Ramesses III's temple at Medinet Habu as offerings for a feast of Nefertum (Nelson 1934: 59). Amenophis III gave 1000 pigs and 1000 piglets to the temple of Ptah at Memphis (Morkot 1990: 331); according to the Nauri decree pigs were bred in Seti I's Abydos temple domains (Epstein 1971b: 341). They are also mentioned in the Third Dynasty Biography of Methen (Sethe 1903: 1-7), and were plentiful enough to warrant the existence of the title "Overseer of Swine" in the time of Senwosret I (Newberry 1928: 216). They were also present in Deir el-Medina (Janssen 1975: 177-78). Pigs are mentioned in medical texts such as the Ebers Papyrus, where their teeth were useful for coughs and boils, their eyes to cure blindness, their blood for "hair in eye" (an unidentified disease), their grease for easing stiff joints and encouraging the flow of blood, their ordure for weakness in the veins, and their intestines to cure worms (Darby et al. 1977: 189-90).

The most conclusive evidence concerning the Egyptian consumption of the pig comes from the faunal remains. Pigs appeared in the Fayyum sometime between 4500 and 4000 BC (Clutton-Brock 1981: 73). Thereafter there is a great deal of evidence for the consumption of pigs in Egypt. In the Early Dynastic Period there were more pigs than cattle or sheep/goat, especially in the delta (Boessenek and von den Driesch 1988: 122). Pig bones were found in Maadi, Merimde Beni Salame, Buto, Helwan and other Lower Egyptian sites (Keimer 1935: 28). The most abundant source of animal bones in Kom el-Hisn were the pig bones. According to Wenke, that they were all domestic pigs is clear from the dentition (1985: 9). In Upper Egypt, pig bones come from Armant, Toukh, and Abydos (Darby et al. 1977: 173). In later periods pig bones were found in Malkata (Chapter X) and Amarna, where there are even pig-sties (Kemp 1984: 47-49), Elephantine, Deir el-Medina, and Tell el-Daba. Pig bones are frequently found at settlement sites, but rarely in funerary contexts. One reason for this is that very few bones from tombs have been preserved and examined, thus skewing the available data. The other is that, as pigs were a cheap, low-status, food (see below), and would not have been considered worthy of being offered on important ritual occasions, especially if a more expensive, higher status beast could be afforded. Thus most joints of meat found in tombs belong to bovids rather than pigs or sheep/goat, although all these types of meat were consumed.

Pigs are very efficient animals to breed for food. They reproduce very quickly and are easy to care for, not requiring much herding far from a settlement and are quite good at foraging for food in the debris discarded by people. "They are not as demanding in requirements for protein and carbohydrates as other domestic animals" (Miller 1990. 128-29), their growth rate also being much faster than that of other animals. This makes them a cheap and efficient source of food, one probably much favoured by the common people who would not have the necessary grazing-land or feed that was required for cattle and sheep. Nutritionists consider pigs "more important in the diet than sheep/goats (and cows). Each pig could provide 1.5 times as much meat as a sheep/goat and each gram of pork contains about 100 more calories than a gram of mutton" (Pellet and Shadarevian 1970). 89% is edible in proportion of it weight (Paul and Southgate 1978: 90-3).

Pigs were not only useful for food, but also for some agricultural work. As Herodotus mentioned (II: 14; Godley 1946: 291), and as tomb paintings illustrate, pigs were used, as were cattle and ovicaprids, to trample in newly sown grain (II: 14; Godley 1946: 291). They also could have acted as the garbage collectors of the ancient world (Miller 1990: 126), much as the *zebaleen*, garbage collectors of Cairo, use pigs even now. Their hair could also be of use. The Dokki Agricultural Museum contains a brush made of wild boar bristles, which was found in the Theban necropolis (the exact location is lost) and dates from the Eighteenth Dynasty (Keimer 1935: 28).

Although pigs were consumed, they rarely featured in representations and have been considered taboo. Sometimes foods are deemed impure due to religious beliefs, e.g. the pig's affiliation with Seth. Spell 112 in the Book of the Dead (Allen 1974: 91)[12] emphasizes the pig's association with Seth and helps explain why the pig might have been taboo. The relevant portion, Re's dictates to gods and Horus' followers, states: "Abominate the pig for Horus' sake, so that he may recover". However, one could argue that if the pig symbolises Seth, then the slaughter and consumption of 'Seth' would be a positive rather than a negative act, although, as Seth was a god, consuming him might be construed as an act against the god. Such sensibilities did not, however, stop people from consuming animals sacred to, or associated with other gods, e.g. cows of Hathor, geese of Amun, etc. In actual practice, it is possible that the taboo might be ignored by the irreligious and by those who could not

[12] All references to spells from the Book of the Dead use the Allen (1974) translation, and will only give the spell number.

afford to live by the taboo[13]. It is also possible that in Egypt pigs were considered impure for certain people (priests) at certain times of the year (holy/unholy days, feasts, etc.) and not others.

Taboos do, however, stem sometimes from pragmatic (health) rather than philosophical or sentimental considerations. This could certainly be so in the case of the pig. Animals can become taboo if they have offensive habits, smells, transmit diseases, or are not ecologically advantageous to rear[14]. Pigs are pungent and have a tendency to wallow in mud and their own excrement (necessary for their sweat glands), which they sometimes eat[15], all fairly unattractive habits. Additionally, though less likely, the pig taboo might exist because the Egyptians feared that they might acquire some of the attributes of the pig merely by consuming it (Webster 1942: 8). The most important point, however, is that pig meat spoils faster than most other types of meat save chicken (probably unknown in Egypt) unless consumed or preserved immediately. The fact that pigs can transmit trichinosis (Millet et al. 1980: 79-84), a parasitic worm causing "diarrhoea during the intestinal stage, by pain, stiffness, swelling (especially periorbital), sweating and insomnia" (Darby et al. 1977: 171) may also have been a factor in its prohibition as food. However, if pork is well cooked then there is no such danger as extreme heat applied for a long time destroys the organisms, and as much Egyptian food (meat) was boiled for a long time, this would have made the animal safe to eat. Of course, were the Egyptians to grill/roast the entire animal as is shown in some reliefs, there might be a danger of the inside being raw and therefore a host for bacteria. Furthermore, pigs compete with humans for water and shade, two prized commodities in Egypt.

Thus, despite Herodotus' reports of the supposed injunctions against eating pork as unclean, the paucity of their two- and three-dimensional representations and their association with Seth, archaeological and textual evidence confirm that the pig was consumed in ancient Egypt.

[13] Taboos are frequently most applicable to wealthier and more important members of the community (Webster 1942: 324).

[14] Although pigs have more fat and meat than other domesticates, they do require water and shade. Both these items are at a premium in the Middle East and therefore "pig farming was a threat to the integrity of the basic cultural and ecosystems of the Middle East" (Harris 1974: 40), and that "pigs are thermodynamically ill-adapted to the … climate (Harris 1974: 42).

[15] None of these are particularly appealing features (Douglas 1970: 2), especially to people who seemed to disapprove expressly of eating excreta (Coffin & Pyramid Texts), and who in some areas would have been short of spare mud, for example in Amarna, although the Workmen's Village is where most pig-breeding took place.

I.6 Fish

Fish are an excellent source of protein, vitamin A, and the minerals Magnesium and Phosphorus (Smith and Walters 1967: 93); in fact, most fish can provide humans with 70% of the necessary protein required for humans (Paul and Southgate 1978: 134-149). They were readily available to all Egyptians as food. Although many species of fish are found in Egypt, as the varieties included both freshwater (Nile) and marine (Red Sea, Mediterranean) fish, virtually all fish represented and found on monuments and in the archaeological record are freshwater types. There is so far one notable exception in the archaeological record. Bones of *Sparus auratus*, a marine species living mainly in the Red Sea have been found in Memphis (Jeffreys 1989: 9) and Kom el-Hisn (Wenke et al 1988: 17). The reason for the lack of marine fish in the archaeological record might be due not only to the difficulty in identifying species by bones other than the head bones, the relatively limited and only recent faunal investigations of settlement sites, but also that unless marine fish were processed at the sea side and then brought to the Valley they would spoil. There is, so far, very little evidence of large scale settlements around the Red Sea which would export fish to the Nile Valley, an area that already had an abundance of fish. The absence of Mediterranean species is more surprising, but this might be due to the dampness and consequent low rate of preservation of faunal material from delta sites, the logical sites for fish from the Mediterranean. The lack of marine species (the Punt reliefs being a notable exception) in the artistic record is more surprising, but perhaps they only recorded journeys made on the river itself.

Fish have been consumed since Palaeolithic times (Brewer and Friedman 1989: 5); their remains are found in the Fayum Neolithic and Fayum A. In the Predynastic period they are found throughout the delta at Tell el-Iswid (van den Brink 1989), El Omari, and Merimde, and in the Nile Valley in Elephantine, Hierakonpolis, and Nag-ed-Deir. In the Early Dynastic period a cooked fish formed part of the funerary repast found in a tomb (Emery 1962: 6). Bones and representations confirm that fish continued to be part of the diet through Dynastic times. Although a regular part of the diet, fish do not seem to have been "farmed". It is possible that the pools in people's gardens appearing in wall reliefs (Beni Hasan 3, TT187, TT90, and TT296, to mention but a few), or mentioned in texts (Anastasi III: 2.2 [Caminos 1954: 74]), as well as the actual sacred lakes in temples, were used for breeding fish in addition to being ornamental and sacred. Representations of fish and fishing become part of the standard set of tomb scenes.

Fish were shown being caught by fishermen, speared by nobles, served to the deceased, kept in storerooms, being eviscerated and dried, and having their roe prepared for consumption. The *Tetradon fahaka* and two species of mullet were venerated in Elephantine in the Middle and New Kingdoms (Brewer and Friedman 1989: 15). In the late New Kingdom and subsequent periods several species were mummified (Louvre: N2898 and N2899d). Fish lent their form to utilitarian objects (slate palettes, dishes), as well as jewellery. They are mentioned in literature (Wenamun), in taxation (Griffith 1898: 97) and offering-lists, and as part of the payment for workmen (e.g. Deir el Medina), and are found in tombs as offerings for the deceased (Schiaparelli 1927: 159-60; Emery 1962: 6). There are several classes of artefacts (nets, hooks, harpoons, net-weights) devoted to their capture.

Yet, despite the prevalence of fish in art, artefacts, literature, and their presence in faunal assemblages and funerary offerings, fish have been considered taboo animals which were not eaten by the Egyptians. As is the case with pigs, this is primarily the fault of Classical authors who listed fish among the animals considered taboo (see discussion in I.5). Diodorus Siculus is especially emphatic about this, while Herodotus restricted the taboo to priests. This might be due to the belief that fish had consumed the penis of Osiris, and, due to deference to the god, fish were not consumed. Primary among the offenders was the oxyrhinchus fish (probably *Mormyrus kannume*), although the Lepidote and the "Phagrus", *Hydrocyon forskali*, were also considered culprits (Keimer 1939: 269). As far as ancient texts go, there is one non-religious instance which supports the fish taboo, and that is the victory stele of Piye (l. 151) where fish-eaters were "an abomination to the palace, and King *Namart* entered the palace because he was clean and did not eat fish" (Lichtheim 1980: 80). However, this is a Twenty-fifth Dynasty text, which is quite late, and one written by a Nubian, therefore not necessarily reflecting Egyptian habits.

This belief was augmented by the absence of fish hieroglyphs in the Pyramid Texts, though fish are not the only creatures avoided therein[16]. The Coffin Texts do not seem to object to fish, but the Book of the Dead

[16] According to Lacau the Pyramid Texts object to the representation of all living things, while Middle Kingdom Sarcophagi show animal hieroglyphs with severed heads/limbs, or omit limbs (Lacau 1913; de Morgan 1895: 106, fig. 247).

contains two spells at least, #64 and #148, that object to the consumption of fish (and goat — a unique example) (Brewer and Friedman 1989: 15). Also, the word *bwt*, the hieroglyph for the fish, *Barbus bynni*, means forbidden or taboo. Some scholars have argued that this is because the fish itself is taboo, while others have argued that the phonogram is being used and it bears no reference to the fish. A final reason for the persistence of the idea of taboo is the lack of fish shown on offering-tables. Though fish are less common than other animals on offering-tables they do occur, and are frequently shown being brought as offerings to the tomb-owner (Table III; el-Bersha 2); apparently it is only in the Memphite Necropolis of the Old Kingdom that they are not present. They appear on offering-tables and storerooms, and are even shown being eaten at a banquet of Horemheb (Fig. 7).

It is quite possible that at certain times in certain areas some fish were taboo for all or specific (e.g. priests) people (Desroches-Noblecourt 1954: 36)[17]. However, the evidence, especially the lists of taxes and tribute, as well as the faunal remains, prove that fish were consumed. It is possible that for offerings some species were favoured over others; "the most frequently listed fish are: *Mormyros; Tilapis nilotica; Mugil cephalus; Clarias Anguillaris; Synodontios*" (Handoussa 1988: 107), while most species caught were used for general consumption.

In the Nile there are twelve families of fish, yielding approximately fifty-six species that were easily available to the Egyptians. The following are those more commonly found in representations and the faunal remains and are organized by family[18]. Marine fish have been excluded due to the lack of representational and faunal evidence. All of the fish listed below are consumed in Egypt today.

Mormyridae (elephant-snout fishes): There are fourteen species of this fish. Most are present in the Nile from the delta to the Sudan. They include the *Mormyrus kannume*, the Egyptian *ḫ3t* (*WB* III: 359) with its distinctive snout (Brewer and Friedman 1989: 48-53) which is occasionaly seen being eviscerated and prepared for storage.

[17] G. Pinch (1992) has kindly pointed out that fish bones were found in the Hathor shrines at Mirgissa (Vercoutter 1970: 318, fig. 16) and Timna (Rothenberg 1988: 241-5; pls. 143-6). There is no doubt, especially in the case of the former, that these were placed as offerings in the shrine. Thus it seems probable that fish were forbidden only to certain cults, and that even during specific times of the year.

[18] All scientific names come from Brewer and Friedman 1989.

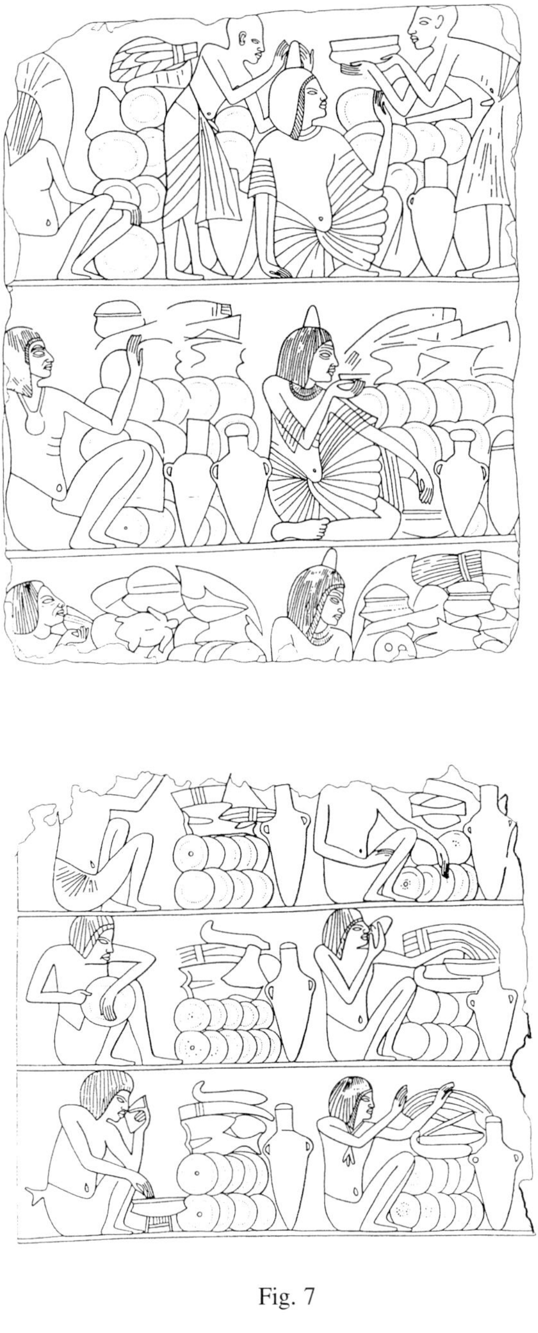

Fig. 7

Characidae (tiger-fish): eight species of this family are found in the Nile. The genus "*Hydrocynus* is thought to be the most likely candidate for the Phagrus of ancient Egyptian legends" (Brewer and Friedman 1989: 54) concerning Osiris.

Citharinidae (moon-fish): Two species of this fish have been identified in Egypt. They are quite large and are available throughout the Nile valley (Brewer and Friedman 1989: 56).

Cyprinidae (carps): There are thirty-nine species found in Egypt (Brewer and Friedman 1989: 57-9), though most are quite small and therefore would not have been a very important source of fish protein to the Egyptians as the effort required to catch a significant number would be quite great. The larger species of the genus *Barbus* were frequently depicted by the Egyptians. The most common was *B. bynni*, one of the six fish hieroglyphs, *bwt* (*WB* I: 453-4)(see above).

Clariidae (suborder: catfish): Though this family has only four species (Brewer and Friedman 1989: 60-63) found in the Nile it is one of the most common in faunal assemblages (Wenke 1985: 9). *Clarias anguillaris* and *C. lazera* live not only in the Nile, but also in the small canals found throughout Egypt, and can be collected by hand (Brewer and Friedman 1989: 61).

Schilbeidae: Two species inhabit the Nile. They were, in the Late Period especially, associated with Mendes (Brewer and Friedman 1989: 64-65).

Bagridae: There are two species of this fish in the Nile, with only one identified representation (Tomb of Ukh-Hotep in Meir) found to date (Brewer and Friedman 1989: 66).

Mochokidae: Although there are four genera of this family in Egypt, only one is common: *Synodontis*, there are four species of this genus, although only two, *S. schall* and *S. batensoda*, seem to be represented. The latter might be the fish that swims upside-down (Brewer and Friedman 1989: 67-69). *S. schall* appears in the archaeological record (Jeffreys 1989: 9; Wenke 1985: 9). The supposed Egyptian name for this fish is *wh* (Faulkner 1976: 66; *WB* I: 350).

Malapteruridae (electric catfish): Only two species of this family are known in Egypt (Brewer and Friedman 1989: 70-71), and both, despite their name, are edible.

Mugilidae (mullets): Three of these species are found in Egypt. They are easy to catch as they inhabit the area near the shore and move in large schools. Not only were the fish eaten, but the roe of *Mugil cephlus* were dried and pressed for consumption (Brewer and Friedman 1989: 72). In Egyptian they were called *ʿḏw* (Faulkner 1976: 51;*WB* I: 240).

Centropomidae (suborder: perch-like fish): One species of this family is known in Egypt, *Lates niloticus*. It, along with the *Tilapia* (see below) and the *Mugil* (see above) is the most frequently portrayed fish in Egypt (Brewer and Friedman 1989: 74-75). It has been found in faunal assemblages throughout the delta and in Memphis, it has been presented as offerings in tombs at Deir el-Medina, and it has been mummified because it was sacred to Neith (among others) in the Late Period. The Egyptian name for this fish is thought to be *ʿḥꜣ* (Faulkner 1976: 46).

Cichlidae: Eight species of the genus *Tilapia* exist in Egypt, though only three are common: *T. nilotica* (*Bulti* in Arabic, and *in.t* [*WB* I: 92] in Egyptian), *T. galilaea, T. zillii*. These fish, commonly represented on a bident spear along with *Lates*, are most abundant in the delta, while the latter are common in Upper Egypt; this pairing is thought to represent the unity of the two parts of Egypt (Brewer and Friedman 1989: 79). *Tilapia* are one of the most frequently represented fish (in both two and three dimensions) in Egypt and also one of the most frequently retrieved in excavations throughout the country.

There do not appear to be any other Egyptian names attached to specific types of fish. There are, however, some words that denote fish in generic terms: *mhit* (*WB* II: 127), *rm* (*WB* II: 416), *ꜣbḏw* (*WB* I: 8), *fttw* (plural) (*WB* I: 588), *nšmw.t* (*WB* II: 340), *sšn.w* (*WB* IV: 294), *ʿwbbw* (*WB* I: 172), *ʿdw* (*WB* I: 237), *pꜣḳr* (*WB* I: 500), *ḥbꜣ* (*WB* III: 62), and *swr* (*WB* III: 429).

II

SLAUGHTER

All mammals, regardless of size, would have been killed in the same way. The method of flaying and jointing would have varied somewhat depending on the size of the animal rather than the species. Small animals would have been suspended from trees, while larger animals would have been killed and processed on the ground. Birds and fish were each killed in a different way. Surprisingly, given the fate of the mammals, the birds are shown being strangled rather than beheaded.

II.1.1 Slaughtering, flaying & gutting of mammals

Information about ancient Egyptian methods of slaughtering mammals (oxen, goats, oryxes, sheep, antelopes, and pigs)[1] comes primarily from tomb and temple reliefs and paintings, along with such paintings that appear on coffins (e.g. the sarcophagus of Ashayet [JE47267] from Deir el-Baheri). This pictorial evidence is supplemented by three-dimensional representations in the form of wooden tomb models of the Middle Kingdom. There are also some textual references to the slaughtering of animals, and to slaughter-yards. Ethnographic studies further supplement the traditional sources of information and are especially useful in interpreting the artistic record. Furthermore, cut-marks on bones bear evidence to the manner in which they have been butchered (Chapter VI).

One of the most common scenes occurring in Egyptian art from the Old Kingdom onward is that of cattle-butchery; almost every tomb in Egypt includes a scene showing the slaughter of cattle. This scene first appears in the late Third/early Fourth Dynasty, and is continued throughout Egyptian history, reaching its acme in the Middle Kingdom.

[1] Although the Egyptians are known to have consumed pork (faunal remains, offering lists, and a few tomb paintings attest to this), there are no known reliefs showing the slaughtering of pigs or of hyaenae. This is also the case for many of the hunted mammals as some of them were extinct by dynastic times, while others just did not figure in tomb paintings; therefore they will not be discussed separately although the statements made on butchery are as relevant to them as to the other types of animals.

In the New Kingdom the sequence is less commonly found in its entirety, but in the Amarna period, and then in the Saite period, it had a resurgence. Scenes showing the slaughtering of goats and oryxes and other antelope are also included on tomb walls throughout Egyptian history.

These scenes generally show the slaughter of animals as funerary and religious offerings, rather than as a routine activity (see Table I). There are occasional scenes showing hunted animals or small ungulates (goats) being slaughtered and jointed in a clearly secular setting — after hunts or as food for people working in the fields (Fig. 5).

There is a great deal of symbolism involved with the wall paintings. As mentioned before, several scholars believe that all scenes in tombs and temples are allegories awaiting interpretation rather than literal representations of the daily activities of the Egyptians or those directly and practically connected with their burial. Though this is possible, one should not exclude a pragmatic interpretation of the scenes. It is likely that the Egyptians decorated their tombs in a way such that their paintings could be interpreted on many different levels, none of which were mutually exclusive. Also, not all scenes depicting butchery need have had the same hidden meaning: the meaning depended to a large extent on context.

Certain butchery scenes are clearly sacred in nature and these are highly charged with symbolism, for example the Opening-of-the-Mouth ceremony (Blackman 1924: 55), or the sacrifice to a god where the scene clearly has several possible meanings. The animal can be seen to represent chaos in the form of the unfettered forces of nature, the enemy of the god, king and *maat*, and by killing it before the god, the god achieves victory over chaos (Otto 1950; Junker 1910: 72). When an animal is slaughtered before the king this too can be seen as a symbolic triumph over the enemies of the Pharaoh. The animal's meat that was consumed "symbolized the destruction of the enemies of the god, and the king" (Fairman 1958: 80). This interpretation would be especially feasible in the case of the offering of oryx and antelopes as, despite being semi-domesticated at various times in Egypt, they were more wild than the usual domestic cattle that were sacrificed. In fact, in Edfu, Denderah, and other Late Period temples, texts make it clear that the slaughter of these animals symbolizes an attack on the enemies of the king (Derchain 1962). In addition to symbolizing the destruction of the enemies of god and king, the ritual might suggest that the slayer (the king, in theory), would absorb the desirable qualities and powers of his enemies by eating the sacrifice (Fairman 1958: 91; David 1973: 272; Otto 1950: 164). The ritual interpretation of killing as a symbolic triumph over chaos would be more appropriate in the later depictions of the killing/hunting of hippopotami and crocodiles (Ghoneim 1977: 185). Additionally these dead animals are offerings of food to the god or Pharaoh (e.g. Offering to Anubis and Osiris in the tomb of Rekhmire [Davies 1943: pls. LXXXVIII and LXXX]), as appeasement, an attempt to buy blessings, or for thanks and praise both in the present and for eternity.

Otto has provided another symbolic interpretation for the butchery scenes found in funerary contexts, specifically those associated with the New Kingdom examples of the Opening-of-the-Mouth Ceremony in which the *ḫpš* is used as a mouth-opening tool[2]. He states that the earliest preserved texts for the ritual are in the tombs of Rekhmire and Seti I. In the scene "the victim — a bull — lies fettered on the ground while the butcher stands over it cutting off its foreleg. (Nearby) a standing woman is depicted. At one side of this group appears the *sm* priest with raised *ḫrp* — sceptre, followed by the lector priest (*ẖry-hb*)" (Otto 1950: 165). Otto interprets the woman as a kite or Isis, while he writes that the "bull is not necessarily Seth — he is Thoth, while the butcher is Horus who gets his eye from Thoth, the eye being represented by the heart and/or the foreleg. The slaughter represents a battle" (Otto 1950: 168). However this is just one possible interpretation, and, as Otto himself admits, it is difficult to understand the ritual meaning of a depiction of butchery when we are alien to the culture (in cultural tradition as well as in time), and because the "Egyptians themselves had often lost all knowledge of their (illustration and text's) real meaning, thus the relations between ceremony and accompanying ritual texts do not remain fixed, so that ... the same ceremony is accompanied, at a different place and time, by different texts, since illustrations and texts were often transmitted through various channels" (Otto 1950: 164) and change through merging, insertion, and variation by interpretation.

These scenes provide invaluable information on the ancient Egyptian practice of butchery, regardless of the symbolism involved and whether the depiction was ceremonial or secular in nature, as the actual method of slaughtering and jointing an animal was probably the same in both contexts, especially since the majority of the meat derived from offerings was consumed by ordinary people rather than the deceased or a deity (Chapter X). The only difference between a sacred and a secular slaughter

[2] The Ceremony of the Opening-of-the-Mouth had a stellar context and the *ḫpš* fitted into this context as the Egyptians compared it to the constellation we see as the Great Bear/Big dipper (Roth 1991: 4, 1993: 57-79).

might be in the intention of the butcher manifested by the words spoken while performing the operation. Unfortunately there appears to be no satisfactory Egyptian formula that consistently appears in the texts attached to scenes, save for, perhaps, the statement: "An offering to N" serves as adequate consecration[3].

According to such representations, the beef that the Egyptians ate was from oxen — castrated male animals — rather than from cows. Occasionally cows were slaughtered, as can be seen in donation lists (e.g., P. Harris I: pl. IV, line 8), though, on the whole, oxen were favoured as meat animals (see discussion in Chapter I). The ancient Egyptian method of slaughtering and jointing an animal can be reconstructed by using all the sources mentioned above, assisted by ethnographic investigation to help interpret the artistic record.

Bringing To The Ground

The first step in the sequence of killing an animal is to bring it down to the ground. The tomb paintings show that the oxen were brought down by lassooing and wrestling the animal to the ground which is then laid trussed on its back. This method is still used today with oxen. A front and a back leg are tied together to make the animal trip and fall down. Then the butchers assistants tie up its remaining legs and hold down the animal. Smaller animals, such as calves, sheep and goats, are merely wrestled to the ground and trussed, or held still by the butcher's assistants during the subsequent operation (Fig. 6).

Cutting The Throat

Once the oxen were laid trussed on their backs being kept still by the butcher's assistants, the butcher held its throat and cuts it. Later representations show the butcher holding the animal's head to the the right side so the left side of the neck was exposed, and cut across its throat, thereby severing the carotid artery (Fig. 8). It is probable that the carotid

[3] A current example of this practice can be seen in the Muslim, Coptic, and Jewish methods of slaughtering. In these traditions specific words and gestures are made over an animal when it is slaughtered. The basic text of what is said over the beast is the same regardless of whether the butchery is sacred (for a special festival) or secular in nature; there are only occasionally some variations in what is said depending on the reason for sacrificing an animal especially. The Muslims say: "In the name of Allah, the Merciful, the Compassionate", the Copts say: "In the name of the Father, the Son, and the Holy Ghost", and the *Shokhetim* (trained kosher butchers) of the Jews say: "Blessed be our Lord, King of the universe who commanded us to slaughter".

Fig. 8

artery was cut in earliest times, but it is most clearly depicted in the New Kingdom. Cutting the carotid artery and then separating the head from the body is still the method used in Egypt now. Some scholars believe that an earlier custom "was to stupefy the victim by a blow on the head with a sceptre (*ʿb* or *ḫrp* sceptre) before actually killing the beast" (David 1973: 320; Otto 1950). Although the sceptre does appear in some reliefs it does not appear in all butchery scenes and might indeed have had symbolic significance in some of the scenes, as Otto (1950) suggests, but even so it is quite possible that the sceptre was used to consecrate offerings by passing it over the animal and not necessarily by clouting it on the head and rendering it unconscious (Jéquier 1921: 182). There does not appear to be any tangible evidence to support the claim of slaughtering an unconscious beast which might have sprung from a desire for humane methods of butchery to be ascribed to the ancient Egyptians.

In modern butchery, it is at this point that one would consecrate the beast by speaking the relevant formula over it (see footnote 3). Not all butchery scenes involve priests, although it is possible that some of the butchers were priests (see below). Most Egyptian bylines to the illustrations do say: A sacrifice to N, but no more than that. This can be interpreted as a label for the illustration, or perhaps as what the butcher says over the animal by way of consecration.

In some of the Old Kingdom scenes of butchery for offerings to the deceased, a *wꜥb* -priest stands near the butchers to inspect the blood and the entrails of the beast to see if it is indeed pure and undiseased, and therefore acceptable to the god. (Quibell et al. 1898: 31). In the tomb of Ptahhotep a man with a bloody hand waves it beneath the nose of a man whose title is *imy-r wꜥb swnw*, which has been variously translated as a form of *wꜥb* priest, a vetinarian, and a doctor (Fig. 9). Once this person has approved the purity of the animal, the butcher continues his work. (Leibovitch 1953: 59). Examining the entrails of an animal and sniffing the blood are methods used by modern-day meat inspectors and contemporary Egyptian village butchers to determine the health and edibility of an animal. This examination of the blood and entrails of the beast is the final step before the butchery proper commences.

In current butchery, after the first cut has been made through the animal's carotid arterey and jugular vein, and blood has spurted out, the butcher makes another cut across the throat to speed up the process, and then waits until the animal bleeds to death before proceeding. This second cut, if it occurred, is not separately shown on the wall paintings.

After the throat was cut, the blood was drained from the body by pumping the foreleg of the animal, thereby forcing the blood to flow out from the severed veins and arteries of the neck. This helps explain why the foreleg is given so much attention even before it is cut off as an offering to the gods — it is a vital component of the "pump" used to empty the body of blood, thus preventing spoilage of the meat (Fig. 6). This is no longer practised in Egypt, although it was in parts of the Middle East and Eastern Europe early in this century (Fabish 1990). Frequently paintings and models feature a man holding a bowl who kneels beside the head of the animal. Various scholars have hypothesized that he is collecting the blood for blood-pudding (Winlock 1955: 25), or, perhaps to have it tested by the vetinarian-priest for purity. After watching the

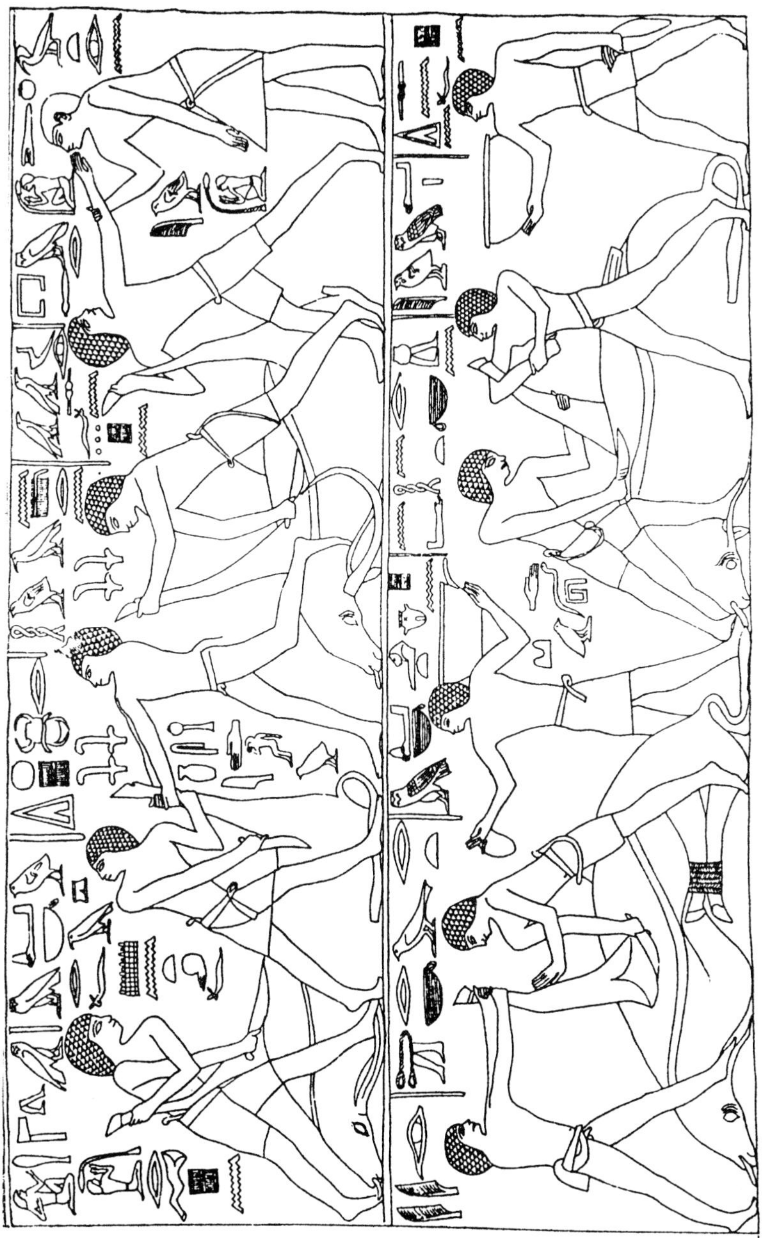

Fig. 9

slaughter of a *gamoosa*[4] performed in modern Egypt in a manner very similar to that of ancient Egypt, it occurs to me that it would be very difficult to collect the liquid from the bleeding throat of a bovid as, after the initial spurt, the blood flows straight down and would be very difficult to catch in a vessel, especially in the large, deeply curved jars depicted in tombs being used for that purpose. It is possible, however, that the blood would continue to spurt out of the body if it were being pumped out by the foreleg, and could then be caught by the butcher's assistant and kept for later use or for testing the purity and health of the cow. A more detailed discussion of the use of blood appears in Chapter VIII.2.

Flaying

After the actual killing of the animal the pictorial sequence becomes more confusing. Sometimes the actual flaying of the animal is not shown, although it is implied as the next vignette in the butchery series shows the skin hanging off the animal. This is where ethnographic information is extremely useful. After the animal has been drained of blood it is skinned. The modern Egyptian method of skinning cattle might not have been used by the ancients, but it is intriguing. Once the animal is quite dead a small slit, about five to six centimetres long, is made in the skin on the inner side of the right hindleg. A stick, generally a portion of the center of a palm frond that has been dried, is pushed into the hole and poked under the skin, between it and the flesh, tearing through the connecting membranes. This operation continues for between four and five minutes with the side-skin and the leg-skin being gradually detached from the flesh. This is only done on one side of the body. Then, one of the butcher's assistants kneels down and blows into the cut made in the skin. He keeps blowing into the skin which starts to puff up like a balloon. Another assistant takes a thick stick, generally a walking stick of the type used by many Egyptian *fellahin*, and start to beat the ballooning animal. They flip the body over to beat the other side too. Then the animal's skin is blown into once again, and the beating is repeated. This blow-and-beat sequence, a method of losening the skin to facilitate flaying, is repeated two to three times. After this, the two forelegs (right one first) are cut off from the bloated body at the carpals, followed by the severing of the head. Tomb decoration is not always clear as to when the head is removed — frequently it is shown to be severed but attached to

[4] *Gamoosas* or water buffalos were not known in ancient Egypt.

the body throughout the proceedings, even when servants are shown walking off with the head. This might be a convention of Egyptian art rather than an exact illustration of what happened. Interestingly, severed heads of animals frequently are found pictured on offering-tables before the deceased, although archaeologically they are infrequently found as a part of the funerary offerings.

The ancient drawings and the modern method of flaying merge at this point. The butcher takes a small knife whose blade is an elongated oval, much like the leaf knives of ancient Egypt (see Chapter III.1.1), and starts to skin the animal. Skinning starts from the right hind leg, above the tarsals, where the small cut for blowing was made on the inside of the leg (e.g. Tomb of Idut, Sakkara). The cut is carried down the inside of the leg, and then another similar incision is made on the interior of the left hind leg. The animal is on its back with its limbs sticking up. In both ancient representations and modern practice the butcher's assistant holds the back legs while the butcher removes the skin from the hind legs and cuts off the feet below the tarsals. Then he continues making an incision for skinning the animal along its belly. The skin is peeled away. When the beast is half skinned it is removed to the tree where it is hung from its hocks and the skinning continues. In ancient Egypt, as in modern Egypt, if the animal is full grown this operation continues on the ground. However, if the animal is young or a small animal such as a sheep or goat, then the operation is performed with the animal hanging from its hind legs. In modern Egypt a large flat metal dish *(tisht)* is placed under the suspended animal's throat to catch the blood. A bucket of water and a cup are placed within easy reach of the butcher. The butcher continues removing the skin, periodically pouring water on to the carcass with the cup, or having an assistant perform this service. In modern Egypt the leaf- shaped knife is preferred for this function as both sides can be used to cut and skin — very useful in flaying (Nimr 1992). Perhaps this is why this knife-shape was favoured in antiquity. Once the skin is removed it is put to one side.

Small mammals (ovicaprids and gazelles) are flayed whilst suspended rather than supine. Once their throats have been cut the carcass is hung from a tree by its hind legs (Fig. 5). Then a slit is made from the anus down, the internal organs are removed, and an incision is made in the skin on the inside of the right hindleg. From here the flaying progresses, now as then, in the same way as for a large mammal. It is considerably easier to flay a suspended animal than a recumbent one, as the author found whilst doing experimental work on wolves and horses, the former being recumbent, and the latter suspended. The ancient Egyptians, like their descendants, did not erect a pulley system to raise the animals to ease the flaying operation.

The Foreleg

Some scholars believe that in certain funerary slaughter scenes there is a point when the leg of a live cow or calf is cut off and presented to the deceased. Junker (1938: 229-31; Vandier 1969: 137-38), maintains that in some Old Kingdom representations the *ḫpš*, foreleg, is removed when the animal is alive as the tail is aloft and the animal appears to be held down by the butcher's assistants. On closer investigation (e.g. mastabas of Ptahhotep, Kagemni, Ti, Mersankh, Akhet-hetep, G 2120, G2150 etc.) this does not appear to be so as the bovine is shown with its head thrown back and its tongue protruding, a conventional manner of showing death. Furthermore, the tail is only occasionally shown aloft; this can be ascribed to artistic variation or the work of a particular atelier. The aid of the assistant is necessary as the animal is cumbersome when dead, rather than merely (and understandably) recalcitrant when alive. A. Eggebrecht, who has also written on the subject of butchery in the Old Kingdom (1973), also disagrees with this notion of vivisection. The belief in vivisection is not limited to Junker however. A. Weigall (1915: 10-12) comments on the occurrence of a tomb scene depicting the amputation of a leg from a living male calf in the presence of its mother. This is derived from the scene which illustrates Chapter One of the Book of the Dead. The text does not really explain this vignette; in fact, it seems to indicate nothing regarding the vignette (Allen 1960). The vignette seems to appear at the Opening-of-the-Mouth-Ceremony as shown in the book from the Eighteenth Dynasty onward. Such a vignette appears in several Theban tombs, virtually all of the Ramesside period[5]. If these representations had a symbolic meaning to the ancient Egyptians it is now lost. As for whether or not this act was actually performed it cannot be proved or disproved. However, the practical difficulties in holding a live calf's leg and cutting it through with one blow of the knife are manifold. Also, the questions arise: What was done with the leg? And the calf? The animal offerings found in tombs are frequently of young animals, but they are generally not as young as the one pictured; furthermore,

[5] Nineteenth and Twentieth Dynasty: TT19, TT23, TT31, TT41, TT45, TT58, TT138, 360A; Ramesside—no specific dynasty: TT141, TT218, TT259, TT296; Eighteenth Dynasty: TT151; Psammeticus I: TT36.

there tends to be more of the animal present than just the leg (except for the tomb of Wah where a leg constitutes the sole food offering [Appendix II: **R15**]).

Although I do not believe that the animal was vivisected, I do think that in many cases the foreleg was removed before it had been skinned. Wah's offering had the skin on it, as do several of the whole forelegs that were used as temple foundation deposits (Arnold 1979: 51-7). The removal of the foreleg before it had been skinned might aid its preservation, for if the leg was to be offered after some hours, then its skin covering would slow down the putrefying process. Furthermore, if it were offered/ placed on a fire with the skin on, it could be removed from the fire, stripped of the skin, and then eaten without any ill-effects being felt (Klebs 1922: 173).

Gutting

The next step after flaying the animal was and is the removal of the entrails. The butcher takes the big butchery knife that was used to cut the throat of the animal, and makes a slit from below its genitals, all the way down its belly. He then pokes and prods the inside of the animal and removes the entrails. The heart was removed and sent almost immediately as an offering to the god or the deceased. In contemporary butchery the butcher takes another small knife and using his hands he frees the entrails and pulls them out of the animal. They are placed on the flayed skin. The lungs, heart and liver (the liver is one of the last organs to be removed) are also removed and the animal is then ready for jointing[6]. In the tomb of Ramesses III (KV35) and Ipy (TT217) the entrails of the butchered animal hang suspended on a line with the other cuts of meat.

Although the paintings and reliefs do not always agree on the exact order of these events, most seem to support the above-stated order. The Meketre model (MMA20.3.10) and other roughly contemporary models show the animals being butchered in the above order (with the exception of the live-foreleg removal), which, on the whole, seems the most practical.

Another source for the study of Egyptian butchery and one much closer in time to the ancient Egyptians is Herodotus. According to Herodotus (II: 39-40 [Godley 1946: 321-24]), the Egyptian method of dispatching the victim was as follows: they cut the throat, and having

[6] An unusual scene depicting the removal of the intestines appears in the tomb of Ankhmahor at Sakkara.

done so, (completely) severed the head from the body. The ox was flayed and the stomach was taken out, leaving the entrails in the carcass. Unfortunately, however, Herodotus' observations concerning the Egyptians date from a time when Egypt had long been open to foreign influence, causing many native Egyptian practices to be either completely changed or, at the very least, modified. Furthermore there is the question as to how much Herodotus actually observed compared with how much of his information is based on secondary sources. If much of what he wrote concerning Egypt and Egyptian customs was based on heresay, one must regard his data with some suspicion and use it with care. Therefore his remarks are not entirely applicable to Pharaonic Egypt. Nonetheless, the mode of slaughter which he describes was probably quite similar to that practiced in the Pharaonic Period, with a few exceptions. It is doubtful that the Egyptians left the entrails in the carcass — several wall paintings show the animal being completely eviscerated after its slaughter, and one tomb painting (TT217, KV35) suggests that some of the viscera was consumed as it is hanging on a line above the butcher's head. This would be in keeping with contemporary practice where even the intestines of the animals are (after cleaning) fried or boiled in a soup and eaten. There are good microbiological reasons for this order (especially the gutting and skinning of the animal) because once the animal is killed and the intestines are cut, then the visceral contents will infect the body cavity unless they are removed (Heath 1990). If tomb decoration is to be believed, the head was completely removed from the body only in late Eighteenth Dynasty contexts, but in many other New Kingdom depictions of the same date it remains partially attached to the carcass. Perhaps keeping the head partially attached to the animal emphasized its identity throughout the sequence of representation.

Not all animals were butchered recumbent and trussed. Ungulates (goats, oryxes, gazelle) and small mammals (hares, etc.) were hung from a tree by their hind legs, their throats cut from ear to ear, and gutted anus-to-gullet. Lastly they were skinned before being divided into the usual Egyptian cuts (Chapter VI). All butchery representations of these animals appear to support this sequence. It is almost certain that this method was not used for larger mammals (oxen, large antelope) only due to their size which would make suspension a practical difficulty. In modern Egyptian villages adult cows/oxen are never hoisted up and suspended from a tree or hook untill they have been cut into more manageable sizes.

There are some unusual early examples of oryx being slaughtered in a different manner. The tomb of Nefermaat and Itet in Medum shows a man seizing a standing oryx by its horns with one hand, while he cuts its throat with a knife held in the other hand (Fig. 10). This sort of slaughtering is shown in only these reliefs at Medum and only seems to apply to the oryx[7]. This is strange as there are contemporary reliefs from Medum (e.g. from the tomb of Rahotep) that show the usual mode of butchering, as described above, being applied to oxen. Whether there is any symbolic meaning attached to this particular mode of killing the oryx, or whether this was an initial way of slaughtering medium-sized animals (these tombs show smaller mammals such as goats being suspended from tree branches to aid in their despatching) is unclear. When this image of butchery was shown to butchers in Egypt and

Fig. 10

[7] Similar scenes of slaughtering an oryx appear in the Late Period, with one such scene appearing in Luxor Temple, dating from the time of Amenophis II. These scenes are thought to have a complex symbolic meaning including killing apophis and providing offerings for the king and gods (Derchain 1962: 27; 30-6). Also, see the above discussion pertaining to the symbolism of butchery.

Britain they all thought that it was practically impossible. According to them the closest one could get to killing a small (sheep, gazelle, or goat) standing animal would be by straddling the body and gripping it firmly between ones legs whilst holding the throat back by the horns or hair. None of the butchers recommended this method of slaughter, although they deemed it possible.

The question of what was done to animals that had been hunted and killed by a spear or an arrow still remains unresolved. It is quite probable that these were dealt with on the spot (e.g., gazelles on trees). If they were larger (oryx, or other form of antelope), or if the hunting-ground was near the hunter's residence, the game was probably brought home and butchered in the courtyard, or if it were a wealthy household, in the butcher's establishment, such as the one figuring in Meketre's funerary equipment. The exact method of dealing with already dead animals destined for the table remains unknown. If animals were supposed to be bled (for religious as well as hygienic reasons) as a part of the slaughtering process, then something would have to be done with the hunted animals. One possibility is that the animal had its neck slit and was drained of its blood after it had been killed by the hunter, and then hung from a tree (depending on size) and then jointed[8]. This would prevent the meat from spoiling, and is what is done currently with hunted animals throughout the Middle East, albeit ostensibly for religious reasons. If the Egyptians did not bleed the hunted animal then the meat would become gamey very quickly, although in the Egyptian climate it is more likely that it would become a health hazard rather than merely "gamey". Also, it is unknown whether the Egyptians appreciated the gamey taste that is frequenly cultivated in certain animals and birds (e.g. hares, etc) in Britain. Unfortunately there are no known representations of the disposal of hunted animals, so one cannot arrive at a firm conclusion regarding the butchering of hunted meat.

II.1.2 Season of death

An interesting question that arises is whether or not there was a special season when there was a concentration of butchery. In countries with

[8] Vandier (1969: 132) briefly discusses the fate of hunted animals. He believes that scenes showing a gazelle hanging from a tree (in Niuserre's sun temple and in Zawiet el-Metin) illustrate the slaughter and jointing of a hunted beast.

firmly defined seasons based on changes in temperature, meat keeps better in the cold, and as animals are fatter in winter, one might suppose that more butchering was performed in the colder months of the year. This drastic fluctuation in temperature does not hold true for Egypt, however, and so this rule cannot be applied to Egyptian butchery, although it is quite possible that the inundation could have influenced the slaughtering of animals, much as the cold weather and lack of food forces Western farmers to slaughter animals in autumn. If animals were grazed, and if they were kept as large herds and not rented out to people in small groups, they would need large areas of land (see Chapter IV.3 for a discussion of grazing versus feeding). Nowadays animals have food brought to them, although in some areas they are taken to the islands in the Nile and left there to graze. In ancient times, during inundation much of the pasture land which might normally have been used for grazing animals would have been under water, and unless the herds were sent away to the oases or outlying areas that briefly became green during the rainy season, there would have been overcrowding of animals and people during this time[9]. Therefore, inundation would have been one sensible time to do a concentrated amount of butchery and to cull the herds, although, if the meat were to be preserved rather than eaten immediately there might by some difficulty in keeping it dry (depending on humidity levels), and thereby necessitating its immediate consumption. Another reason for slaughter during the inundation might be that the animals were in danger of drowning or acquiring hoof-rot, which is caused by too much constant dampness. It is possible that this held true for secular butchery; sacred butchery would be governed at all times by the dates of festivals (or deaths). Unfortunately, in view of the problems surrounding the reconciliation of the civil/religious calendars with the solar, it is impossible to assess how far, if at all, any festivals involving meat offerings can be tied to any specific time of year. The extant evidence suggests that feasts were fairly evenly scattered throughout the year.

The other question concerns the age of slaughtered animals. "There is a theoretically optimal slaughter point for each individual animal when most of its muscular development has been completed and before it starts to deposit large quantities of fat" (Pomeroy 1962: 53) which differs from animal to animal. However, animals are not necessarily slaughtered at an optimal point. Ideas of what an optimal point for

[9] However, as there is little demographic information about ancient Egypt one cannot necessarily assume it as a major factor in a decision to cull.

slaughter differs between societies (for example, the Amarna and other representations of overfed cattle that are unable to walk (Fig. 11) — a far cry from todays ideal lean animal), and there might not have been always a need for meat at a particular animal's 'optimal point'. It is therefore probable that animals were killed as and when the meat was needed and the Egyptians deemed the animal to have attained the appropriate size.

Two- and three-dimensional representations show a fully-grown ox being slaughtered. Fully-grown is a relative term, although an optimum age for slaughtering a cow or bullock for meat (in pastoral societies) is at the age of four years (Dahl and Hjort 1976: 167). However, the majority of the joints of beef from funerary contexts (laid out in Appendix II and Fig. 42) examined show that the animals were under the age of three and a half years, as their leg bones were not fused (see Silver 1969 for ageing). The Malkata bones show that some full grown four to five year old animals were slaughtered, but several animals under the age of three and a half were also slaughtered — unfortunately the sample is too small to make a generalization about the age of animals.

Most of the pigs in Amarna were slaughtered by two years of age (Kemp 1984: 157), while most of the Malkata pork was over two years old. Modern farmers tend to kill their animals before they reach the age of two. The optimum slaughtering age for a sheep or goat is eighteen months (Dahl and Hjort 1976: 207). The royal funerary offerings consisting of sheep/goat have unfortunately not been examined unbandaged, but the predynastic offerings from Nag-ed-Deir contain the remains of several immature animals (details of the author's study is laid out in Appendix II: **R45-65**). Most were under eight months of age, while some were between nine and fifteen months old.

It seems that animals were killed at a variety of ages. This might have been due to necessity, or have been a symbol of status (see Chapter X).

Fig. 11

II.2.1 Aves: slaughter & plucking

According to two-dimensional representations the Egyptians did not slaughter poultry as they did mammals. Representations show that, once the bird was captured (if wild) by whatever means (net, throwstick, trap), it was killed by strangulation[10] (Fig. 12). With the exception of one relief showing Akhenaten strangling and offering a duck to the Aten (MMA 1985.328.2), all representations of bird strangulation are domestic or quotidial in nature, although ducks and geese appear as offerings for both dead humans and deities. There is symbolic value ascribed to the killing of a duck as birds represent chaos and the desert. The title of an Old Kingdom butcher emphasizes the connection between the bird, the desert and Seth: "The Master Butcher of the Slaughterhouse of the Palace, He who is Privy to the Secret of the Darkness, Who Opens It (the darkness) in Strangling the Bird of the Desert Regions" (Fischer 1978: 57), a connection that is seized upon in the Ptolemaic and later, as seen in the Edfu texts (Fischer 1978: 57). However, the scenes showing the preparation and killing of poultry are primarily representations of everyday activities. All the two-dimensional representations of this scene come from tombs and not temples. It is interesting that pigeons, doves and, in fact, any birds other than geese and ducks are never shown being killed and consumed although they were eaten[11].

As described above, a man, seated on the ground or on a stool (a New Kingdom development), twists the neck of ducks and geese to kill them (Fig. 12). It appears that though for mammals it was necessary to empty the body of blood, this was not the case for poultry, although wringing the neck of the bird causes the arteries and veins to become clogged with blood. This is unheard of in modern Egypt where the bird's throat is cut to make sure it bleeds. According to poulterers this is not only due to Islamic dietary laws, but also because they fear disease in the blood that would remain in the bird if it is strangled. This non-bleeding is practiced nowadays in many Western countries when game birds (which include

[10] If the bird was killed by an arrow or any other weapon, then it would probably just be eviscerated and plucked without anything being done to the neck. This contrasts with the treatment of hunted mammals.

[11] Pigeons and doves have been found as mummified offerings (see Appendix II) and commonly appear in the offering-lists. Quails have also been found as tomb fare and are also listed by Herodotus (II:77) as food (Godley 1946: 365). The force feeding of cranes and other birds is also illustrated in several tombs, notably that of Ti (Épron & Daumas 1939: pl. 8).

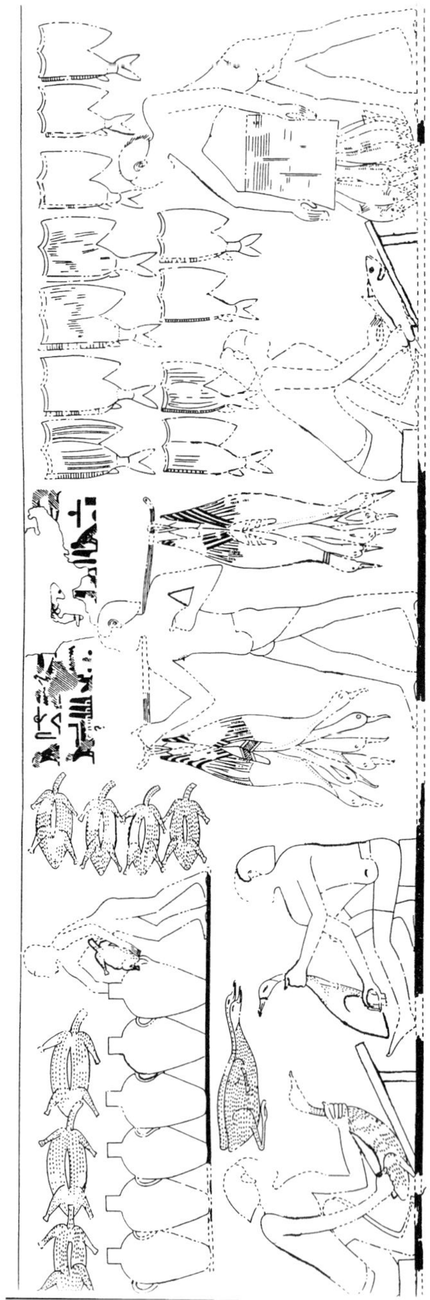

Fig. 12

duck and geese) are either shot or strangled and then hung up 'to age and settle'[12] (Fig. 13); these birds though, are not eviscerated[13]. This can be quite dangerous if the birds are not cleaned thoroughly prior to eating and if they are not well cooked as "ducks … killed without bleeding should be cooked the same day (or well cleaned and thoroughly cooked prior to consumption) to avoid the development of dangerous toxins" (Montagné and Gottschalk 1961: 357). Egyptian poultry did have its internal organs removed as the birds shown hanging up after being killed, or on offering-tables, have slits in their ventral surfaces whence the viscera was removed. Immediate gutting would be necessary for birds in a hot climate as otherwise the meat would putrefy and become a breeding-ground for bacteria which would have been difficult to dislodge, even by cooking. In the hot climate of Egypt it is possible to leave a bird for up to six hours (depending on the weather) after death without doing

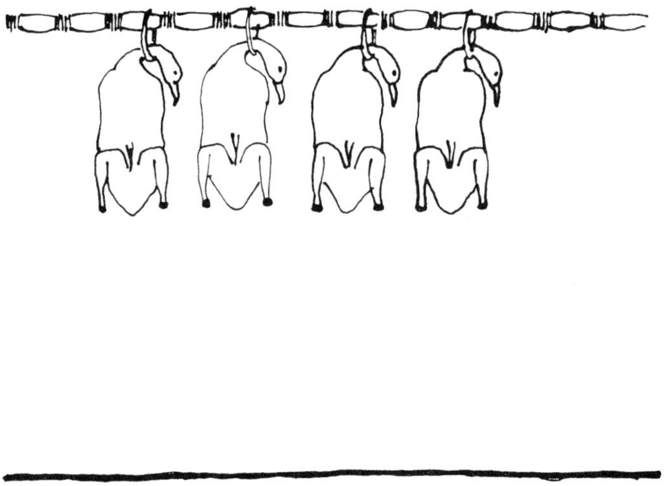

Fig. 13

[12] It is advised that mature game birds are hung up for six to seven days, and between three to four days if they are young. Pigeons and quail are eaten fresh (Hume & Downes 1963: 254, 263).

[13] In some European countries the viscera is left in the bird's body after it is dead, the neck is tied with string and the body of the bird is beaten with sticks. This releases the enzymes and microorganisms in the intestines which tenderizes the meat and gives it a distinct flavour. Some time later (one to two days) the viscera is removed from the body through a small slit in the ventral side of the bird, and the meat is consumed after being well cooked (Heath 1990).

anything to it in terms of cooking or processing as long as it has been eviscerated. After that it becomes smelly and the chances of bacterial infection are very high (Heath 1990). The actual act of evisceration is not shown — there is occasionally some action on a table with a knife and the bird, but the only visual evidence of evisceration is the slit shown in the ventral surface after the bird has been killed.

However, the birds that were put into tombs as food-offerings (Appendix II), do not seem to have been killed by strangulation. S. M. Goodman who studied these birds (ducks, geese, pigeons) writes: "One might expect to find hematoma associated with strangulation. This was not the case with the birds I examined. In a few examples there was a considerable concentration of blood in the chest cavity, which in my experience seems characteristic of slaughter by having the throat cut" (Goodman 1990). Perhaps Dr. Goodman is mistaken in his conclusion and the bird was strangled and then beheaded immediately, as for consumption, thus causing the concentration of blood in the chest cavity. Alternatively, perhaps because of some ritual requirement the birds may have been killed with a knife, unlike those shown in the reliefs[14]. If this was indeed the case, there are no representations to illustrate the process. Unfortunately, it is unknown if the sacred species mummified in the Late Period, e.g. ibises and hawks, met their end by strangulation.

Poultry remains buried as food-offerings in tombs show that although the birds were eviscerated, most of the internal organs were replaced after the interior cavity had been cleaned; a similar mode of presentation to that of modern grocery stores where kidneys, livers, and the heart packaged separately within the body cavity. Whether this was because some of the innards, such as the liver, were edible, or because they were 'mummified' as food-offerings for the deceased and therefore not treated as birds consumed in daily life, or because this was how the Egyptians would routinely 'package' birds for market, as is the case in modern poultry processing is, unfortunately, impossible to determine. Certainly cooked (roasted or boiled) birds would have had empty internal cavities, and processed birds show ventral slits indicating that some, if not all, the viscera were removed.

After the neck of the bird was wrung, most reliefs show that the bird was plucked (*ḫ3d* [Montet 1925: 66]) before being processed further. The head poulterer strangles and eviscerates the birds, while an assistant sits nearby and plucks them. The modern practice of immersing the

[14] This would be true for any religion following the Mosaic laws, but if this were the case, why would the illustration show strangulation even in the case of the Aten offering?

feathered bird in a pot of boiling or hot water to facilitate plucking was either not practiced in the interest of preserving fuel, or was simply not depicted. After plucking, the feet, wing tips, and sometimes the heads of the birds were frequently cut off. This is consistent with the birds studied and listed in Appendix II and the work of earlier scholars (Wilkinson 1837: 173). However, it is not an immutable rule, as two of the roasted spatch-cocked ducks (see Chapter VII.3) in the Louvre still preserve their feet, whilst they are without both head and wing tips. This is also the case for the birds in the amphora from the tomb of Kha (TT8; **R19**) in Turin. Such discrepancies can, however, be attributed to carelessness and a desire for speedy procurement of funerary arrangements; in the ideal world of the afterlife such omissions would not be expected to occur, as they never do so in the pictorial representation where the feet are always severed. Indeed, in many of the representations the head is shown as being intact after the feet and wing-tips have been removed.

II.3.1 Pisces: killing & scaling

Evidence of catching and killing fish comes mainly from two-dimensional representations that commence in Predynastic times (Table III), and achieve their traditional form by the time of Snefru (Montet 1925: 441). These scenes show that, for most fish, once they were caught (by net, trap, or angling) there was very little to do in the way of 'slaughtering'. They were generally pulled ashore/aboard the boat, and left to die by asphixiation. Occasionally when a large fish was captured it was hit over the head to stun/kill it — a fine three-dimensional example illustrating this action is shown in the Meketre boating model (MMA20.3.6), and two-dimensional example illustrating the same action are seen in the mastaba of Seshemnefer in Giza (Junker 1953: 139, fig. 60), and the Fifth Dynasty mastaba of Idut at Sakkara (Fig. 4).

There seems to be no symbolic meaning to the group-fishing scenes unless it is a generic symbol of human control over the forces of nature and chaos, as well as the permanent supply of food for the hereafter. The ubiquitous scenes showing the tomb owner spearing fish are not being considered here as they do have a variety of symbolic meanings (Gamer-Wallert 1970: 124-31). The fact that fish, fishing, and fish-processing for consumption are depicted shows that fish were not considered taboo. This idea is further supported by examples of actual fish recovered from funerary contexts (see above).

After the fish was caught, "it was laid, belly down, on a block or a flat, sloped board. It was then cut in a downward stroke along the vertebral column, after which the viscera was removed...the head and vertebral column...was often left intact. In some cases the fish was opened from the belly instead of down the back" (Brewer and Friedman 1989: 12). The fish were left for further processing once they had been slit and emptied of their viscera. Frequently scenes that show the evisceration of fish include the removal and processing of the roes (mainly from *Mugil cephalus)* as a part of the scene (Fig. 48). The fish do not appear to have been scaled, although this is difficult to determine from the representations. Certainly the dried fish found as funerary offerings in Deir el-Medina (JE315; JE44038; Bruyère 1937: 107) and in Kha's tomb (Schiaparelli 1927: 159-160) and other funeral contexts (BMEA36191) were un-scaled, although dried and perhaps salted. It is possible that some fish were scaled and that this part of the processing of the fish did not figure in the representations[15].

[15] Fish were probably generally scaled as fish scales have been found in middens and other places on archaeological sites.

III

TOOLS & FURNITURE

III.1 Tools

Butchers, poulterers and fishermen all had one implement in common: the knife. Tool-kits used by butchers were basic, consisting of only a knife, a knife-sharpener and perhaps a vessel containing water for ablution. Poulterers are represented only with knives although they probably had sharpeners as well. Fishermen had knives similar to those of butchers and poulterers, with an additional knife-type that might be better called a scraper.

III.1.1 Knives & scrapers

According to all two and three dimensional representations of butchery the butcher despatched the animal with a knife. The shape and the material of this knife, however, changed over time. The earliest known knives were of flint. In Sakkara and Helwan, flint knives were thrown into burials on top of food offering from the Predynastic and Early Dynastic period (Saad 1951: pl. vii.a; Emery 1949: 98, pl. 41B). Flint knives were closely followed by copper knives from the First dynasty onward (Helck 1982: 110-13), which were also used for poultry and fish, although the butchery knives showed more variety in shape.

The shapes of the different types of flint knives from the Predynastic until the New Kingdom are shown in (Fig. 14). Unless flint knives come from an excavated context, precise dating is virtually impossible: the following information has been derived from knives that are part of museum collections or occur in excavation reports, and whose origins are fairly certain.

The most elegant and beautiful knives date from the Predynastic and Early dynastic periods. There are a few different shapes of knives dating from these period, but the ones that are most typical are carefully flaked, with slightly curving backs, rounded or pointed tips and gently curving blades (Fig. 14C, G, I); these quite heavy, very sharp knives would need

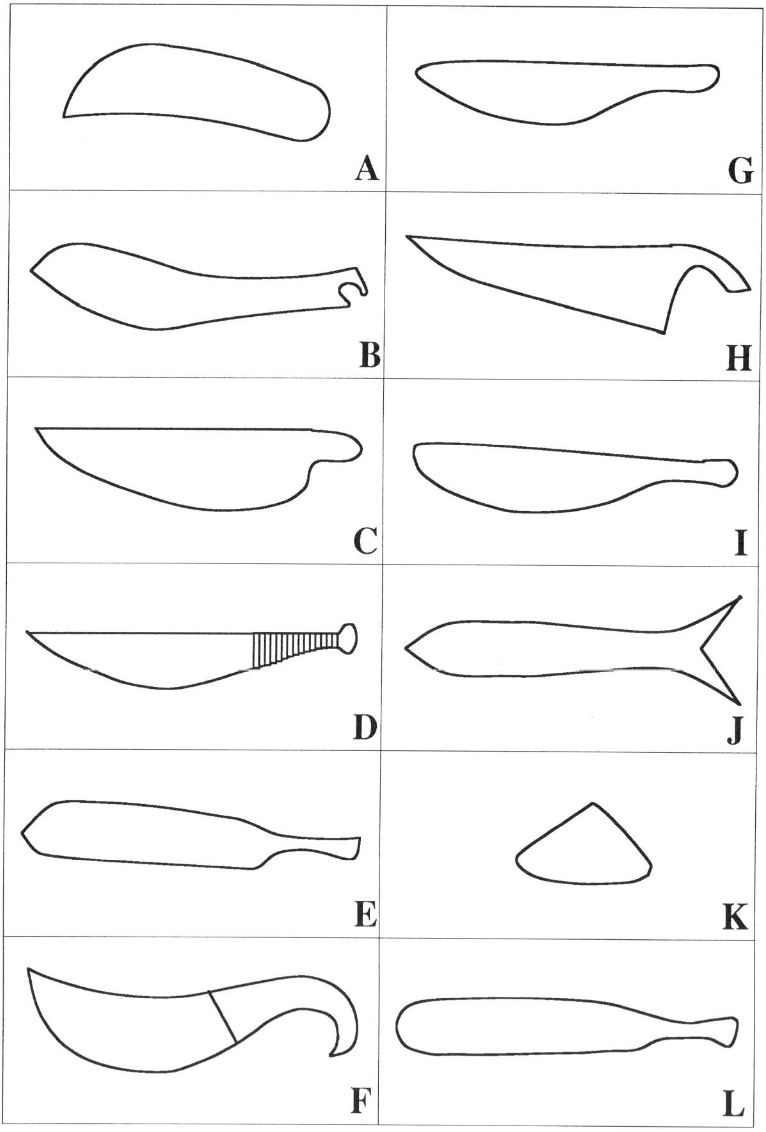

Fig. 14

a firm and steady hand to wield them. The Old Kingdom knives are smaller and less beautifully worked than those found in the Predynastic. The majority have straight backs and pointed tips (Reisner 1931: 230, pl. 18), and are less commonly found in tombs than those of the earlier period. The handles of knives dating from all these periods are made of the same piece of stone as the blade; there is no evidence of a leather or other organic binding in the handle area. Other knives that might have been used in butchery are the "Gebel-Arak" type, and the "Fish-tail" knife (Figs. 14J).

There are fewer examples of Middle Kingdom knives than those of the Old Kingdom. They certainly continued to exist as their manufacture has been represented in the tombs of Beni Hasan (Griffith 1896: 32-38, pl. IX.4 & X.2), and some physical examples have also been recovered (Garstang 1907; Petrie 1891: pl. 13.6; Petrie 1890: pl. 16). There are two very common types: one straight backed with a slightly rounded tip and a sharply curving blade with handles wrapped by leather thongs, the other straight backed with a curving blade and no differentiated area for the handle. It is possible that the extremely large knives pictured in tombs and found were used to make the first cut along the animal's neck and that smaller blades were used for the rest of the procedure, much as is the case today. Some experimental work done on flint knives by P. Kelterborn indicates that most of the knives can be used as butchers' knives, with the exception of a knife similar to the first type described above, but having a thin handle. According to Kelterborn "the thin handle is out of proportion with the sturdiness of the blade. This would indicate purely ceremonial use" (1990: 601).

Fish and poultry would have been eviscerated using smaller knives, although they would have been of the same shape as the ones commonly used in butchery (Fig. 14C, D, I, J). In representations, the shape of the knife used by poulterers remained consistent throughout Egyptian history. It had a straight back and a curved cutting edge: the handle could have been leather bound for a firm grip. Fish were cut with two sorts of knives. One is triangular in shape and could conceivably have been used to scrape away the scales as well as to gut the fish (Fig. 14K). The other is more usual in shape (Fig. 14C, D, I, J), and quite possibly made of flint even in the New Kingdom where both metal and stone tools would have been available (see below). It is possible that the New Kingdom representations showing what appear to be flint knives merely follow tradition (Lythgoe 1922: 64). For fish-gutting, flint knives might have been preferred to metal knives since water would not damage them as it would metal, although, brass and copper alloys do not rust in the same way as steel. Fish processing certainly would involve more contact with water than mammal butchery, and it is possible that a flint implement would be of more use than a metal one. Also, fish was a lower status, easy-access food which might not have been considered to 'merit' the expense of a metal tool, and if the fishermen were independent workers then it is highly likely that the cheaper, more accessible (generally) flint would have been preferred to metal.

Further evidence for knife-use comes from a method of testing flint (and in the future, possibly metal) knives in museum collections. M. Christensen and P. Walter of the Louvre have tested forty-six Predynastic Egyptian knives, most from a funerary context, from the Museum of National Antiquities at Saint-Germain en Laye, using SEM coupled with a Si(Li) detector, along with macroscopic and microscopic examination. This technique identifies the chemical composition of the polish and residue of the worked material on the knife, thereby identifying the substance that the knife was used to cut (see Christensen and Walter 1991 and Christensen, Walter and Menu 1992 for a full explanation of the techniques used). Their research has shown that several of the knives in the collection, notably the "Gebel-Arak" and "Fish-Tail" types (Fig. 14J), had been used to cut meat (Rebeyrol 1992; Christensen and Walter 1991: 168). This evidence would indicate that knives (and perhaps other objects) found in a funerary context were actually used and not just produced for the grave. It is especially interesting that the tail portion of the "fish-tail" type of knife was used to cut meat as this knife has been thought to have been more of a ritual rather than a practical object. This scientific method used by Christensen is a very useful way in which to establish a typology for use. Hopefully more examples of knives with greater varieties of shapes will be examined so that such a typology can be determined. It is interesting that so far no examples of the pictured butchery knives (Fig. 14C, D, G, H, I) have shown any evidence of use-wear, although they are heavily patinated (Christensen and Walter 1991). This confirms Kelterborn's belief that the knife was impractically shaped for butchery. More recently Christensen has examined bladelets of flint which show evidence of having been used to cut meat (Christensen 1992).

Flint was also found in New Kingdom contexts, although to a lesser degree than before (Petrie 1890: pl. 16; Petrie 1891; Frankfort 1927). The flint knives are smaller than those of preceding eras and less elegant, although equally sharp. The most common shape has a straight back and a curved blade, and is pointed at the end. Thus, mainly due to the paucity of flint found in New Kingdom contexts it is thought that by this time flint knives had probably been replaced by copper knives (Griffith 1896: 32-38, pl. IX.4 and X.2; Eggebrecht 1973: 115), or by alloyed (Tin, bronze, etc) copper knives (Ogden 1992).

The evolution of the shape of metal knives is harder to determine than that of flint as very few knives come from an excavated context, with the exception of the finds from Amarna (Frankfort and Pendlebury 1933: pl. XLVII; Pendlebury 1951: pl. LXXVI.8-10). Perhaps this is due to the fact that metal could be melted down and reused, and was considered a valuable commodity that was removed from graves and reused. The knives that do survive are generally daggers or surgical tools rather than knives used in butchery. From the remaining examples metal knives do not appear to have been very varied in shape, some being identical in shape to their flint counterparts (Petrie 1891; Petrie 1890). There seem to be two basic shapes with minor variations: the straight backed with a curved blade, rounded tip, and wrapped handle, and the pointed and blunt leaf-shaped (Fig. 14D, E, G, L). The leaf-knife appears in butchery contexts in the New Kingdom (Tylor 1894), while the former could have been present prior to that, but is not distinguishable from the flint knives due to its similarity in shape. Colour might clarify the representation, as would the presence of a flint knife sharpener (see below).

Another type of metal knife is thought to have been used in meat production: a meat chopper (Pendlebury 1951: 107; pl. LXXVI.8). The faunal remains from Malkata (notably the vertebrae) provide clear evidence that meat and bone was sometimes chopped through. This can be done by Pendlebury's meat chopper as well as by a large sharp knife or an axe, both of which were available to the Egyptians.

There has been some debate over whether butchery knives were of metal or flint throughout Egyptian history, and not just in the New Kingdom. Practically it would have been possible to use both types of knives in butchery. It is not necessary to assume that one technology superseded the other: the flint knives were easy to make, cheap, and readily available, so they probably coexisted with metal knives as there was no need to abandon an efficient type of tool and technology even if an additional technology and material (metal) became available. The author has found through experimental work using flint and stainless steel knives for skinning and dismembering horses and wolves that a flint blade is more efficient and easier to use for butchery and flaying than a metal one as it does not lose its cutting edge as quickly and therefore does not require as frequent sharpening. Furthermore, sharpening a metal knife in antiquity would require using a sandstone whetstone, and would have been more difficult to do than retouching a flint blade (see below for retouchers)[1]. Jointing animals and cutting off pieces of meat is very easy with flint blades and requires virtually no retouching while even for these activities modern metal knives need frequent resharpening.

Further experimental work was performed by the author using a replica of an ancient Egyptian knife, kindly made by G. Killen. The knife is of cartridge brass, a combination of 70% copper to 30% zinc, a combination that is not identical to the metal composition of ancient Egyptian knives (they were generally of a metal similar to tin bronze, an alloy of copper and tin), but similar, although softer. The cutting-edge of the knife was hardened by hammering, and then the sharp edge was honed. Shin meat, which is quite tough, was cut using this knife. The knife was also used to make cuts on a rib. The same cutting was done using stainless steel knives, and small flint blades. Initially the metal knife cut very well and proved to be more efficient than stainless steel knives and even the flint. However, it did become less sharp after use. The flint blades, which were of ancient Egyptian origin, kindly donated by Dr. D. Phillipson, were quite small, and had not been touched since they were excavated, which would account for the fact that they were not as sharp as the metal knife. Furthermore, they were quite small, and due to their size were not overly efficient when cutting thick wedges of meat as they would become buried in the meat. They were, however, sharp enough to cut the meat, and were very effective when cutting smaller pieces of meat. When I have experimented with freshly flaked blades the flint has been very sharp. The flints and the Egyptian replica knife were all sharper than the stainless steel knife used; the latter became dull much faster than the other two. Cutting through bone with all three types of knife was difficult, although it is possible that I did not have the strength that this operation might require. This experiment showed that the metal knives the Egyptians had were quite adequate for the task at hand, but would require some sharpening as the beast was dismembered. Providing the flint blades were of a reasonable size they could efficiently cut through the meat, and tend to require less sharpening

[1] It has also been suggested that metal knives require hammering and pounding (Ogden 1992) to be sharpened and strengthened; these two activities would take time and energy that a butcher could ill afford whilst he was in the middle of slaughter. This could, however, be done prior to the start of butchery, as was done by Killen on the knife that he made.

than the metal knives. Both types of knife are not very efficient for cutting through bone and hide.

The absence of archaeological material has helped fuel the metal versus flint debate. Until recently archaeological evidence — or rather the lack thereof — has supported the idea that flint knives were only used in the Old and [early] Middle Kingdom. This lack could merely reflect excavation method and the excavator's preferences in what is included in any publication, or be a result of the types of site excavated. Tombs of the New Kingdom, unlike their Old and Middle Kingdom counterparts, do not contain flint knives as part of their funerary equipment. In fact, they rarely contain domestic knives at all; bronze daggers and more elaborate knives (i.e., rarer and more valuable higher status items) are found in the tombs. This has contributed to the erroneous belief that by the New Kingdom flint was almost entirely replaced by metal. However, recent excavations of settlement sites (e.g. Memphis [Jeffreys 1986: 7], Tell el-Daba [Bietak unpub.]) have recovered a great number of flint artefacts, showing that even into the the New Kingdom many more flint tools were used in households than was previously supposed[2]. Also, tools used to cut up pieces of meat did not necessarily have to look like the knives — they just had to have a sharp cutting edge. Many flints following this description have been found in settlements, especially those of the New Kingdom when flint had become a predominantly utilitarian material having been displaced by metal for most high status uses.

This supposition contradicts what other scholars have written about the use of flint in the post-Old Kingdom eras. It is believed that after the Old Kingdom all butchery was performed using metal knives except in the case of religious ritual. It is possible that flint knives, due to their antiquity and association with the earliest religious sacrifices, had ceremonial and religious meaning (Eggebrecht 1973: 115) which ensured their continuing use into the New Kingdom as they provided a symbolic return to the time of the ritual's initiation (Eggebrecht 1973: 115; Garstang 1907: 107). Tomb reliefs do not support this dichotomy between sacred and secular, although examples of clearly secular butchery (e.g. goats being slaughtered in the fields as seen in the tombs of Meir and Beni Hasan) in the Middle Kingdom clearly show the use of flint knives,

[2] Earlier excavations did find evidence of flint in the New Kingdom as Frankfort reports that in Amarna he found "a large number of flints" (1927: 212) in conjunction with cattle bones in an area which might have been used for butchery in the house of Panehsy, the superintendent of the Aten's cattle.

while an overwhelming number of New Kingdom examples show the use of metal knives in a ceremonial context (e.g. the Opening-of-the-Mouth ceremony), thereby refuting the idea that flint was used purely for ceremonial butchery after the Old Kingdom. Tomb reliefs of the Old and Middle Kingdoms show the butcher using flint knives. They have been positively identified as flint from their colour (when the paint survives), and by the presence of the retouching tool that could only have been used for stone implements (see below) (Griffith 1896: 32-8). Furthermore, New Kingdom butchery reliefs show knives that are similar in shape and colour to those shown during earlier periods, along with metal knives which are not only a different colour (when painted) from their stone counterparts, but also a different shape — the leaf. These knives appear in examples of sacred as well as what is possibly secular butchery.

Of course, one should not discount the idea that economics had a bearing on the material chosen for a knife. Flint was almost always readily available and cheap. It was a traditional material used by the majority of Egyptians in all walks of life (for artisans, tools, weapons (Vila 1970), food production, agricultural tools) and doubtless there would have been many people who could have easily manufactured the required tool from flint. Metal was more expensive, more valuable, less readily available, and more difficult to work. It is likely that it was used more frequently by the upper classes, and possibly more for display than actual work, thereby making it a more likely candidate for ceremonial use than the humble flint.

Thus both archaeological and representational evidence supports the idea that, throughout Egyptian history, both materials were used to manufacture knives and that, as far as one can discern, both materials were used for both sacred and secular butchery.

III.1.2 Knife sharpeners

The butchers in Old and Middle Kingdom reliefs are shown with objects in the shape of the hieroglyph *sšm* tucked into their kilts or in their hands. These objects are conveniently termed knife sharpeners, frequently translated as 'whetstones' into English. The hieroglyph would be better translated flint(re-)sharpener, as it is in German (*WB* IV: 292), rather than whetstone which suggests a stone implement used to sharpen a metal tool, a possibility in the Old Kingdom, although it is far more probable that flint was used then. In the representations at Beni Hasan, as well as most Old Kingdom representations from the Memphite necropolis, they take "the form of a long rod, which is coloured black in (one painting), and green in (another) ... In the time of the Old Kingdom these sharpeners were represented as pierced at the end and furnished with a loop for suspension, as in the hieroglyph *sšm*. Such are commonly found in the paintings of that period, but there are no loops on those depicted at Beni Hasan ... Possibly the sharpener was made of greenish or black basalt; but the green colour reminds one of jade"[3] (Griffith 1898: 36). It is possible, too, that these sharpeners were made of the mid-ribs of date-palm leaves which are quite hard when fresh — this would satisfactorily explain the green colour of the sticks, though not the black (Miller 1991). Unfortunately no such sharpeners have yet been found in Egypt, which might be due to the ephemeral nature of the material from which they were made. Knives were sharpened by using the stick to strike the knife, thus re-chipping the blunted edges whenever necessary, providing a newly flaked sharp cutting edge. The legend above the scene reads: *pḏi ds*, translated as sharpening the knife, with *ds* meaning flint in addition to knife, thereby underlining the point that a flint knife was being sharpened. These representations appear only in the Old and Middle Kingdoms, thus supporting the idea that during that time flint knives were used for butchery. However, there are a few instances when such retouchers are shown in the New Kingdom. The butchers in Deir el-Baheri's South Hall of Offerings illustrate such an instance (Naville 1901: pl. CVII). Firstly, this might be a deliberate hearkening back to ancient periods when only flint was available, thereby suggesting that flint knives possessed a ritual significance (see above); secondly it might have been purely an artistic decision with no religious significance; and finally, it might indeed mean that flint had a ritual significance, but that it was only shown on the walls rather than actually used by the temple butchers.

Several scholars have called these sharpeners whetstones (e.g. Petrie 1898: pl. XII; Hassan 1944: 270). This might be due to the lack of a better name that is more illustrative of the use and material of the object. The name 'whetstone' implies that the knives used were of metal rather than flint, a fact which would considerably effect the manner and material of their construction. Metal blades can be sharpened in two ways: by

[3] There is no evidence of jade in ancient Egypt. However, there is evidence of green slate being used as a honestone (BMEA36728). It would be possible to use slate to chip and sharpen flint.

stropping on a leather strap, or by a stone. If this were the case, the sharpeners would be made of a leather strap tied to the butchers' waists, upon which the metal knife was stropped sharp. Sandstone could also be used to sharpen metal knives (Killen 1992), but it would have to be a piece that would fit into ones hand so that one would have enough pressure to sharpen the knife, something the long 'whetstones' would not do efficiently.

Objects described as whetstones or honestones have been recovered from Egypt. Several such stones have been recovered from Amarna: #507 in P.43.2 (Pendlebury 1951: 102), #82 (Ashmolean 1921.1161) in O.48.16 (Peet & Woolley 1923: 28), #514 in T.34.1, #507 in P.43.1, #216 in U.35.17, #33 from N.49.51 (Excavation Notes 1924-31), amongst others. #82, examined by the author, is triangular and 23.5cm long, and 2.0cm wide. It is made of white sandstone with a black vein running through it and black spots. Other whetstones dating to the New Kingdom have been recovered from el-Lahun and Medinet Ghurab. Two small rectangular (maybe broken?) white sandstone whetstones from the Eighteenth Dynasty tomb of Maket (Petrie 1891: 22-3; Ashmolean 1890.826 and 1890.827)[4], and three triangular white sandstone whetstones come from Medinet Gurob[5] (Ashmolean 1890.928, 1890.978, 1890.1059).

None of these whetstones showed any striations indicating use and wear under the microscope, but Ash.1890.826 was tapered at the end in a way that indicated use. All of these could have been used, certainly some of the edges and sides were worn, although there were no definite lines indicating this. Perhaps examination of these objects using the techniques perfected in the Louvre (see above) might yield some positive results. Some of the whetstones from the Ashmolean, notably 1890.826 from the tomb of Maket in Kahun and 1890.1059, 1890.928, and 1890.978 from Medinet-Ghurab, showed bits of a green substance, much like aged metal, on them, implying that they had been associated with a metal, either bronze or copper, thereby supporting their identification and use as knife sharpeners. Other hones/whetstones, dating probably from the New Kingdom, have been found in Thebes (BMEA15770 and

[4] Ash.1890.826 measures .113m in length, .020m in width, and .013 in thickness, while 1890.827 is .081X.020X.012m.

[5] Ashmolean1890.1059 (.185X.048m), dated to the reign of Seti II, is triangular and blackened by burning, Ash.1890.978 (.180X.054m), dated to the time of Ramesses II, is also triangular, and Ash.1890.928, (.125X.038m) dated to the era of Amenophis III, is burnt black and broken, but would have been triangular, like the other two.

BMEA36728. BMEA36728 also has some green marks that indicate proximity to metal. It is of metasedamite, like BMEA15770, and shows signs of use. This would argue for the use of metal knives for most activities, including butchery, in the New Kingdom. Close study of Old and Middle Kingdom butchery scenes that contain knife sharpening show a tool answering to Griffith's description, thereby identifying the depicted knives as being of flint. If this is indeed the case, then there are no pictorial examples of the sharpening of metal knives. When metal knives were used it is far more probable that they were sharpened on some convenient stone (a threshold, corner of building, or large stone) rather than by stropping as stropping is, according to many butchers, less effective, especially on larger knives. There are no knife sharpening scenes associated with poultry and fish, although they must have been sharpened in the same way as butchery knives.

III.1.3 Fish scaling dishes

The term 'Fish-dish' has been applied to the class of shallow oval fired pottery vessels dating to the Middle Kingdom, that has been found in both funerary (Petrie 1891: pl.5; Petrie 1890: pl. 13) and settlement (Jeffreys & Giddy 1989: 5, Bietak 1979: pl. 8a) contexts. The inner surface of the dish is incised with designs of fish, water plants, and occasionally animals. Generally the centre of the vessel is slightly convex. The origin of the name, "fish scaling dish" cannot be traced through the literature and is believed to have been passed into it orally (Bourriau 1991). The name of this dish-type implies that it was used to descale fish, though this seems to be neither practicable nor represented in the artistic record. It is unclear whether fish were scaled at all, as the majority found in funerary contexts (as food) preserve their scales. All representations showing the processing of fish never show a vessel involved with this practice. The only vessel in evidence might be an amphora in which to put the completely processed fish for storage. Practically it would be extremely difficult to scale a fish in a shallow dish, or in a vessel of any type, as the sides would get in the way of the strokes necessary to remove scales. Furthermore, the scales would get caught in the incisions and would be very difficult to clear out. There is no evidence of use in these dishes as the incision is still raised, frequently with jagged pieces of clay; had the dishes been used these would have been worn away. Also, if these dishes were indeed used to scale fish, why would they only appear in Twelfth Dynasty contexts? Fish were consumed throughout Egyptian history, not just in the Middle Kingdom.

There have been various theories as to other uses of the fish dish. Bourriau suggested that these dishes were a mould for bread (Bourriau 1981: #117), although she revises her opinion and returns to the definition of fish dish in a later publication (Jeffries 1988: 2; fig. 3). She has also speculated that this might have had some role in the house altar (Bourriau 1991). In a similar vein, Bietak suggests that these dishes are a type of offering bowl (Bietak 1979: pl. 8a). Petrie hypothesized that they were used for serving food (1890: 25). All the dishes examined by the author (Louvre E11267, UC18748-9, UC18746-7, UC16776-8, UC7591-2, UC18455, and UC18812, among others) show no signs of use or wear, so perhaps they were for some ritual purpose, though even then some signs of wear should appear on the incisions.

III.2 Furniture

The furniture used in processing these animals was also basic and consisted only of tables and seats. Most of the information concerning these comes from two dimensional representations, although some similar items have been recovered in the archaeological record. Due to the limited numbers of of representations of poultry (Table II) and fish (Table III) scenes only a few types of seats and tables are shown as having been used for their processing. The shapes of the furniture used in their processing is similar, perhaps due to their likeness in size, while the furniture used for mammals is not only larger, but also, in some cases, structurally different.

III.2.1 Seats

Initially, regardless of whether it was fish, bird or mammal, the Egyptians used to sit on the bare ground or reed mats while processing meat. Fishermen are shown thus seated throughout the Old and Middle Kingdoms with a few exceptions. In the tomb of Niankhnum and Khnumhotep in Sakkara, as well as that of Raemka (MMA08.201.1), the fish gutters are seated on high-backed, legless reed seats (Harpur 1987: 193, fig. 99, table 7.81; Fig. 15). These are very unusual, and though they might have supplied workers with back-supports, did not permit kneeling or crossed legs. These seats are not used in any of the other

Fig. 15

scenes showing poultry or mammalian processing. In Old Kingdom representations the poulterer is commonly seated on bare ground, and occasionally on a mat (Tomb of Ti). The mat is used in a scene that shows the strangling and plucking of a bird, not its evisceration and preservation. There are very few Middle Kingdom scenes related to the processing of poultry. That of Pepiankhrib (Blackman 1924a: pl. 8), is very similar to Ti's, and shows a man seated on a mat, plucking a bird. The other such Middle Kingdom scene in the tomb of Djhutihotep (el-Bersheh 2), shows the complete processing of a duck. Here the poulterer is seated on the ground without the benefit of the mat. Butchers start off by being thus seated and then, move to a small block that acted as a stool of a sort occasionally used by fish-eviscerators. This block continued in use throughout Egyptian history with a few other types of stools augmenting the repertoire of seats. There are also some examples from the New Kingdom which show poulterers (Rekhmire TT100) seated on such blocks.

Stools appeared in the late Middle Kingdom, where they are used primarily in the context of beef processing, and feature prominently in many New Kingdom paintings where they are used not only for beef, but poultry and fish too. Stools were three- or four-legged with square, round, or semi-circular seats. Several examples of these stools have been found, notably in Deir el-Medina (Bruyère 1934: 48, fig. 21; Killen 1980: 44-5) and el-Amarna (Peet and Woolley 1923: 62, fig. 7, pl. xiv.6,8). Three-legged stools have solid seats that are either straight or slightly concave, and are made of wood (e.g. Royal Scottish Museum, Edinburgh 1956.107, BMEA 2481) or limestone. Four-legged lattice stools belong to the New Kingdom (e.g. Egyptian Museum, Turin 8511-12, Brooklyn Museum, New York 37.45E, Nardowe Museum, Warsaw 141433) and appear in some meat-processing scenes, notably TT217 and TT79. TT217 also shows such a stool being used in bird-processing, the only such example known to the author (Fig. 16). Some elegant examples of four-legged stools come from the tombs of Kha and Tutankhamun. They appear in the New Kingdom and are made from "four straight legs of square-sectioned timber with stretchers holding the bottom of the legs firm, while the seat rails are mortised and tenoned to the top of the legs. Between the stretchers and the seat rails is the lattice of braces on all sides" (Killen 1980: 46). Four-legged stools without elaborate latticing were also made. For a thorough discussion of Egyptian furniture see Killen (1980) and Baker (1966).

III.2.2 Tables

There is some diversity in the types of tables used for processing different types of animal meat. The initial type of 'table' used consisted of a solid rectangular slab made of either wood or limestone. This type of table, equivalent to a modern butcher's block, was used throughout Egyptian history, and is frequently mentioned in different contexts throughout the Book of the Dead (called a "slaughtering-block") (Allen 1974), and is thought to have formed part of a butcher's titles (Fischer 1960). Modern butchers certainly prefer wood for meat-processing, as stone would damage the knives when meat is chopped or cut. This might also have been true for the ancient Egyptians, especially if they were using stone knives which are more brittle than their metal counterparts. Wood could have also been used for poultry processing as the legs, neck, and wingtips of the birds were chopped off. However, this would not have been the case for fish as they were merely eviscerated and not chopped in any way. Some of the boards were thicker on the side where the processor sat and tapered off, resembling a triangle with the apex

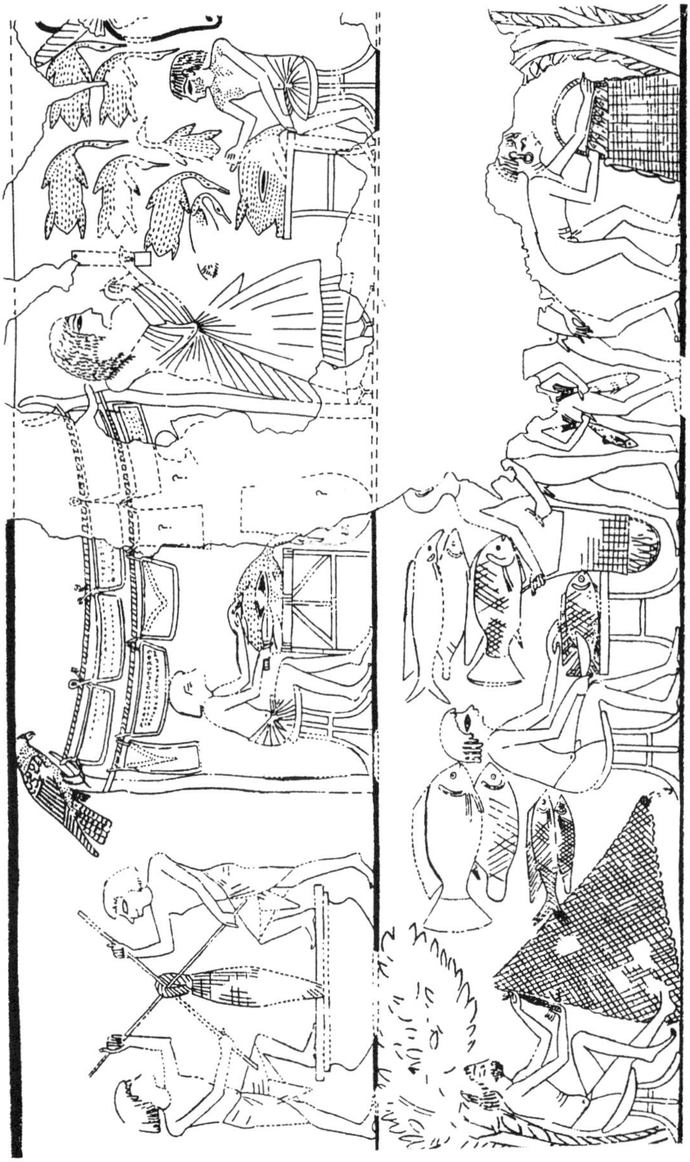

Fig. 16

pointing away from the processor. This is unlike the slant on tables used for fish and poultry which were angled down toward the worker (see below). Early representations of fish processing show that flat, rectangular slabs of wood or limestone were used. As with the meat boards, these boards/slabs acquired in some cases four legs, and in others two solid pieces of wood at either end as supports, and became low tables (e.g. from Tarkhan, Grave 527. Ashmolean 1912.603). There are several representations of both these types throughout Egyptian history, with some real examples made of wood and stone being found in Early Dynastic tombs from Tarkhan (Petrie 1913: 25).

The type of table used by the poulterers is stable throughout Egyptian history. It consisted of a sloping rectangular[6] board held up on one side by either a pair of legs, or else by a long single 'leg' (Fig. 12). This could be made of wood or even of limestone. Although both materials are practicable, the former alternative is more feasible as it is easier to fabricate and carry. An almost identical slanted table is the most commonly represented table in fish-processing scenes (Fig. 12). It differs from the poultry table only in that its legs are placed at the extreme end of the table rather than being slightly under the edge of the table. This table could have been constructed of wood or stone or even a combination of the two materials, although stone might have been preferred for fish, then as now, since the fish remain cooler and the water, blood, and scales from them are not absorbed by the table, thus preventing the spread of bacteria. A slanted table from a First Dynasty tomb from Sakkara has been found. It is of limestone, 67.94cm X 10.79cm, and has four holes where legs can be fitted, although only two legs are present (Emery 1949: 113). It might represent the simple types of tables used by everyone of that period, and although the nobility later adopted grander designs, the poulterers and fishermen continued to use this simple model. It is due to the humble nature of the work and workers that no furniture associated with them or their occupation has been found in a funerary context as no such worker's tomb (to my knowledge) has ever been discovered. If it had, it probably would have been of so humble a nature as to have passed without comment. Certainly no such furniture has as yet been found at a settlement site, so it is impossible to reconstruct precisely the type of table used.

[6] One of the ends could be rounded, although this does not appear to be so in the pictorial record.

A latticed table, similar in construction to the latticed stools, except topped with wood, is shown being used by a butcher in TT217. There are only two known examples of such a table, both from the tomb of Kha: one constructed of reed and rush, and the other of wood (Killen 1980: 66, 68). The top of the latter example would be unsuitable for work of any sort as it is made of wooden poles; any work table would have to have a solid top. There is a single example of a poulterer shown working at such a trussed table (Fig. 16).

One very unusual representation of a high table is also known from the Old Kingdom Sakkara tomb of Nebimakhti (Harpur 1987: fig. 98), where a fisherman is shown processing his catch. Unfortunately, the scene is damaged so that the whole table is not visible and it is impossible to determine whether it was a large solid block or a table with legs, or even a latticed table. The only determinable feature about it is the fact that it is quite high and requires the worker to sit on the high square seat mentioned above. It was probably made of wood. In some late New Kingdom scenes one finds fish laid out on four-legged wooden tables with the fisherman gutting the fish on a proper three-legged stool (Fig. 16). Such representations are rare, especially since this daily-life scene is not well represented in the New Kingdom (Table III).

Three-legged wooden tables, popular in Egypt from the early New Kingdom until Hellenistic times (Killen 1980: 67), were also used in meat production as is shown in the unusual representation on the Amarna talatats in the Luxor museum. An actual example of such a table can be found in the British Museum (BMEA 2469), which is reported to be extremely stable and strong enough to support the weight of a small ox despite only having three legs (Killen 1992). This particular example shows no evidence of ever having been used for meat processing as the top is unscored and smooth. Other tables pictured in scenes of meat-processing might have had either three or four legs — it is difficult to determine which due to the Egyptian conventions of representation. It is more likely that these are three-legged tables as the few four-legged tables known from Egypt have stretchers between the legs (e.g. from the tomb of Kha, now in the Egyptian Museum, Turin 8258), although the one pictured in the tomb of Menkheperresoneb (TT112) seems to be slanted which, although possible to manufacture and use, might not be as stable as a level tri-legged table.

IV

LOCATION

The next question to address is where the animals were slaughtered both in sacred and secular contexts. Information concerning slaughter-houses or/and slaughter-yards used for mammals, birds and fish comes from two- and three- dimensional representations as well as from excavated contexts: in the case of mammals, notably temples.

IV.1 Abattoirs and temporary structures

As ethnographic evidence and the study of contemporary slaughter areas has shown there are a few things that the slaughtering area should have for the sake of hygiene and good health regardless of whether an animal is being killed for food or for a religious reason. As fresh meat of any sort can be contaminated by soil, dust, water, manure, flies, other foods, inadequately cleaned butchery equipment, and water activity on the surface of slaughtered meat (Mossel et al. 1975: 226), the slaughter area should protect against these as much as possible. Perhaps the ancient Egyptians, like their modern counterparts, did not stress these requirements as much as Westerners today would, although modern Egyptians would be conscious of health considerations. A slaughter area should provide a separate area in which to keep the animals before killing them, so that they may be carefully tended, fed and rested before slaughter, and, more importantly, protected from the smell of blood as it makes animals nervous and traumatizes them, causing the glycogen level to fluctuate, thereby adversely affecting the quality of the meat since not only does its colour, water holding capacity, texture, and succulence all deteriorate with increased glycogen levels, but also spoilage becomes more likely (Lawrie 1976: 431). Thus, ideally, a slaughter-area should have space for air to circulate, protection from the dust, and a place where the butchers could wash themselves, their instruments, and the meat.

The amount of information derived from two-dimensional representations changes from the Old Kingdom to the New Kingdom as butchery scenes are more prevalent in tombs dating from the Old and Middle

Kingdoms. In the New Kingdom the scenes become increasingly rare and are abbreviated, with an accompanying change in the physical context of the scenes. The majority of New Kingdom (not TT217, and a few others) butchery scenes are clearly religious in nature and are specifically associated with a ceremony that presumably took place at the mouth of a tomb, probably in the forecourt that was commonly located in front of Theban tombs of the New Kingdom. By contrast, poultry and fish processing scenes were always clearly secular activities and are limited in number throughout Egyptian history: these scenes do not vary significantly with regard to architectural detail over time.

Virtually all Old Kingdom tombs, mainly from the Memphite necropolis, contain scenes of butchery. These scenes rarely show a large establishment in which animals are butchered. The only indication of an architectural setting comes when the animal is shown being jointed, after it has been killed. These seem to show the activity taking place in an enclosed area by their bordering of columns. This implies a separate enclosed area where the slaughtering and the immediate processing of the meat was performed. Such a structure also appears in scenes that depict the killing and processing of waterfowl, and can be construed as a temporary structure placed within a courtyard or any open space near where an animal was butchered, strangled, or plucked and processed. It would probably have consisted of a line and poles that were strong enough to support joints of meat suspended from it, and a separate canopy, perhaps of reeds and wood, to protect the meat against sun and dust (Fig. 17). It is possible that they just had the former without the latter. The meat-hanging line could probably be raised and lowered to allow the butchers/cooks easy access to the joints (Vandier 1964: 258-60). Frequently the meat was shown being boiled (see below) in a pot near the jointing location, with a bird being roasted on an open fire adjoining the cooking pot. An unusual and elaborate example of the poultry processing enclosure is the bower shown in TT217 (Fig. 16). Reed structures of a similar nature have been found in the courtyards of houses near the North Palace at Ballas. The excavator also believes that these structures were probably for the preparation of food, either for immediate consumption or for longer term storage. This interpretation of the reed structures is supported not only by the artistic evidence, but also by the presence of grain silos, ovens, and ethnographic information (Lacovara 1990: 4).

Middle Kingdom scenes showing butchery tend to have an architectural framework defining their perimeter. Most scenes show the butchery being performed in an area bounded by columns, with a meat-hanging line

Fig. 17

suspended between the pillars and being covered by a light roof of reeds and wood, very similar to the ones discussed above.

As mentioned above, New Kingdom butchery scenes are fewer than in preceding periods. The majority are concerned with the sacrifice in front of the mummy in the Opening-of-the-Mouth-Ceremony when the foreleg is offered to the mummy after its mouth is opened. This is perhaps a reference to the constellation the foreleg sometimes represents and the divinity it confers (Roth 1992). It may also refer to the Book of the Dead. Most of the other New Kingdom representations do not have an architectural perimeter indicated by columns. A notable exception to this comes from the tomb of Djhutnofer (TT104). The scene in question shows his house (Fig. 18). In a side building, or a part of the house removed from the main living quarters, up a series of stairs, and on what seems to be the topmost floor, is a man seated on a stool in front of a table filleting or further dividing joints of meat. A double "drying line" is suspended above his head. From it hang triangular pieces of meat as well as thin strips of biltong, the first such example before the Amarna period (see Chapter VIII.1). Another man, carrying a *ḫpš*, is ascending the stairs to the chamber. This would indicate that butchery was done within the estate and would argue that the slaughter and main jointing took place in one area, while further divisions of the joints, perhaps for preservation, took place in another, more interior space.

It has also been suggested by Yvonne Harpur that for the Old Kingdom examples, at least, the location of the scene in the tomb suggests the actual location of butchery. She writes,

> "if fishing and fowling scenes were placed near the tomb entrance on account of their outdoor setting, it is conceivable that similarly-located butchery scenes were depicted here for the same reason. Entrance thicknesses showing men leading oxen into the tomb, and interior thicknesses showing men overthrowing an ox, imply an indoor sacrifice, which no doubt took place before the false door[1], but this does not necessarily mean that the dismembering of the carcass took place here as well[2]. Perhaps out-

[1] The tethering-stone (see notes 4, 6 and 10) in the tomb of Mereruka would imply that a sacrificial beast was tethered and offered in the chapel. This would be difficult from a practical point of view as the room is pillared, and wrestling a fully grown ox to the ground and slaughtering it would require more space for manoeuvering. If it were a calf the operation might be performed more easily in the constrained space. It seems most likely that the animal was brought into the chapel, presented as an offering, and then removed outside and killed or perhaps turned over to the funerary priests to dispose of as a sacrifice at a later date.

[2] However, killing an animal and then dragging it out to dismember it is equally difficult, especially if the animal is large as many false doors with related butchery scenes are located in the interior of the mastabas, and would therefore involve considerable manoeuvering of the carcass through several doors and rooms.

Fig. 18

> door butchery was a more practical alternative, or it may relate to a different ceremony than the one in the offering room. On several entrance thicknesses at Sakkara, dating between (dynasties) V... and VI, butchery is depicted next to the statue-dragging scene. The combination of these activities suggests a sacrifice made before the statue of the deceased, prior to its passage through the chapel and enclosure within the serdab" (Harpur 1987: 57).

Thus, although the placement of such scenes within Old Kingdom mastabas might suggest an actual location of sacrifice, this is by no means certain.

Three-dimensional representations, i.e. models, also provide some information concerning butchery areas, though none shed any light the methods of processing poultry and fish. The majority of late Old Kingdom and Middle Kingdom models merely show a butchery scene involving two or three butchers, a bovid being killed, and occasionally a man busy over a bowl of what might be fat or blood (see Chapter VIII). The entire tableau is set out on a plinth. In these cases there is no indication of a tethering stone or of any architecture containing these activities. Perhaps the plinth and the lack of walls is meant to suggest that this was an open-air activity. Frequently the butchery tableau is adjacent to a brewery and/or bakery. The fact that all these activities are shown taking place on one platform suggests that either they were activities that took place in one area, or that they are grouped together for convenience as they are all domestic tableaux. In any case, even in these group tableaux, there is no sign of any architecture, which might lead one to suppose that all these activities took place outdoors. Although this is a possibility, there is a considerable amount of archaeological evidence that contradicts such a theory as several specialized brewery-bakery complexes have recently come to light (Mark Lehner's work on the Giza Plateau; Kemp's work at Amarna). However, it is conceivable, though doubtful, that the sites where these large installations have been found catered for several hundred people and therefore had a more permanent architectural setting. By contrast, smaller domestic areas for brewing and baking (most large houses must have had the latter even if not the former) might not have required such formal, permanent structures and so light reed shelters rather than large mud brick structures could have sufficed. Of course, the fact that these models did not provide an architectural setting for the activities depicted might be due to model making conventions and/or convenience and economy[3].

[3] For a discussion of models see A. Tooley (unpublished).

One set of models does not fall into the 'open-air pattern' discussed above. The most complete extant representation of a (secular) butchery area comes from the funerary equipment of Meketre (MMA 20.3.10), and to my knowledge is unique in its degree of detail and the inclusion of a slaughterhouse that was one of the structures on Meketre's estate. His slaughterhouse model shows what might be taken to be the Egyptian ideal slaughterhouse. It also shows how practical considerations regarding the hygienic butchering of animals affected the architecture (see above). The slaughterhouse is the tallest of Meketre's models: a large rectangular room, with a sloping ceiling that extends upto a clerestory-like opening over a mezzanine under which are three rooms. Gilbert takes the model to be a scale model, with the scale determined by the size of the people, and calculates its dimensions as: "height at rear 3.41 m; height at front 3.86 m; length 5.70 m; width 4.35 m; interior area (approximately) 20.0 sq. m" (Gilbert 1988: 78). However, Gilbert (1988) and Winlock's (1955) belief that the model is to scale is debatable.

The model shows butchers holding down and slaughtering a bovid and apparently collecting its blood in a bowl. It is interesting to note that there are no tethering stones in the model; indeed, none of the models examined contained tethering stones. A man on one side of the room roasts a duck/goose, and another man near the three rooms under the mezzanine, possibly storerooms for processed meat, cooks something in a bowl, that Winlock has suggested is blood (Winlock 1955: 24). These activities are observed by an overseer. The mezzanine level is clear of people and is the site for a line, very similar to the meat-hanging lines shown in Middle Kingdom wall paintings, from which hang joints of meat. This building would indeed have been an ideal place not only for butchery but also for processing meat, as the improved air circulation caused by this design, with its monopitched roof and mezzanine level, would have aided the cooling and drying of meat. Due to the building's height, and "with proper insulation, and a sloping roof channelling warm air out of the front windows, Meketre's butcher shop might have been quite cool by comparison to outside temperatures" (Gilbert 1988: 78-79). This would have delayed the spoilage and the breeze would have discouraged the settling of the ubiquitous flies. There is also an area in the building that is exposed to direct sunlight. This is very important as "sunlight (Ultra-violet light) can be an effective disinfectant as long as the area is thoroughly cleaned and is allowed to dry while exposed to direct rays from the sun" (McNitt 1983: 48), and would have disinfected the area where the butchery itself took place. A large bowl of water, possibly one of the bowls shown to the side of the building (?), as ethnographic evidence attests, would easily have provided all the water necessary for flaying the animal and cleaning knives and hands. It is also interesting to note that the painted threshold was of wood. It is quite possible that this was so not only in the model, but also in reality as the more usual stone thresholds found in other buildings would cause the animals to jar their hooves, resulting in pain, trauma, and a fluctuation in the glycogen level, thereby effecting the quality of meat adversely (Lawrie 1976: 431).

The Meketre model is one of the only extant representations of an entire slaughterhouse from ancient Egypt; its components are very similar to slaughterhouses and curing areas found in large farms or wealthy estates throughout the world and throughout history (e.g. Mount Vernon in Virginia, Wimpole Hall in Cambridge, and many properties in modern Egypt). The model is also unique in that it represents a particular individual's slaughter-yard — secular identification that is not always clear in tomb decorations and models. Most of the archaeological information concerning slaughter-yards comes from temples and is therefore clearly sacred in nature, thereby making Meketre's model all the more valuable as a source of information regarding slaughter-yards that might have been common on Egyptian estates — a part of Egyptian life about which relatively little is known. It is possible that such slaughterhouses were part of the outbuildings in the larger houses in Amarna such as that of Panehsy (Frankfort 1927: 212). Certainly evidence of external kitchens, storage areas and animal pens (Peet and Woolley 1923: 49) has been found, and it is not improbable that slaughterhouses formed a part of these complexes, but have not been identified as such.

Since butchered offerings played a major part in funerary and cult ritual, information about the slaughterhouses and slaughter-yards that existed as part of temple complexes (P. Harris I, pl.14, line 8), should provide us with more information. Indeed, some temples had a "Department of Meat" (Spalinger 1975: 232). Information collected from these sources is also important as butchering an animal would have similar sorts of requirements regardless of whether it was sacred or secular in nature. Ramesses III's Medinet Habu calendar reads: "I have made festive thy regular offerings with bread and beer, while cattle and desert game are butchered in thy slaughterhouse" (Nelson 1934: 10). Slaughter-yards in temple precincts are pictured on several tomb walls in Amarna. Furthermore, temples are more extensively and more commonly excavated than settlement sites and ought to yield considerable amounts of information regarding these areas. However, despite the large part slaughter obviously played in temples, there is a remarkable paucity of concrete evidence for slaughter areas in temple sites.

Theoretically almost all temples would have had individual slaughter areas regardless of whether or not they were dependent on other temples for the majority of their meat offerings[4], although some constraints could be caused by space and practicalities. However, there are only a limited number of temples where excavators and/or scholars have located slaughter-yards. Some of these are indeed slaughter areas, while the identification of others remains ambiguous.

The earliest temple where there is thought to be any evidence of a slaughter area is the Abydos Temple of the First Among the Westerners (Ricke and Schott 1950), dating to the Archaic Period. In this temple first excavated by Petrie (1903), Ricke and Schott (1950: 45) identify a *pr* shaped building as the temple's slaughter court. They base this identification on the thickness of the wall and the mention in the Pyramid Texts of the existence of slaughter halls. Petrie identified this area (E) as a set of store rooms rather than a butchery yard. He writes: "the great building at E was 42 feet wide inside with walls 8 feet thick…a space of 21 × 42 feet, with an entrance on the north. This must have had some supports for the roof, and probably the 42 feet was divided into three, five, or seven spaces. Both three and seven spaces imply unlikely spans, and five spaces with spans of eight feet each across the width, and three spaces with seven ft. spans in the depth, seem more suitable" (Petrie 1903: 8). This was later re-studied by Barry Kemp (1968: 152) who thought that there was insufficient evidence concerning the building (Petrie's E, his J) and that it was impossible to discern its use or date. There is really nothing in the shape or thickness of wall to determine the function of this building; even if it had divisions, an hypothesis based on minimal evidence, it does not mean that this area was for storage or for stalls where animals were kept prior to slaughter. Certainly nothing in its shape points to its use as a slaughter yard. This is not to say that it could not have been used as such — any open space is potentially a slaughter area. However there is no indication that this is one while another court-yard is not.

From the evidence of the Pyramid Texts and the presentation of food offerings in the daily *prt-ḫrw* ritual in Old Kingdom Pyramid temples, as well as references to these establishments in the Abu Sir Papyri (Posener-

[4] Although the funerary temple of Neferirkare was dependent on his Sun Temple for most of its meat offering, it contained a slaughter area of its own. See discussion in text.

Kriéger 1976) it is believed that slaughter yards were a definite part of the Old Kingdom pyramid complexes. In most temples it is hypothesized that the slaughter area was located in the main pillared court. This idea is based on the fact that the courts contained altars which have been thought to be for offerings, as well as drains which might have been used to remove the blood of the animals sacrificed and offered. Temples with such drains are Khufu's Pyramid Temple (Lauer 1947: 257-8; Ricke and Schott 1950: 45), and those of Khafre (Hölscher 1912: 56-7), Raneferef (Verner 1991), and Sahure (Borchardt 1910: 75-6). Other temples with holes rather than drains are the valley temple of Menkaure (Reisner 1932: 25-6) and Pepi II (Jéquier 1940: 23). However it is unlikely that these pillared courts were the site for sacrifices as neither the presence of altars or drains and basins is a convincing enough argument for such an identification. According to Malek (1988) these altars were points of redistribution of offerings rather than altars on which offerings were placed during a ceremony for the king. The fragility and brittleness of their material, calcite, would certainly preclude their use as cutting block. Furthermore, there is no reason for the slaughter to be effected before the altar — certainly if this act were performed as part of the ritual it would be very bloody and undignified and make further activity before the altar a messy affair. The drains which frequently flow into basins have been thought to be used not only for blood removal, but to remove water from embalming installations (Hassan 1943: 69-102), or as a symbolic attempt to imitate a sacred lake (Malek 1988: 32). It is more likely that these drains and basins were connected with the removal and drainage of rainwater rather than blood (Malek 1988: 32), especially since the blood of slaughtered animals tends to splatter and not flow in a regular manner that allows for effective drainage. One could argue that water was sloshed on the floor to aid the drainage of blood, but there is no support for this theory. There is a question as to where the water came from, and if it would be sufficient to warrant such an elaborate drainage system. Unless several animals were being sacrificed simultaneously (see below) there would not be a great deal of blood and drains would be superfluous[5]. More to the point, in the Fifth Dynasty the temple documents of

[5] Ethnographic work has shown that the blood for a calf does not fill half a *tisht* (a metal dish) — only approximately three to four pints flow from it after slaughter and during butchery; an adult would yield more. This is because "animals bled while lying flat...would drain less efficiently as a result of poor gravity flow" (Gilbert 1988: 80). Potentially 16-17kg of fresh blood can be removed from a steer (Gilbert 1988: 80), depending on the weight of the steer as blood constitutes 10% of the total body weight (Jackson 1992).

Neferirkare (Posener-Kriéger 1976: 507-8) clearly show that the abattoir was separate from the pillared court (*wsḫt*) which was used for offerings (Spencer 1984; Eggebrecht 1973: 124-8).

However, the temple archives do suppose that an abattoir existed in funerary temples. In the temple of Neferirkare it is identified as Room 11, a long semi-covered columned room to the north of the sanctuary with a separate entrance to the outside of the temple proper (Posener-Kriéger 1976: 507). This would make a better slaughter area as it is convenient to the altars, sanctuaries and other offering/redistribution areas, it is not directly located where blood would inconvenience performers of rituals, and it provides a separate entrance for the animal. It is interesting to note that it's axis is not the same as that of the main temple, thereby lessening its religious focus. However, in this temple the abattoir was not well used as most of the meat was supposed to come from the sun temple of this king, although festival slaughter and slaughter of fowl were supposed to take place in the funerary temple (Posener-Kriéger 1976: 519). In non-festival times room 11 supposedly served as a depot for the meat from the sun temple. It is possible that further dismembering of the carcass was done here. The only alternative location for the abattoir would be in a mud brick structure outside the temple proper — a structure that has not appeared in the archaeological record, although another one, from the funerary temple of Raneferef (see below) has been found.

Borchardt agrees that Old Kingdom funerary temples had slaughter yards, and in the case of the Pyramid Temple of Niuserre he locates the area not in the broad or pillared hall, but behind the side entrance to the complex, conveniently placed for the offering areas, with a separate entrance, and decorated by scenes of butchery. It is a small hall with a stone floor, its walls decorated with scenes of animals being slaughtered (1907: 70, figs. 49 and 56). He bases the identification on the wall reliefs (see discussion below) and the parallel example of the similarly situated chamber in Neferirkare's funerary temple. Although this room is a possible candidate for a slaughter area, the room seems too small for actual butchery, especially as the doorway is quite narrow and would be tight for an ox. Jointing however would have been feasible here.

The funerary temple of Raneferef contains what its excavator, Verner, has identified as a slaughterhouse. Not only does the architecture argue for this identification, but the presence of embedded tethering-stones and the quantity of animal bones (mainly cattle), along with flint knives and flakes support it (Fig. 19). According to Verner, this rectangular building was constructed on a north-south axis, its entrance (1.5m) on the north,

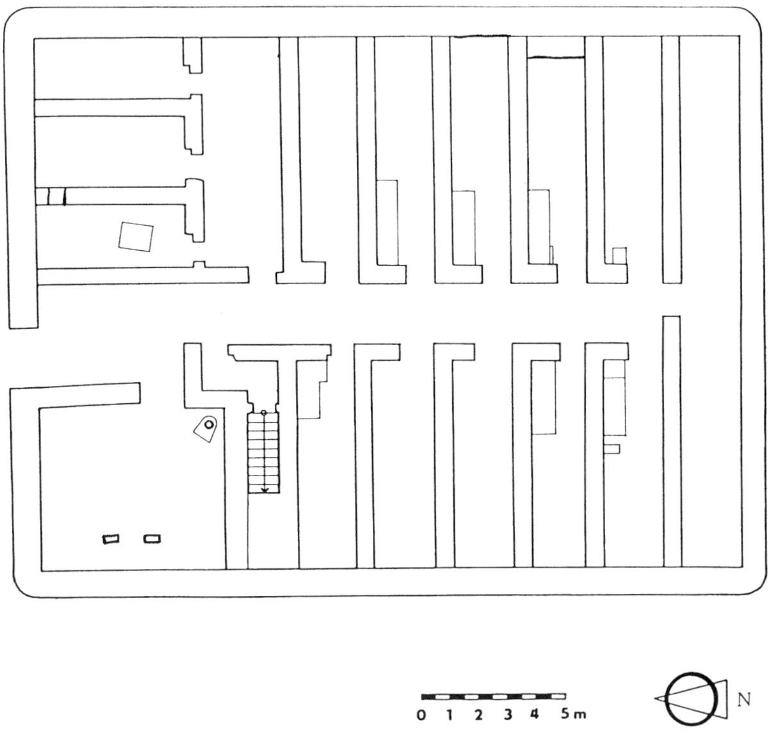

Fig. 19

with an area of 15 × 27 m. Depending on windows and doors, this axis would permit the wind from the north to blow through the complex, removing foul smells and blowing away flies. The outer face of its enclosure wall (1m thick), and all four outer corners of the structure were rounded, and the wall was plastered and white washed, as were all the rooms. The rooms had irregular clay floors. The building was of two stories with butchery on the ground floor and (presumably) the drying and processing of meat occurring on the upper floor. The main butchery chamber was in the north-west corner where two cone-shaped tethering-stones were found embedded in the floor along the west wall, and one was found nearby[6]. This room was unroofed — as mentioned above this was helpful and hygienic as the air circulation would remove the smell

[6] Initially identified as an anchor by A. Nibbi (1979: 44), an opinion she has subsequently altered.

of animal excrement and blood (Verner 1986: 181), and the ultra-violet light would have an antiseptic effect. The three rooms in the north-east corner of the building, one of which contains a mud brick plastered and whitewashed butcher's block which has now lost its wooden top, were where the joints were cut. Most of the primary jointing must have been done in the slaughter-room as the doors of the three north-eastern chambers are too narrow to permit the whole body of an ox (Verner 1986: 182) to pass through; these chambers show signs of burning and fireplaces, thus it seems that they were not only used to joint the animal, but also to cook it (Verner 1986: 183), whether for immediate consumption or for long-term preservation. The other rooms in this complex might have been used for storage, processing, or other food processing activities; it is doubtful that they were used as temporary stalls for the animals as there are no tethering-stones and the doorways are narrow.

This slaughterhouse compares quite favourably with that of Meketre. It has different areas for slaughter, jointing, drying and storing of the meat. It also has several of the practical attributes, like exposure to light and air, mud floors, wide entrances, and tethering-stones associated with slaughter areas.

Cult temples of the Old Kingdom also boast the presence of slaughter yards. Out of the six Sun Temples known to exist, those of Userkaf, Neferirkare and Niuserre are known to have contained slaughter areas; nothing can be said about the other three as there is no extant evidence. It is believed, mainly because of textual evidence, that the brick temple of Userkaf contained a slaughterhouse (Posener-Kriéger 1976: 523), although the archaeological evidence is too scanty to support or disprove the belief. More substantial evidence comes from the sun temple of Niuserre where Borchardt claims to have found two slaughter courts. One is on an east-west axis in the court north of the altar, and the other is in the north-west corner of the enclosure (Fig. 20).

The larger long and rectangular slaughter area in the north-east is on a slightly raised platform surrounded by a balustrade with a gap for an access ramp via which the animals were led to their doom. The floor is of stone which is cut by shallow runnels, supposedly to carry away the blood of the slaughtered animals or any cleansing water (see above). This was deposited in a series of ten circular calcite basins set into the east end of the pavement. There is no outlet for the liquid, so it was presumably removed manually. There are no tethering-stones (von Bissing 1905: 46-8), although, if the models are any indication of reality, they are not necessary in a slaughter yard. The smaller north-western butchery

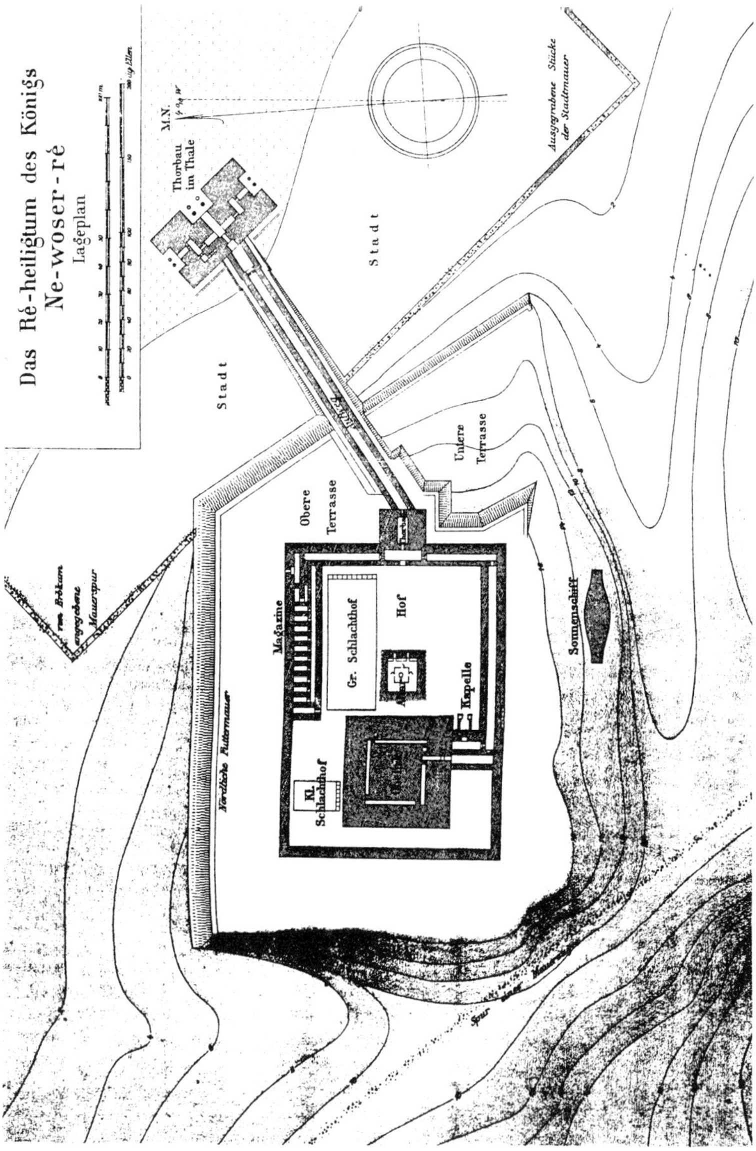
Das Rê-heiligtum des Königs
Ne-woser-rê
Lageplan
Thorbau
im Thale
M.N.
Stadt
Stadt
Ausgegrabene Stücke
der Stadtmauer
Obere
Terrasse
Untere
Terrasse
Magazine
Gr. Schlachthof
Hof
Kapelle
Kl.
Schlachthof
Sonnenschiff
Nördliche Futtermauer

Fig. 20

area with seven basins was similar to the larger with a small difference in that the basins are fed by *three* holes, not one, each at the termination of a runnel (von Bissing 1905: 51). In light of the evidence from the funerary temple of Raneferef it is difficult to believe that these buildings were slaughter courts. As discussed above, the presence of drains and basins is not a convincing argument for identifying these two areas as slaughter-courts. The practical problems of emptying the basins of blood and water also arise. The fact that the building was of stainable stone rather than mud brick, and the lack of tethering-stones, argue that these structures were used for some other cult rituals. If this area was, indeed, associated with butchery, it is possible that meat was consecrated here or that the structures acted as symbolic slaughterhouses where meat was displayed and cleaned, but there is no reason to assume that these buildings were the site of the actual slaughter (Eggebrecht 1976: 129ff). The orientation of these "abattoirs" is noteworthy in that, like most of the others discussed so far, it lies north-south, while that of the main temple is east-west, perhaps to take advantage of the northerly wind to provide ventilation.

This is not to say that the sun temple had no slaughter court. Written evidence from the funerary temple of Neferirkare states that most of the meat offerings came from his sun temple and were further redistributed from his funerary temple (Posener-Kriéger 1976: 519), thereby arguing that the sun-temples, or at least this particular one, contained abattoirs. It is quite probable that the sun temple abattoirs, like that of Raneferef, were of mud brick and consequently have not been found as they have been destroyed or perhaps ignored. Unfortunately, there are no extant remains of Neferirkare's temple so it is impossible to investigate this avenue further.

Evidence concerning slaughter-courts from Middle Kingdom temples is scarce. This is mainly because very few large monuments of the period remain. The temple of Mentuhotep II in Deir el-Baheri does not have an area that can obviously be called a slaughter yard although there are four foundation deposits containing sacrificed bovids (Arnold 1979: 51-7). The pyramid complexes dating from the period do not reveal any particular area which might have exclusively been used for slaughtering, although it is likely that they did contain slaughter houses as there were no adjacent sun temples to provide the meat offerings. It is always possible, of course, for the Egyptians to have slaughtered a sacrificial animal in a side court or room, or even outside the walls of the complex, and equally possible that the Residence took over the job, or that some other cult temple was responsible. Unfortunately, not enough is known about temple ritual and economy to determine the exact location and source of the sacrifice.

Despite the large number of standing temples from the New Kingdom, only four claim the inclusion of slaughterhouses. Perhaps the lack of evidence is an accident of preservation, excavation and identification which can be corrected once excavators become aware of the constituent elements of a slaughter area.

The two Theban temples that claim slaughter-rooms are the funerary temples of Seti I in Gurna, and Ramesses III at Medinet Habu. The areas defined as slaughterhouses are similar in both. Rooms 5 and 6 in Medinet Habu (Fig. 21), just off the columned hall, have been identified as the slaughterhouse. These two connected rooms are quite small, room 5 being 2m wide and 3.5m long, and room 6 being 1m wide and 3m long. A portion (1.3m × 3.5m) of room 5 is roofed, and room 6 is entirely roofed. The floor is of mud and might have once been plastered. The doorway to room 5 is only a meter wide, and the entrance is very near columns. It does not seem probable, however, that these rooms were used as slaughter houses (Nelson 1934: ix). The doorways are too narrow to permit the entry of a plump and wide-horned animal and the access to these doorways is obstructed by the columns of the hall. The space within Room 5 is limited and would make wrestling down and slaughtering even a small ox well nigh impossible; moreover, it lacks a tethering-stone which might have assisted in positioning the animal for slaughter. It is difficult to imagine the possible use of Room 6 in butchery other than storage or perhaps jointing, although the lack of light would seriously endanger the butcher's fingers. The main reason for identifying these rooms as an abattoir are the wall reliefs. The majority of reliefs on the walls of the two rooms show the slaughtering of oxen, *rnnwt* -cattle, hornless cattle and oryx (Nelson 1934: 33), and the offering of the entire animals or parts thereof to Amun, and taking the idea that the wall decoration determines the function of the room, this room has been labelled an abattoir. However, it has been argued that "the reliefs do not in most cases describe the rite that took place where it appears" (Vassilika 1989: 2) and frequently "scenes interject which have no relation to the function of that room" (Vassilika 1989: 2)[7]. Thus, basing the function of a room on the wall scenes is too simplistic.

[7] For a thorough discussion of decoration principles see E. Vassilika 1989, Chapter I, and for Medinet Habu specifically see D. Arnold (1962: 90).

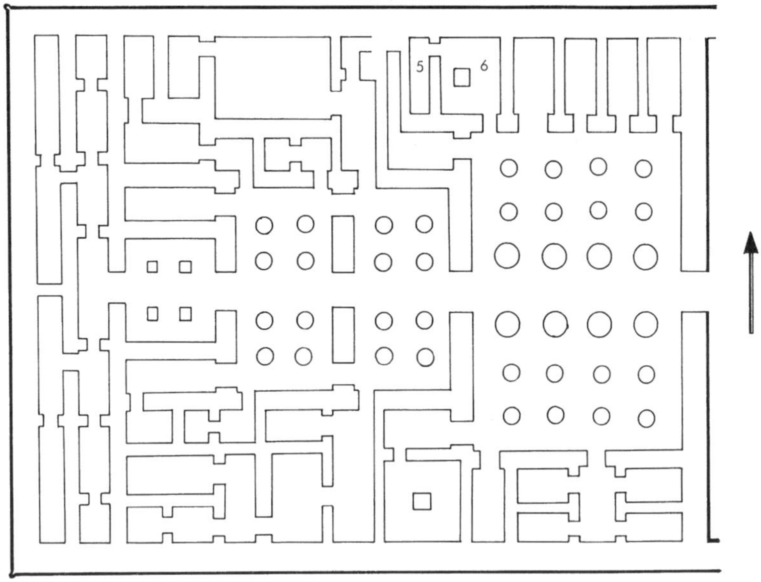

Fig. 21

Furthermore, there is some debate as to whether the location of this so-called slaughter area is tenable as it is situated in the heart of the temple close to the sacred focus. All the suggested Old Kingdom abattoirs were to one side of the sanctuary and had their own entrances. This pair of rooms is very much a part of the temple proper with no outside egress. This is problematic as it raises the question as to whether butchery was considered a sacred and sufficiently pure activity to take place in the heart of the temple. Slaughtering an animal, even for a ritual sacrifice, is messy due to the animal's blood and excrement, noisy until the animal's throat is cut, and undignified until the initial struggle in securing the animal firmly so its throat can be cut is over. This is not what one might expect of a carefully choreographed ritual. The presence of excrement if not blood in a temple strikes one, due to later texts and literary allusions, as a violation of temple purity and something to be frowned upon. It is more than likely that the actual slaughter and initial jointing of the animal was performed in an abattoir, possibly one of mud brick, near the outer wall of the temple away from the sacred focus (Ghoneim 1977: 167). The joints were then taken to the temple (perhaps to Room 5 and 6) and offered to the deity at the appropriate moment in the ceremony. The wall decorations showing the slaughtering of beasts would ensure a symbolic continuity without any of the physical problems associated with its profane aspects. It is useful to see how another religion which does believe in ritual sacrifice resolves this problem. In Islam ritual sacrifice on feast days is obligatory, if affordable. This activity is often associated with the mosque from where the meat is frequently distributed to the poor. The presence of both blood and excrement is forbidden in a mosque, so the slaughter takes place in a courtyard or open area next to the mosque (if it is a wealthy mosque this *midan* is a part of the mosque but not where the main ritual activity takes place) so that it is associated with the sacred building without profaning its interior. It seems likely that the Egyptians would have had some similar arrangement of separation in their temples so as not to offend proprieties.

The other Theban temple reputed to contain a slaughterhouse is that of Seti I in Gurna. Rooms XXXIX-XLI (Fig. 22) in this temple are thought to have been the equivalent of Rooms 5 and 6 in the temple of Medinet Habu (Ghoneim 1977: 167; Arnold 1962: 90). They have again been identified as abattoirs only due to the scenes on the wall. These two tiny rooms are also located in the centre of the temple behind the chapel of Ramesses I, have no separate entrance, and due to the same reasons as discussed above, seem to be storerooms for meat or temple goods, or rooms for some cult purpose other than slaughter.

It is possible, although there is (so far) no remaining evidence, that the mud brick buildings surrounding New Kingdom mortuary temples such as the Ramesseum contained slaughter areas, and perhaps even stables for the rearing and fattening of sacrificial beasts.

The major cult temple containing an abattoir is that of Seti I at Abydos. This sandstone room (Figs. 23, 26) is beyond the King List corridor and forms a separate complex built on to the main temple and, like the abattoirs of the Old Kingdom funerary temples, is convenient to the sacred focus although separate from it. The structure consists of one large partially covered room whose roof is upheld by seven columns, the slaughter room, with three rooms leading off from it. It has a large doorway (about two metres wide) to the outside of the temple in the east wall (Fig. 23). There is a door connecting the slaughter room to the main temple, although there is a high screen wall in front of the doorway so that any activities of the slaughter yard would be hidden from view from the main temple. Behind the screen wall are three large storage (?) jars set into a mud-brick bench with a drainage hole at their base. These might be a later addition to the room. Some of the lower registers of the walls

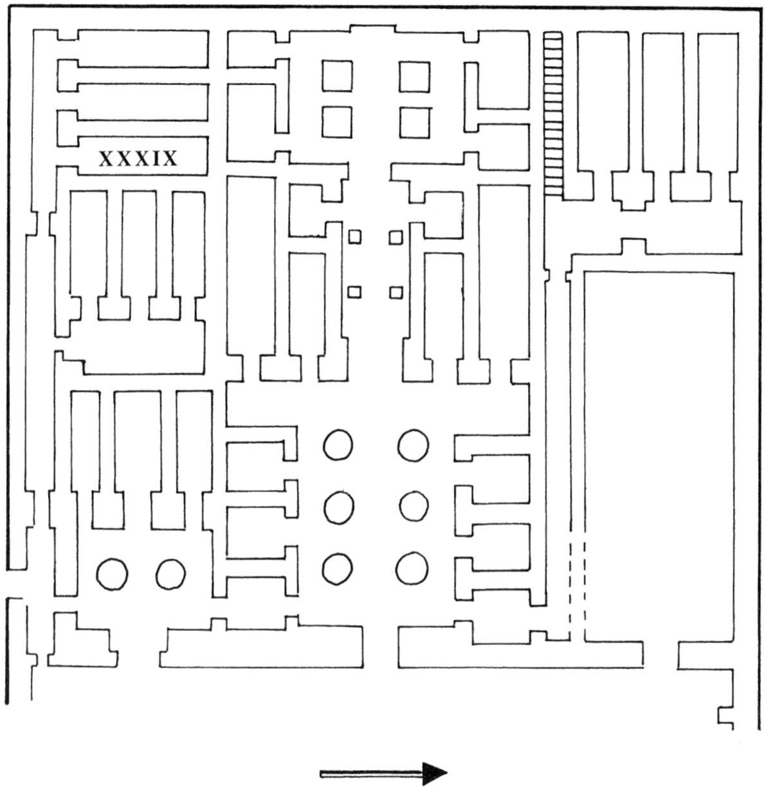

Fig. 22

are decorated with scenes of butchery, although most are covered by meat and other offering scenes. As discussed above, decoration does not indicate room-use, and therefore cannot be relied upon as the sole means of determining the types of activities that took place in a room[8]. This area is more credible as an abattoir than any of the other New Kingdom examples mentioned thus far. It has a separate entrance suitable for cattle, a large area, a stone floor that could be cleaned of blood, an open area to lassoo, hold down and slaughter a cow, and a possible location for water that might be necessary for flaying and washing of implements and

[8] The walls of the first court (Mariette's A) in the temple of Ramesses II at Abydos are covered with scenes of butchery (Mariette 1880). Although an open courtyard is a very feasible location for butchery, the drawings do not necessarily mean that this is indeed so.

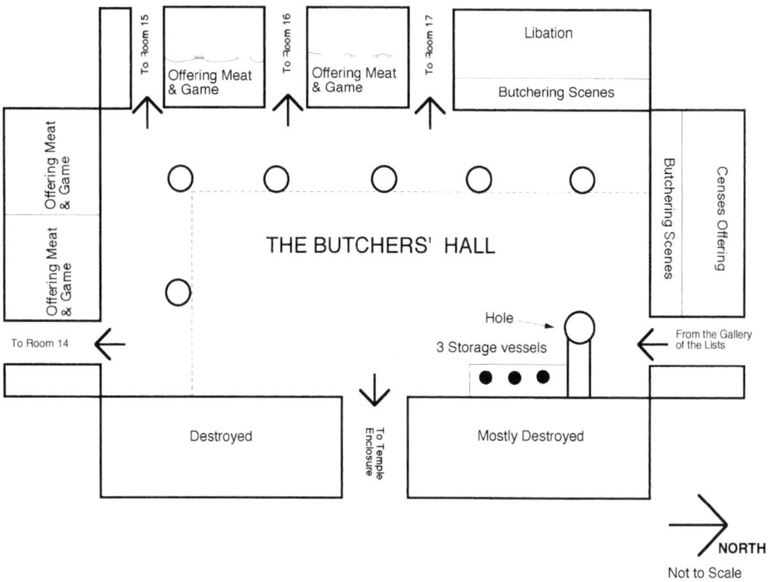

Fig. 23

hands, especially if there were many animals killed there. The three rooms in the back, reminiscent of the three rooms in the Meketre model, could be used for the storage and processing of the meat. No tethering-stones have been found here, but had they existed they might have been reused elsewhere. If they had left any evidence of their existence here in the shape of holes in the floor, it is quite possible that these too have vanished over time as the temple was reused.

Tethering-stones are not necessary in a slaughter yard although they do provide an easy indication for identification as they are frequently pictured in Amarna tombs as 'furnishings' of slaughter yards, as well as in the temple of Seti I at Abydos (Naville 1930: pl. III), and also have been found in front of the offering-table/false door in some Old Kingdom mastabas (e.g. Mereruka in Sakkara [Duell 1938: pl. 126], Sheshemnefer IV at Giza [Junker Giza 11: 108])[9]. A tethering stone, but of a slightly different design has been found in the tomb of Pepiankh (D2) at Meir. It is shaped like the letter 'I', the hole for the rope is located in the upward stroke (Blackman 1924a: 27; Fig. 24). The tethering-stone might be

[9] See notes 4, 6 and 10.

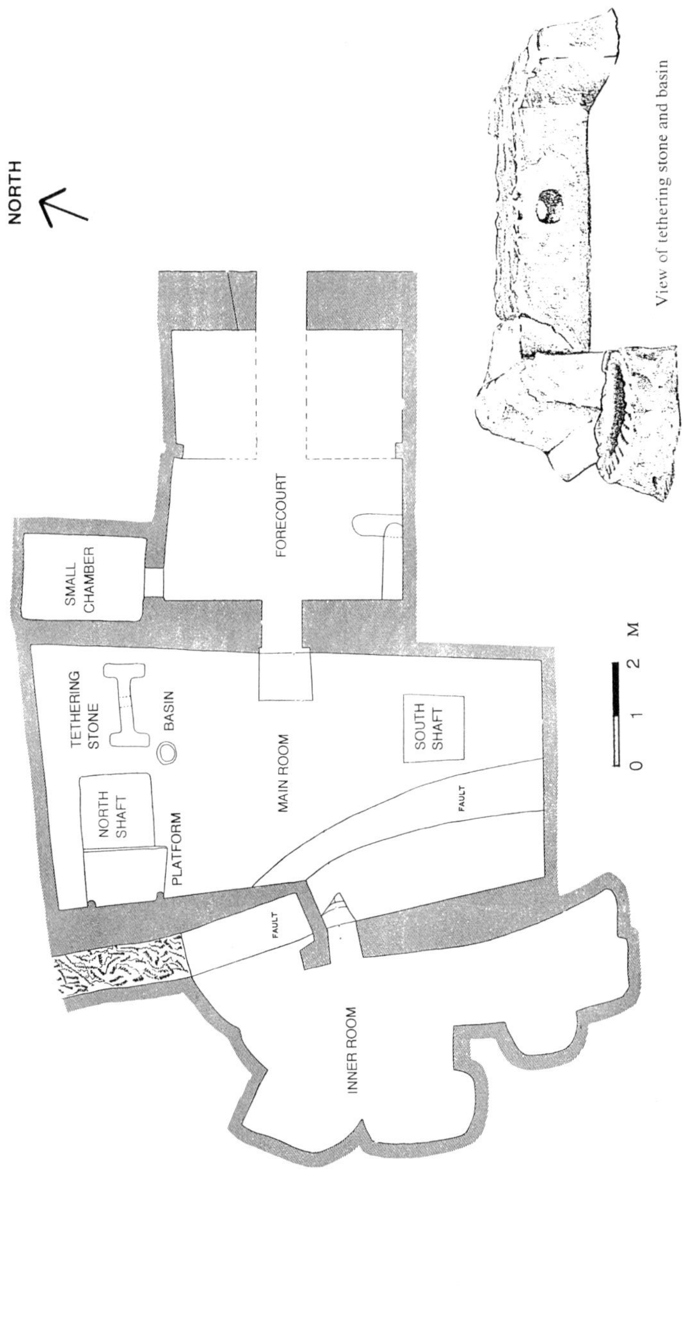
NORTH
SMALL CHAMBER
FORECOURT
TETHERING STONE
BASIN
MAIN ROOM
SOUTH SHAFT
NORTH SHAFT
PLATFORM
FAULT
FAULT
INNER ROOM
0 1 2 M
View of tethering stone and basin

Fig. 24

necessary in small areas with little room to manoeuvre the animal to the ground[10]. It is of note that, though such a stone was depicted in the Seti temple, none has been found.

Further evidence that the Abydos temple was used for ritual slaughter comes from some inscriptions within the temple which tell of the building of a sacred slaughterhouse by Seti (Naville 1930: 5-7). Also, several mud-brick structures of a more prosaic nature are located near the abattoir. These structures might have been related to aspects of daily life in the temple, or have served some specific purpose in meat production, i.e. as cattle stalls, storage areas for food, etc. The inscriptions, shape, dimensions, separate entrance, and proximity to the outside wall of the temple are all arguments that favour identifying this complex as a slaughter house, or at the very least, a room where meat was jointed, offered, and/or redistributed.

A great deal of information concerning butchery comes from the Amarna period. Tombs as well as temple reliefs are rich in the depiction of butchery, while archaeological evidence from the period also provides insights into the construction and location of abattoirs. Tomb reliefs (such as Meryre's, Ahmose's and Panehsy's) show a one-roomed slaughter court to the left of what has been deduced to be the entrance of the *pr-hʿy* and the *gm-pꜣ-itn* compounds, while the supposed *ḥwt-bnbn* is immediately flanked by a slaughterhouse that stood in the courtyard (Leprohon 1988: 48). The slaughter-yard has a little entrance, separate from the main entrance to the courtyard, and in the tombs of Meryre (Davies 1905a: 11) and Panehsy (Davies 1905b: pl. 19) was built against the north side of the temenos wall, while the depiction in the tomb of Ahmose (Davies 1905c: pl. 30) locates the building more in the centre of the north courtyard. The animals are shown headless and trussed, lying next to tethering-stones (in which case what was the role of tethering-stones if the animals were already trussed? Perhaps they provide a bit of narrative? Did they indicate the unlikely idea that the cattle were fattened there? See below). The lack of divisions in the enclosures suggest the meat was processed here once the animal had been killed or perhaps the animals were slaughtered, butchered, and offered immediately. According to these scenes, temple enclosures (here defined as areas within the largest temenos wall) also contained spaces near the storerooms where live animals were force-fed and kept prior to

[10] It is also possible that the tethering-stone was used to secure the animals in the tomb briefly, and then remove them for slaughter outside the tomb. The presence of a trough, mis-identified as a bowl for blood, in Meir D2 would support this theory.

butchering. The reliefs complement the evidence for cattle-rearing in the North Palace where tethering-stones and feeding-troughs have been found (Newton 1924: 295-6).

A butcher's yard has also been found in the Great Aten temple enclosure (Pendlebury 1951: pl. I). It is situated between the sanctuary and the *gm-pꜣ-itn*, off the central axis like the one depicted in the tomb of Ahmose (Davies 1905c: pl. 30), but not against the north wall. This large enclosure was bounded by thin walls and entered over a stone threshold at the east end of the south wall. One or two pierced tethering-stones were found on the surface. The building's yard was never cleared though the deposit must be very shallow. If any other slaughter-yards similar to this one are located against the north temenos as described in the tomb scenes, they are unknown as, unfortunately, neither the German nor the British expeditions to Tell el-Amarna have closely examined this section of the temple, so there is no archaeological evidence against which these representations can be checked. There is also an enclosure on the north side of the sanctuary of the Small Aten Temple, visible in aerial photographs, but still buried under a Pendlebury dump (Kemp 1992). It, too, might be an abattoir, but this can only be determined by excavation.

South of the Great Aten Temple lies the house of Panehsy, Overseer of the Cattle of the Aten. One of the out-buildings on his estate, S.40.1 (Frankfort 1927: 212; Pendlebury 1951: 28-9), as well as the area of the entire estate, contained remains of cattle bones and horns, and flint-chips (Frankfort 1927: 212), material identical to that found in the Raneferef complex. A whetstone (#26/82, [Pendlebury 1951: 27]) was also found in the surrounding area. S.40.1 is located forty metres south of the temenos wall (Pendlebury 1951: 28), quite in line with the "Butcher's Yard" located within the temple. There is some evidence of a break in the temple wall at this point, possibly for a door (Kemp 1992), which would provide direct access to the slaughter-area from S.40.1. The building is made of mud-brick, over 70m × 60m, with stone thresholds, and a flooring of mud everywhere save in the south (Fig. 25). "Apart from a projection to the east it is divided into four divisions by screen walls built between the square brick piers which supported the roof" (Pendlebury 1951: 28), both of which are whitewashed. The entire building was not roofed. "Some five metres in front of the building proper runs a wall pierced by doors corresponding to the entrance into the four main divisions" (Pendlebury 1951: 28). Some of the walls show evidence of having been stone lined, suggesting an extensive use of water, an idea

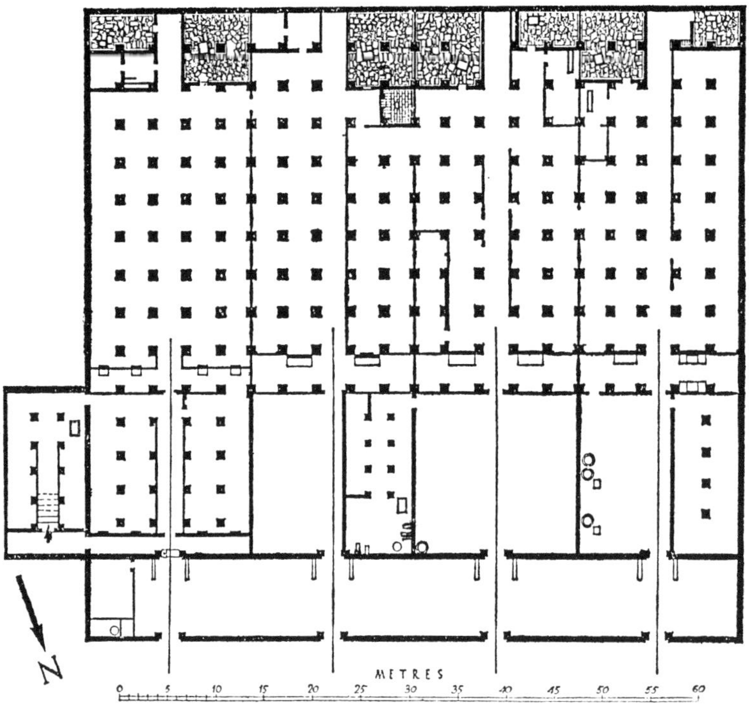

Fig. 25

which is further emphasized by the presence of stone troughs (Pendlebury 1951: 28-9). There is also considerable evidence of burning, as well as several ovens fond in the building. The doorways measure approximately two metres across, and the areas between the pilasters three metres. It has been suggested that this was a purification area for the priests (Pendlebury 1951: 29). However, given the dimensions and structure of the building, as well as the presence of troughs, it is also possible that a part of this structure served as a cattle shed prior to their slaughter, and other portions (perhaps the ones with ovens) were used for cattle processing (and the preparation of other offerings) after slaughter. The proximity of the building to the butcher's yard and the separate entrance to that part of the temple would make this very convenient. Furthermore, the latter would obviate the need to lead the animal through the temple proper, and prevent any defilement of the temple by sacrificial beasts caused by 'mishaps' with nervous bowels. The title of

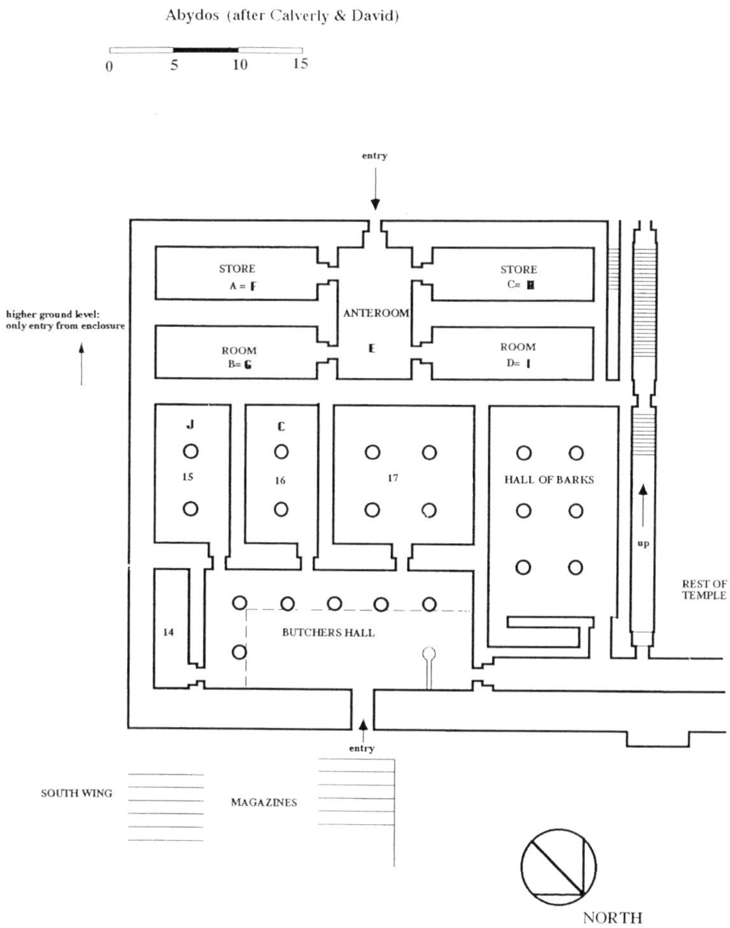

Fig. 26

Panehsy, and the large number of cattle remains also strongly support this idea. The location of temple magazines north of this structure also leads one to think that animals might have been processed here, and that any unconsumed but preserved food would have been taken to the magazines for immediate storage.

Further information concerning slaughterhouses from Amarna comes from the Chapels and their environs (Kemp 1984: 13). Several areas near the chapels have been identified as animal pens (Kemp 1984: 13;

Kemp 1987: 47) and cooking areas. It is possible that some of these spaces in the chapel annexes were butchery yards where pigs and goats, the commonly-herded animals in the village, were slaughtered prior to offering, consumption, and preservation. Indeed, the tethering-stone found in the courtyard annexe of the Main Chapel (561/450) might indicate butchery as well as animal husbandry (Kemp 1984: 24, 33), although they point to larger mammals rather than the small goats and pigs, the latter species being loath to being tied up. Kemp has suggested building 540/541 as another butchers yard for pigs (Kemp 1986: 72-75). Given the evidence of open spaces with a great deal of organic deposit (much of it is coprolitic) and charred areas for fires necessary in pig processing and preserving it is quite possible that animals were brought to these buildings and kept there until they were butchered in the courtyard and consumed or processed in the environs of the Chapel. The gypsum floor and wall plaster of these chapels would indicate that most, if not all, of the bloody work would have been done in the courtyard as otherwise the blood would have permanently stained the floor if not the walls as well.

In conclusion, there are very few ancient Egyptian buildings that can be positively identified as slaughter areas. Temples or large villas with dependent buildings are the best sources of information archaeologically. Both these building types would have a fairly formal structure to conduct the activity in as it would be performed regularly. The former would also have had need of a formal structure due to the ritualistic and religious nature of the act of slaughter. It is interesting to note that most of the spaces in temples described as slaughter areas have a different orientation to that of the temple. From an examination of the evidence, the attributes accorded to abattoirs are: a large open space, partially or completely unroofed, preferably with mud floors, possibly a tethering-stone, and access to at least a pot (a litre or even two) of water (more if processing is taking place or it is pig-slaughtering).

IV.2 Open ground

Although one associates an abattoir with some sort of (even temporary) structure, one should note that in modern Egyptian villages when only four or five animals are being killed, the abattoir consists of an open space, sometimes even the side of the road, with a pot of water being brought from the communal well or pump which can be some distance from the slaughter site. Tethering-stones (solid stones around which the rope can be tied) are sometimes used, although now the animals are frequently tied to a telegraph pole as they await slaughter. Any debris is neatly and speedily removed by scavenger dogs, although most of the blood soaks into the earth and is gradually scuffed into oblivion during the course of the day.

Certainly fish in ancient Egypt were processed in the open as there does not appear to have been a special building or even shelter where they were processed. In representations the fish are almost always shown being cleaned and eviscerated by the banks of the river/canal where they are caught except for when they are shown being processed on the boat. This would have been practical as the fish could be deposited here immediately after being caught, and be processed on the spot, thereby avoiding spoilage. Furthermore, the fisherman responsible for cleaning the fish would have access to water to wash his knife, table/block, and/or the fish if necessary. There are some representations of the evisceration and cleaning process being performed on the boat where the fish were caught (Klebs 1934: 88). This would be practical if the boat were large, or if the fishermen were on the water for the greater part of the day, demanding quick processing of the fish on board the ship to avoid spoilage, and to shorten the time taken in the production and processing of dried fish.

It is also possible for poultry to have been processed in the open without the benefit of a shelter as is shown in some tomb scenes. A poultry yard, *mḫwn* (Barguet 1962: 18), has been found in the temple of Karnak south of the sacred lake. The extant structure was made by Psammetichus, built to replace a construction built in the Eighteenth Dynasty, and includes a courtyard, some pen-like installations of mud brick for the birds, storage areas for grain, and a chapel dedicated to Ernutet (Barguet 1962: 18). Birds were bred here; whether they were killed as sacrifice for Amun in Karnak is only conjecture. The killing of poultry requires few facilities and can be undertaken in any home. It is therefore difficult to define and isolate an area where this activity might have taken place.

IV.3 "Storage" of live animals

While it is clear that the fish that were eaten were either caught or bought, and the fowl came directly from the market, fowl yards or hunts, there is a question as to where the cattle destined for slaughter were kept and how they were transported to the slaughter areas. Tomb reliefs show herds of cattle belonging to the tomb's owner, but only individual animals being offered, implying that animals were generally kept in a herd and isolated prior to slaughter. It is possible that animals were rented out to private individuals by temples, to be reclaimed when it was time to slaughter the animals although how common a practice this was is unclear.

The funerary temples of the Old Kingdom were at some distance from areas under cultivation, as were some of those dating from the New Kingdom. Thus, cattle put out to pasture would be kept in the cultivation, and brought to the temple when it was time for them to be slaughtered[11]. However, if feed and water were brought to the animals then it is quite possible that they were kept in a mud brick enclosure nearer the temple, though no such structures have been discovered. Representations of animals being tethered and fed are ubiquitous on tomb-walls, and this is further illustrated by Meketre's model of a cattle stable (MMA20.3.9), as well as being referred to in texts (Sallier I: 4.8 [Caminos 1954: 307]). Tethering seems to have been used only in situations when an animal was being fattened, or was awaiting slaughter. It is also possible that the animals were kept by the cultivation until the time of sacrifice approached, and that they were then moved to the temple precincts where they would have been tethered and fattened before slaughter. This pattern might hold true for all the temples that are at some distance from the cultivation, and those that slaughtered once or twice a week. However, this might have proven to be have been more difficult in the case of the sun temple of Neferirkare where an ox a day was slaughtered (Posener-Kriéger 1976). This suggests that a large (and hungry) herd of cattle had to be readily available and regularly restocked (from other herds belonging to the temple or perhaps to the Pharaoh's residence) in order that the temple should be effectively supplied with regular offerings.

Textual evidence about cattle in the estates, and cattle in the field (e.g. Leiden 3486.2-6:4 [Caminos 1954: 491]), points to the fact that in ancient Egypt, as in modern Egypt, animals are both put out to graze and given feed. This depends on the availability of grazing land, the safety of the area, and the time of year. As to how the animals would have reached the slaughter-yards, this would depend on the location of each. It is doubtful, however, that the animals were herded up a causeway, through a sacred precinct, and then butchered, as they might accidentally have defiled the temple. It is more probable that they reached the slaughter-area by a more private route, perhaps through the desert.

[11] See Ghoneim (1977) for a complete discussion on herding and rearing cattle.

V

BUTCHERS, POULTERERS & FISHERMEN

Pictorial, textual, and archaeological evidence show how animals were dispatched, and where this was done. However, they do not provide much information about the people performing the act other than illustrating the basic tasks of killing, eviscerating, flaying, scaling, plucking, and jointing. Information concerning poulterers and fishermen is especially scarce, while information concerning the butchers themselves comes from their titles and references made to them in jar labels (Hayes 1951: 102).

Butchers

The most common Egyptian word for butcher was *sftw* (Faulkner 1962: 225), "one who slaughters or makes sacrifice", followed by *sšm.(ty)*, and the variant, *ḫrp imy sšm,* "Director of Those With Whetstones" (Fischer 1960: 177), a name given to butchers because of their retouching tool[1]. The title *ḳbḥ nmst* can be interpreted as "butcher of the slaughter-house" (Fischer 1960: 171), although it has a connotation of "Master Butcher", and is amongst the titles of butchers who were of high enough rank and wealth to merit their own tomb inscriptions (Fischer 1960: 177). However the job of the *ḳbḥ nmst* seems essentially the same as that of the *sftw* and the *sšm.(ty)*. Furthermore, there is nothing in their clothing or accoutrements to differentiate them from the other butchers. However, it is clear from the titles listed above that there was a hierarchy of butchers as some of them were awarded titles such as "Master Butcher", "Director of Those with Whetstones", and "Director of Butchers" (Fischer 1960: 179), thereby suggesting different levels of butchery knowledge, much as exists amongst butchers today.

These titles indicate that the butchers were quite a large group of skilled workmen. A butcher (or butchers) might have been attached to an estate (e.g. Meketre) or have been a shared servant among a group of estates. Evidence from this comes from jar labels found in Malkata which list the names of butchers and the estates to which they were attached, most of which belonged to nobles active at court bearing the

[1] See above for discussion of the word *sšm*.

title of "Chamberlain" or "King's Scribe" (Hayes 1951: 102). Obviously these butchers were of some importance or skill as their names appear on their products. They could also have been independent workers who sold their services to a village or group of villages, depending on the size, population and wealth of the settlement. Small animals that were occasionally butchered for a specific event might have been killed by their owners, but it is more likely that then, as now, a professional butcher would come to perform the operation. Temples probably had a butcher, or more probably a group of butchers of their own. Butchers are listed as members of temple personnel on a writing-board now in University College, London (Gardiner 1947: 64). These butchers might have served the temple alone or might have sold their services on the side to inhabitants of local settlements, much as the Deir el-Medina workmen, for example, made coffins in their spare time. These individuals were probably butchers rather than priests, although the two are not exclusive, especially when involved in temple duties. According to P. Bulaq 11, an account papyrus recording the transactions between an unknown seller and merchants, there were meat shops that were butcherless as the merchants purchased cuts of meat (perhaps from the temples) to sell to private buyers. This would suggest that the temple had a monopoly on butchers, though this is arguable.

Priests were definitely involved in butchery as is attested by several butchery scenes (Montet 1910: 49-50). A *ẖr-ḥb* might oversee a scene of butchery (Meketre's model shows a man holding a *ḫrp* scepter who might be a priest or an overseer) or take away the jointed meat (Montet 1910: 49-50). However, these priests did not actively participate in butchery — they just oversaw or transported the goods. A priest who did participate in butchery was a *snw (wʿb)*, or veterinary priest (Chassinat 1905: 223-8; Fig. 9). He checked the animal's blood and entrails for purity and acceptability for consumption (see Fisher 1976: 99 for further discussion). These priests occur rarely in scenes, so it is unknown if they checked the meat for hygiene regularly or only for certain festivals. It is quite probable that most ancient butchers, much like modern butchers, could judge the edibility of meat themselves. Veterinary priests would be present if there were any doubts on the health of the animal, or as a more formal 'quality control'. In the New Kingdom, when butchery scenes are clearly associated with the funeral ritual rather than food availability in the hereafter, priests are present during the slaughter, and in fact, might perform the act themselves (Nelson 1934: pl. 173-4).

Presumably butchers would have learnt their job through experience, as apprentices, their knowledge being handed down from one generation to the next. They would then have risen in the ranks (if in a temple, etc.). Butchers, like other workers, probably kept the work in the family and some positions became hereditary (Fischer 1960: 178).

Most of the Egyptian butchers that one knows about from titles, statues, and tombs were obviously men of some substance to possess these obvious manifestations of wealth. They were also of considerable social significance as the titles "Acquaintance of the King" and "*wꜥb* priest of the King" suggest a connection to the court and the king himself (Fischer 1960: 169). The titles: "Great of Cattle", and "Overseer of the Rendering House" (Fischer 1960: 169) also suggest that they were powerful as they had control over the food supply of the court and the temples. Naturally this social status would not be accorded to all butchers, and it is uncertain that the people with these titles actually performed any butchery at all, although they clearly had dominion over this activity.

Some information on Egyptian butchers in the very late period of Egyptian history comes from Herodotus. In his discussions he mentions pig-herders and pig-butchers as being outcasts. Undoubtedly this does not hold true for the bovine and caprid butchers in the Old, Middle and New Kingdoms as they figure prominently in tombs and are mentioned in texts. The only other substantial work relating to butchers (and butchering techniques) is that of A. Eggebrecht (1973) who does not really address himself to the question of the social class of the butcher.

In modern Egypt there are travelling butchers who work on market days in different villages in a certain district, e.g. Sohag, Asyut, Luxor. There are also butchers who are particularly skilled, and are hired (by others) for major festivals (birth, *Eid*, sacrifices). Mosques favour specific butchers, either because of their skill or for personal reasons, who then become permanently affiliated with the mosque. Unlike mosques where sacrifice is a rare event, the Egyptian temples would have had permanent butchers in their "Department of Meat" for their daily slaughtering.

Poulterers

While it is likely that some priests were butchers, it is probable that the poulterers were cooks or other servants rather than priests. None of the individuals have priestly or other titles, and it seems that the activity is an everyday one with no religious overtones. It is possible that butchers also killed poultry as is suggested by the title of the butcher "The Master Butcher of the Slaughterhouse of the Palace, He who is Privy to the Secret of the Darkness, Who Opens It (the darkness) in Strangling the Bird of the Desert Regions" (Fischer 1978: 57), although this conclusion is not assured. Wringing the neck of poultry is a simple operation and one that anyone could and can perform. Akhenaten holds a bird up to the Aten in offering and strangles it (MMA1985.328.2). Furthermore, birds are not as large and impressive as the bovids, and so require fewer 'slaughterers'. There was apparently, no testing of birds for purity — perhaps because if they are seriously diseased it is obvious, while salmonella and other bacteria appearing after slaughter cannot be recognized. Although birds are offered to both deceased and deities alike, killing and plucking them was not generally given as much prominence in representation as the butchering of mammals, and are activities never depicted in temples.

Fishermen

It seems clear that fishermen were responsible for the capture, cleaning and preservation of all fish. There is very little evidence to show whether these people were independent individuals who made their livelihood by fishing, peasants who were thus employed part time, or servants of wealthy men who fished full-time in the Nile or areas allotted to them. Tuthmosis III's decrees at Aswan suggests that, in Elephantine, a body of full-time fishermen existed as they were pressed into Pharaoh's service to dig a canal (*Urk*. IV: 814-5; Breasted 1906: 259, §649). This would indicate that fishing could be a full-time occupation. Like the bird catchers, fishermen do not seem as culturally important as butchers. In the Ramesside era there is evidence, however, of fishermen owning their own boats and renting them out to temples or even private individuals such as merchants (Castle 1992: 248-9). This would provide the fishermen with an additional source of revenue. Unlike butchers, there are no titles associated with fishermen, and, thus far, there are no surviving tombs belonging to fishermen.

VI

JOINTING

Jointing applies only to mammals as fish were eviscerated, opened, and left to dry whole, while poultry was also preserved whole (see below). The sole example of poultry being jointed in any way comes from TT217, a tomb noted for its unusual reliefs (Fig. 16). The scene shows what seems to be a poulterer seated at a table cutting slices off a waterfowl and hanging them to dry. The bird is cut into semi-rectangles. The only part of a bird that might be thus sliced is the breast (see Chapter VII.1), though carpals and tarsals are removed.

After mammals are slaughtered, flayed, and gutted, they are jointed. Each country has different methods for dividing up an animal for consumption, and the evidence for ancient Egyptian practices comes not only from the artistic record, but also from faunal material from tomb and settlement sites. The former category includes the following: wall paintings where joints of meat are taken away from, or hung above, the slaughtered animal; soul houses and offering-tables with pictures of joints commonly offered, and, in the unique example of the offering-table of Princess Neferuptah from Hawara (Fig. 27; CG28792), where some of the joints are labelled[1]; models, notably those of Meketre (MMA20.3.10, JE46718), providing three-dimensional representations of the meat; boxes in the shape of joints of meat (examined by the author and discussed in detail in Appendix I). The latter category encompasses excavated faunal material (some with butchery marks; Figs. 29-30) from tombs (Appendix II; Figs. 31-36, 42-43, 45), temple, and settlement sites (Fig. 44), and mummified joints of meat (Appendix II; Fig. 43)[2]. Textual evidence is also useful in naming some of the joints. As a lexicographical study regarding the names of joints is beyond the scope of this work I have relied mainly on the work of A. H.

[1] As will be seen below, some of the names attached to the depictions (e.g. *ḫpš*), are clearly mislabelings, and therefore cast some doubt on the reliability of labelling on the table. Nonetheless they have been mentioned wherever appropriate, with an accompanying discussion as to their validity.

[2] A possible allusion to this type of offering might occur in Spell 28 of the Book of the Dead, where it says: "A wrapped leg is what they found and buried" (Allen 1974: 39).

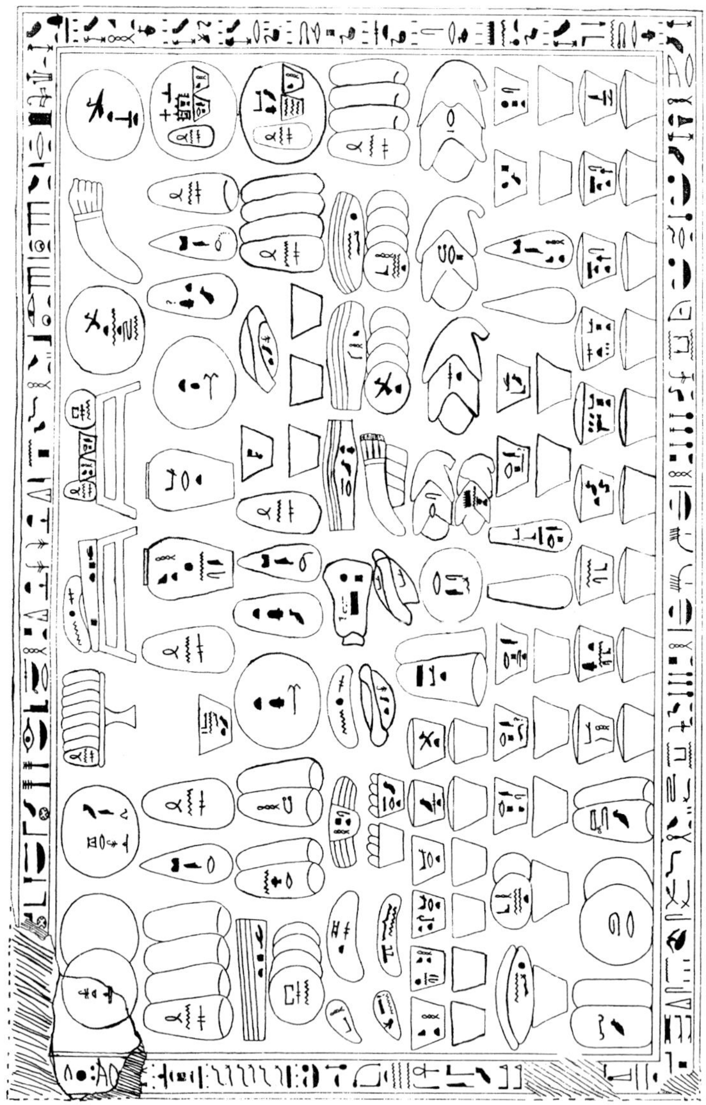

Fig. 27

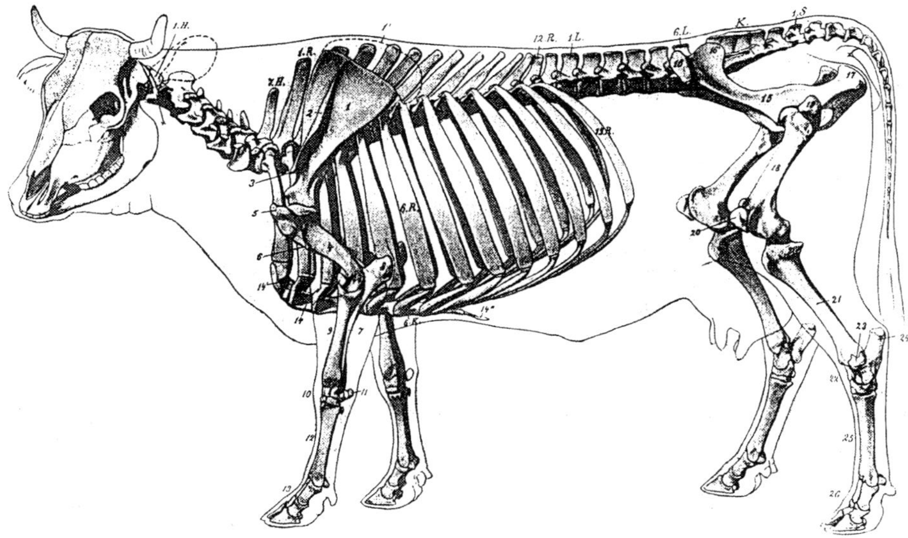

Fig. 28 — Skeleton of Cow

1.H., Atlas; *7.H.*, seventh cervical vertebra; *1.R.*, first thoracic vertebra; *6.R.*, sixth rib; *12.R.*, twelfth thoracic vertebra; *13.R.*, last rib; *1.L.*, first, *6.L.*, last lumbar vertebra; *K*, sacrum *1.S.*, first coccygeal vertebra; *6.K.*, sixth costal cartilage; ×, wing of atlas; *1*, scapula; *&1′*, cartilage of scapula; *2*, spine of scapula; *3*, acromion; *4*, humerus; *4′*, lateral condyle of humerus; *5*, lateral tuberosity of humerus; *6*, deltoid tuberosity; *7*, ulna; *8*, olecranon; *9*, radius; *10*, carpus; *11*, accessory carpal bone; *12*, metacarpus; *13*, phalanges; *14*, sternum; *14′*, manubrium; *14″*, xiphoid cartilage; *15*, ilium; *16*, tuber coxæ; *16′*, tuber sacrale; *17*, tuber ischii; *18*, femur; *19*, trochanter major; *20*, patella; *21*, tibia; *21′*, lateral condyle of tibia; *22*, tarsus; *23*, distal end of fibula; *24*, tuber calcis; *25*, metatarsus; *26*, phalanges. The bones of the fore limb are placed too high in relation to the thorax. (After Ellenberger-Baum, Anat. für Künstler).

Gardiner in his *Ancient Egyptian Onomastica* (1947a, b), the *WB*, and W. Barta's (1963) study of offering-lists for the names of the joints.

By using this diverse evidence one can deduce how an ancient Egyptian butcher would divide up a carcass. However, one cannot necessarily determine the order in which this was accomplished as there are no Egyptian manuals with step-by-step illustrations. According to reliefs, the first limb to be removed is the foreleg, which, once having been used as a pump handle to expel the blood from the body, is cut off, seemingly at the scapula, and given as an offering. Frequently it is followed by the head, which, despite being the first part of the animal to receive a cut, is not the first to be severed from the body. This remained the case until the late Eighteenth Dynasty when, after the removal of the foreleg, complete decapitation occurs. The precise order of subsequent disarticulation is unclear.

In modern Egyptian butchery the first limbs to be cut off are the two forelegs, the right one first. However, in modern Egypt, unlike in Pharaonic Egypt, the leg is cut below the carpals. Then, as in ancient times, the head is removed, followed by the hindlegs below the tarsals. The remainder of the sequence cannot be definitely thought to reflect ancient practices, although it is useful in providing comparisons and ideas for reconstructing the ancient methods.

The modern butcher next cuts both the forelegs mid-humeri. Then, using a medium-size knife, he makes a small shallow cut halfway down the back of the animal, extending the cut along the back down to the neck. Then he makes a cut from the pelvic area, and using a cleaver hacks down the vertebral column to meet the previous cut (see below). Once there he removes most of the spinous processes of the thoracic vertebrae, and hacks down the pre-existing cut to the neck, dividing the animal into two halves. Once it is divided, he removes more of the tops of the vertebrae. None of the vertebral meat is lost during this operation. Then the butcher proceeds to joint the carcass, one half at a time. The neck is removed from one half of the animal and is sent to join the forelegs that were previously removed. The lower ribs of the animal are cut off with a medium-sized knife, followed by the forequarter (scapula and humerus top). Then the upper portion of the ribs, the part attached to the vertebrae, is cut off, leaving a quarter of the carcass still hanging. After that the butcher attacks the bottom of the haunch, i.e. any remaining vertebrae, and the pelvis followed by the femur and tibia, the latter being used in conjunction with the fibula to suspend the animal from the tree. All these large chunks are placed next to the butcher who sits at his block.

At the block the butcher starts with any piece that might come to hand. The first thing he does is to remove as many bones as possible from each piece of meat. The meat is frequently peeled off the bone with no marks being made on the shaft of the bone from either knife or cleaver, both of which are used in the final jointing. The proximal and distal ends of bones are not always removed from the meat, and, when they are, they are generally without large parts of the ends as they have been chopped off in the effort to separate bone from meat. The ribs are hacked away from the attached vertebrae and removed, leaving only meat. As much meat as is possible is removed from the vertebrae; sometimes the vertebrae are chopped up in the process. The humerus, radius, and tibia are divested of most of their meat. Many of the bones are sold separately for soup or as food for pet animals. The meat is hacked up in a fairly random manner, and most of it seems to be of the same price. The legs (carpal and tarsal down) and the head are the cheapest part of the animal, and the liver is the most expensive. The meat from the rump is generally agreed to be the best, although many people feel that the shoulder meat is its equal. The rump meat, depending on the butcher, is often sold at a slightly higher price. The tail is not generally eaten, although some very poor people do buy it for soup. In some shops in Minya, Sohag and Cairo where I watched butchers work, people would buy meat by pointing to the particular portion they wanted, or else they would tell the butcher how much they wanted to spend and he would cut off portions of the animal for them, depending on the amount.

The jointing of the animal described above is as observed in Middle Egypt, with the animal in question being a baby *gamoosa* (water-buffalo). Its size might have influenced the types of joints procured as the animal was suspended from a tree for jointing. Unfortunately, most of the meat available in the butchers' shops in Middle Egypt and Cairo was from young *gamoosas*, so I was only able to study the jointing of one mature animal, which was also suspended, so the jointing method was much the same as that of the calf. If the animal had been recumbent, as it was in ancient Egypt and is in some parts of Egypt today, it is possible that the order of jointing and the shape of the joints, notably the vertebral ones, might have been different. It is possible to butcher a supine animal in the same way as one suspended, but it is more difficult, as I found when dismembering a recumbent wolf, which, despite its size, was harder to joint than a large suspended horse, both of which were dismembered by the author in Britain.

Although the order of division cannot be determined, it is possible to identify the discrete elements into which the animal is divided. All the evidence mentioned above is used to establish the jointing preferred by the ancient Egyptians. Figure 28 can be referred to throughout this section for illustrations of the joints and muscles mentioned in the text.

Head: *ḏ3ḏ3*; this term is used for an animal's head.

The head of the animal was severed from the body as mentioned above, and formed a joint of meat in itself. It features in most tomb and temple depictions of slaughter and offerings, and frequently appears on offering-tables and soul houses. Its popularity might be due to the fact that in art, it is the obvious way to represent the whole animal, rather than a remark on the delicacy of the joint. The Old Kingdom tomb of Medunefer at Balat does contain several ox-heads of all ages as offerings (**R37**, **R38**). The Middle Kingdom tomb of Mentuhotep II (Carter 1901: 201-2) contained the heads of bovids, **R4**, as offerings, while seven of the predynastic tombs in Nag-ed-Deir contained horns (goat and oxen [Appendix II.ii]), and some fragments of skulls as part of the funerary offering. Foundation deposits of temples frequently contained ox-heads[3]. These include Mentuhotep's temple in Thebes (MMA22.3.80-84), Hatshepsut's temple in Deir el-Baheri (Hayes 1959: 86), and Abydos (Petrie 1903: 11, 20). However, there is only one example of a head in New Kingdom mummified offerings, a calf's (?) head from the tomb of Queen Isitemkheb (**S268**). It is possible that no heads have been found from the New Kingdom due to the paucity of undisturbed or partially disturbed tombs, although it is odd that apart from **S268** the head is not found amongst the mummified offerings as they encompass most of the other joints from an ox. Perhaps the head was given to the priests or funeral workers or even the butcher in payment, but this is impossible to determine with any certainty.

The faunal material from Malkata (laid out in Fig. 44) shows very little evidence of cranial material in all portions of the sites, with area E, the palace midden, having a concentration of cranial remains, thereby indicating that headmeat might have been consumed by the upper classes, and so contradicting the conclusions made above. Of course, this information might be skewed due to taphonomic reasons as well as the method of excavation and recovery of the bones.

[3] There is also some pictorial evidence showing kings placing ox-heads in foundation pits in temples (Weinstein 1973: 7 n.22, 8). For a complete discussion and corpus of foundation deposits see Weinstein 1973. As Pinch has suggested (1992), it is possible that the heads found in tombs and temple deposits were apotropaic objects rather than food offerings.

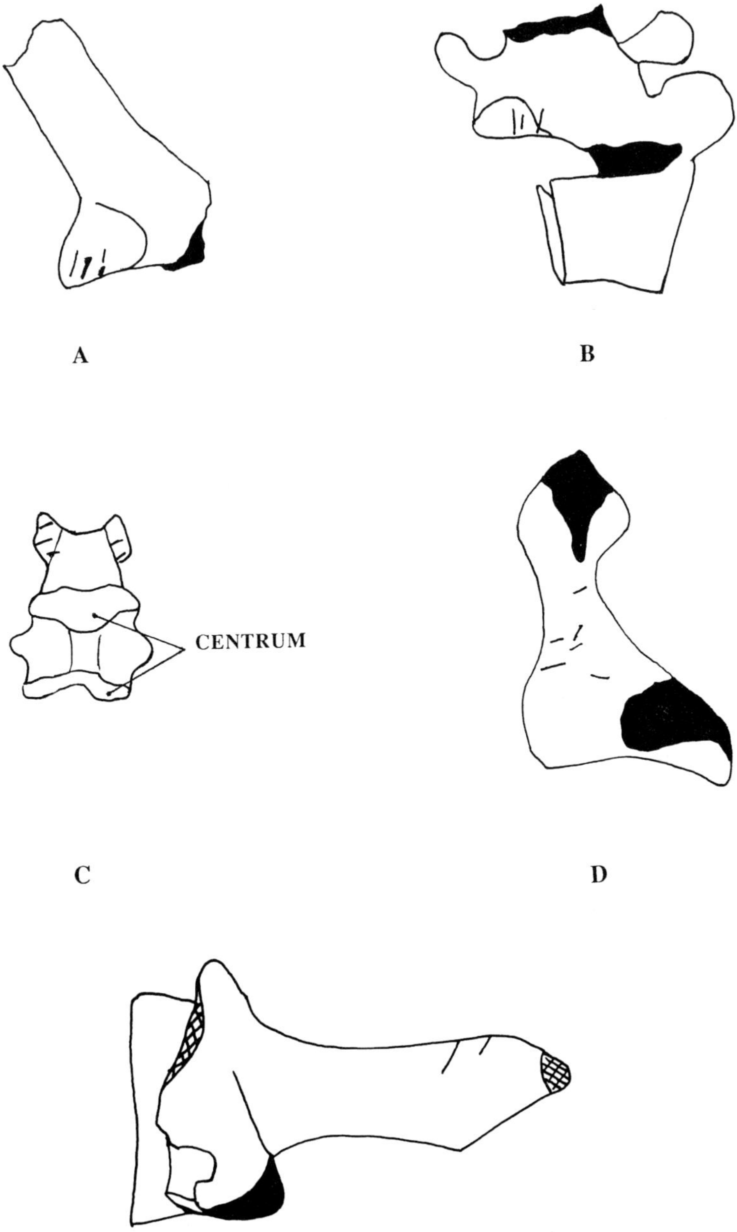

Fig. 29

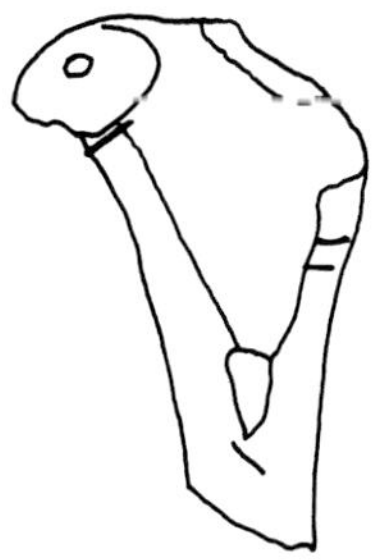

A

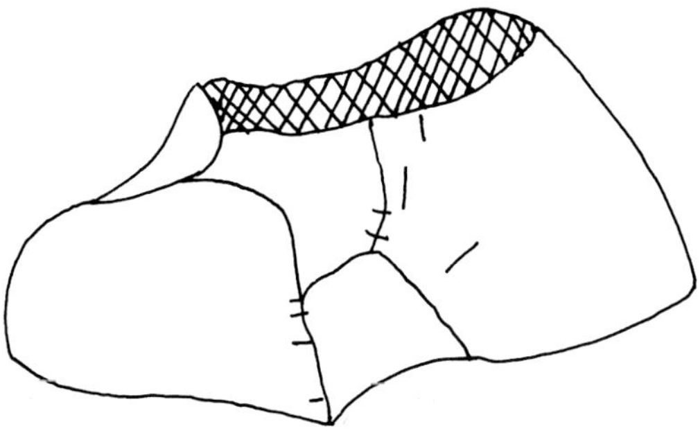

B

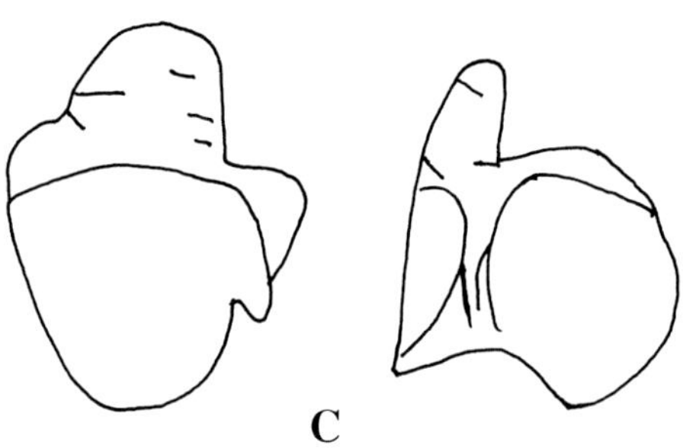

C

Fig. 30

Face meat is not a highly prized piece of meat in modern Egypt, or for that matter, in Europe, the US, and the Indian sub-continent. This might be one reason it was given as an offering, or at least shown as such. It is the most representative part of an animal, and does not yield a large amount of meat, thus making it both an economical part of the animal to offer, especially if a burnt offering was made, and a highly symbolic one. Brain was and is consumed, although today it is not a very high status food. It is possible that in earlier times it was. Some bovid crania dating from the Twenty-Sixth Dynasty were split along the median basicranial axis to permit access to the brain (Moustafa 1964: 256). There is no similar evidence from the Malkata skull material.

Neck: *nḥbt*; *ṯst ʿḥʿy*

The neck, here defined as the seven cervical vertebrae, is not a cut recognised in either two- or three-dimensional representations, although the word for it does appear in texts (*WB* V: 400). The cervical vertebrae do figure in mummified offerings (**S66**, **S67**, **S71**). Faunal remains from Malkata, Nag-ed-Deir, and other sites also show that vertebrae were consumed — especially in the case of the former, where some vertebrae show evidence of butchery (Fg. 29)[4].

Back: *pds*; *iꜣt*; *ꜣt ʿhʿy*; *ꜣt šspt*; *sꜣ*.

Mummified joints show that the Egyptians separated the vertebrae into units between four and eight, and removed the ribs from the side. Faunal remains from Malkata also correlate this evidence with nicks on the ribs (costal area) and vertebrae showing where the ribs had been removed. Mummified joints show wrapped thoracic (**S63-65**, **S68**, **S70**, **S404**) and lumbar vertebrae (**S69**, **S72**, **S187**, **S188**) (Figs. 31-32). This method of jointing where the entire rib is isolated in not used in modern Egypt or Europe, where the vertebrae are chopped through the centrum when the carcass is divided in two. There are a few examples from Malkata where vertebrae have been found intact, as well as cut through the centrum. This might suggest that both types of vertebral jointing were used, perhaps differing according to the requirements of the cooking process to follow.

[4] Throughout this work I have used Binford (1981) as well as my experience with butchery for associating butchery marks with different types of jointing and filleting activity.

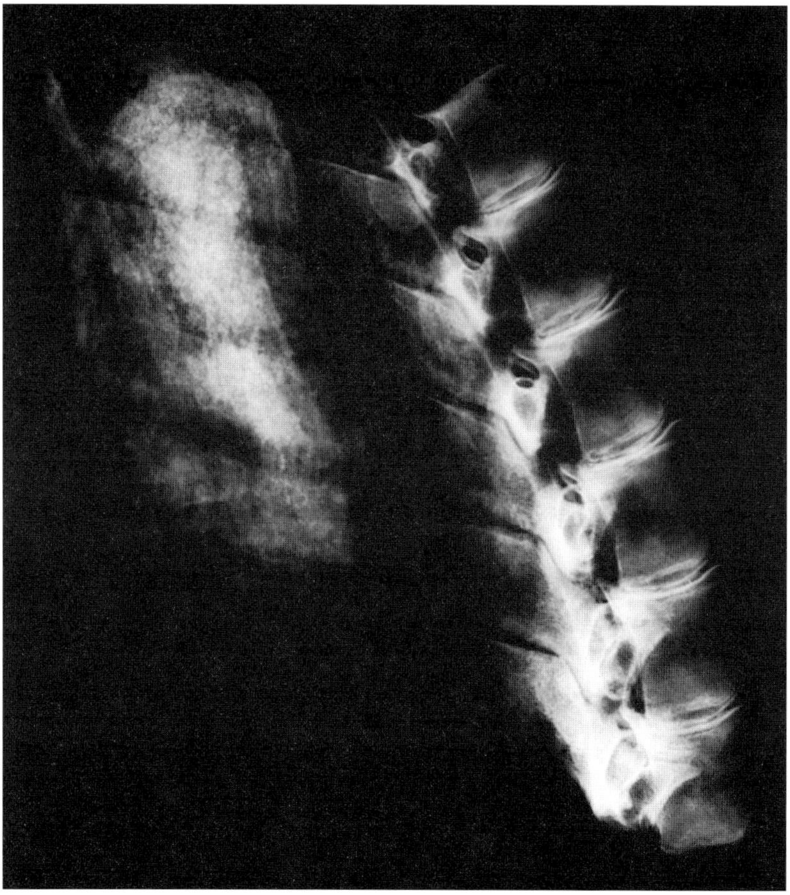

Fig. 31

The entire back was used, even the sacrum, as a mummified sacrum (**S69**) was found. Some painted limestone models of parts of an ox, labelled ribs, might actually represent the sacrum (MMA64.66.8).

<u>Shoulder</u>: *ḏnḥ* (?)

Several scapulae of cows (Boessneck, J. and A. von den Driesch 1982: 120-21), oxen, sheep/goat, and antelope have been found as offerings in tombs from all periods of Egyptian history. There are five examples that come from the Predynastic graves at Nag-ed-Deir (Appendix II; Fig. 45), two examples (in association with humeri) from the Middle Kingdom tombs in Dahshur, and nineteen examples of single scapula from

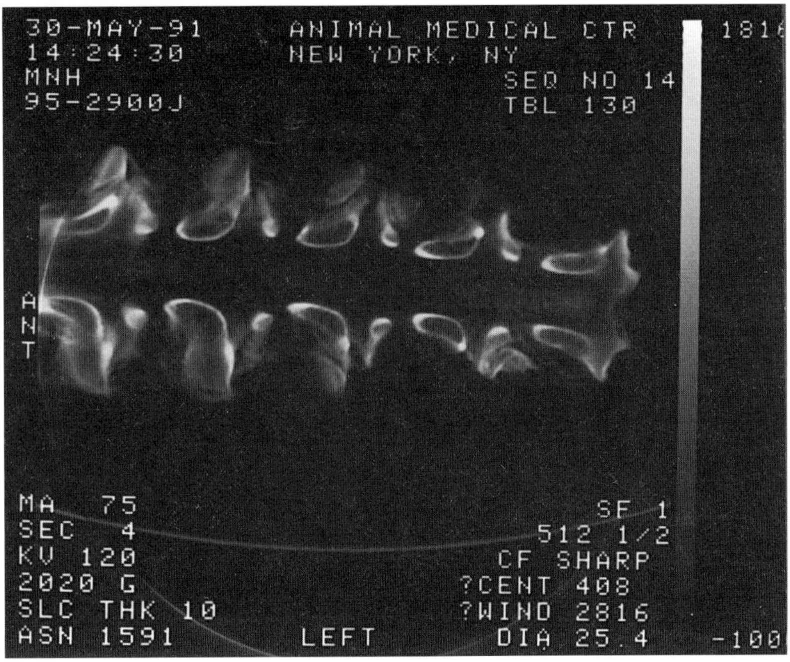

Fig. 32

the New Kingdom victual mummies (Appendix II; Figs. 42-43). They also form a large part of the faunal material recovered from Malkata — 4.2%[5] (Fig. 44). They do not feature obviously in the art, except in the fact that they are part of the foreleg joint that is removed as a whole.

According to Kamal (1909: 10-3), an alternative word for shoulder, *swt*, features on the table of Princess Neferuptah (Fig. 27 [see below]), although this is a doubtful translation. Appearing on the offering-table of the Princess Neferuptah (Fig. 27) is a joint marked "*ḫpš*", the word traditionally associated with foreleg (see below). The joint itself looks very much like a scapula. To the author's knowledge this is the sole example of such a joint being so marked, and raises the question as to whether the artisans labelling the meat offerings on the table did so with a view to convention, or whether their labelling was somewhat random, an hypothesis that would lead one to conclude that using the offering table to identify cuts is unreliable (see note 1).

[5] This is out of clearly identifiable bones (901) in whose case fragments and unidentified bones do not make up the majority of the percentage. All percentages from Malkata mentioned herein are based on this assumption.

Fig. 33

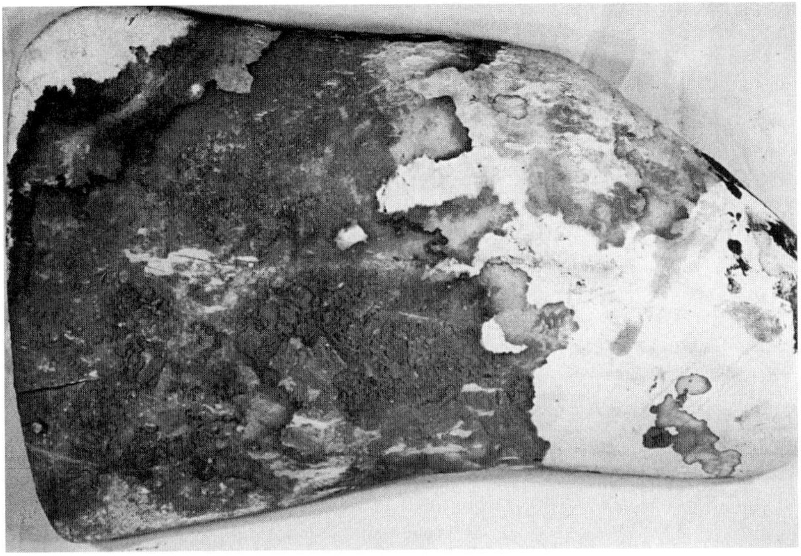

Fig. 34

There is one example of a triangular limestone box that probably represents a scapula (or a scapula and a humerus), **N1** (Appendix I; Fig. 37). The triangular shape is virtually identical to that of the wooden boxes found containing scapulae (see listings in Appendix II, Figs. 34-36, 42-43); furthermore, **N1** has a slender raised ridge running down the centre of the box. Presumably this mimics the fold found on the bone.

Joints including or consisting solely of scapulae are very rich in meat and are, in modern Egypt, regarded as more worthwhile and expensive cuts.

Sternum:

This is the breastbone of an animal. Verhoeven has identified an offering-joint (Fig. 38) as a sternum (1984: 114-15). This object is frequently depicted being boiled in a pot along with other joints of meat in tomb representations of the Old and Middle Kingdoms, though it is unclear if it is indeed a sternum. A small model of a painted limestone joint that resembles this two-dimensional object (MMA64.66.8) has also been found, although it could in reality represent the sacrum (see above).

This two-dimensional representation does resemble a sternum more closely than anything else in an animal skeleton, although it could conceivably represent part of the vertebral column with the upper spines not shown clearly. This would have little meat attached. The fragile nature

Fig. 35

of the bone is responsible for its absence in the archaeological record. (Verhoeven 1984: 114-15).

The victual mummies, however, attest to their consumption. Two examples of wrapped sterna come from the tomb of Tutankhamun (**S197**, **S198**). The hieratic on the boxes containing these joints seems to have been a "*ḳ*", which might have stood for *ḳnḳn*, joint of some sort (Gardiner 1947b: 244), *ḳʿḥt*, a part of the leg (*WB* V: 19), *ḳbḥt*, leg (Gardiner 1947b: 255), or, most likely, *ḳ3bt*, breast (Gardiner 1947b: 241; Cerny 1965: 17) (see below for further discussion). There are also several other instances of sterna in this class of object: **S73**, **S74**, **S75**, **S76-7**, **S78**, **S79**, **S80-2**, and **S83.**

In modern Egypt this is not a joint that is cut off separately and consumed.

Chest: *iwf n ḥ3t*; *ḳ3bt*

Although the sternum can technically be considered a part of the chest, it is dealt with as a separate anatomical element. The chest would constitute any meat from the ventral side of the animal. The term *iwf n ḥ3t* (*WB* I: 52), means literally meat from the animals' foreparts, rather than meat from the ventral side. The word *ḳ3bt* is a more definite word for chest meat, and appears in lists of meat in both funerary and daily

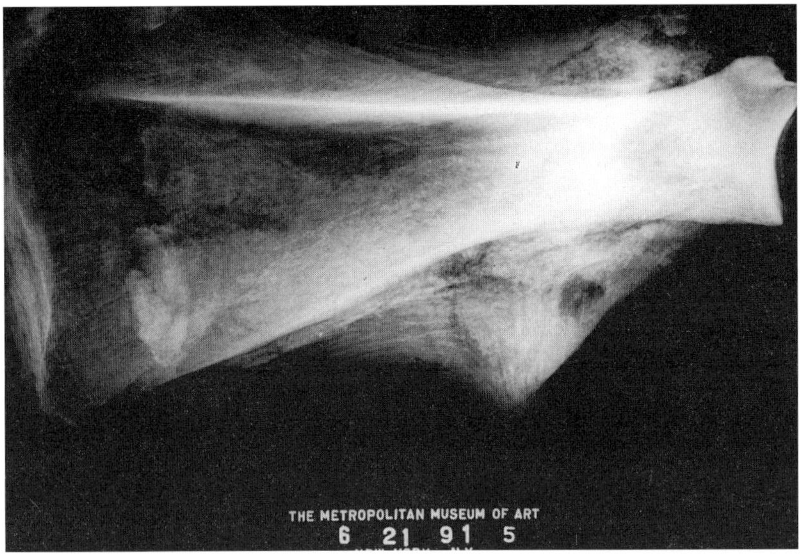

Fig. 36

life texts (Pendlebury 1951: 172-3). Cuts from this part of the animal will be rich in meat rather than bone, and are therefore considered "good" cuts in the modern world. It is possible that some of the "steaks" (**S395-403**, **S405-413**, **S369-377**, **S379-383**, **S386-388**, **S391-393**, to mention but a few) originate from the chest.

Ribs: *spr*, *ḏrw* side of ribs (Gardiner 1947: 244-5; *WB* V: 602), *grgy(t)* wing/prime rib (Gardiner 1947: 244-5), *spḥt* (Gardiner 1947: 244-5; 19).

Egyptians frequently separated the ribs from the vertebrae as cuts on the vertebrae and several examples of mummified ribs (Appendix II) attest. Ribs, in groups of four or more, or sometimes an entire side (*ḏrw*; *spḥt*), are shown on offering-tables or in the hands of offering-bearers. Some of these divisions are confirmed by the victual mummies as well as the appearance of ribs as offerings in Nag-ed-Deir, Sakkara, Dahshur, and Lisht (Appendix II: Part II). Ribs also feature in the foundation deposits of temples (Weinstein 1973), including Sesostris I's pyramid complex (MMA32.1.92), and Mentuhotep's mortuary temple (MMA 22.3.82; 22.3.132; 22.3.192).

Several limestone boxes shaped like ribs (generally in groups of four) have been recovered from Memphite tombs of the Old Kingdom (**N8-11**). Some of these are painted ochre to imitate the colour of flesh.

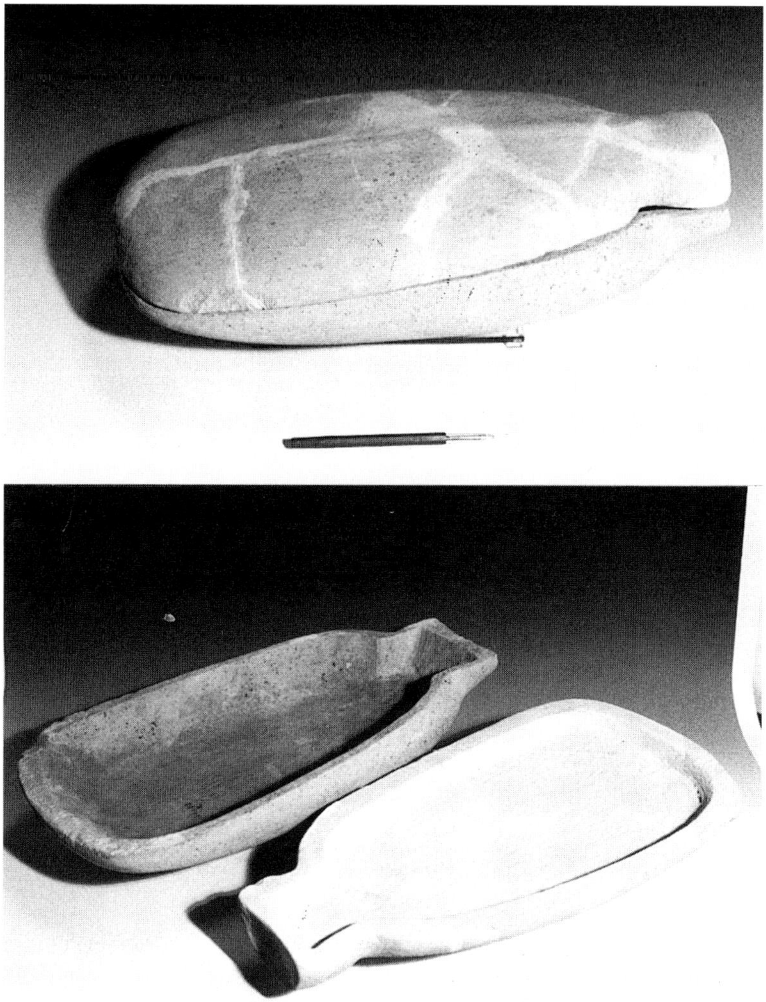

Fig. 37

Rib fragments, aside from longbone fragments, are the most commonly found bone in the archaeological record. Although this is due to their shape, which lends itself to excessive fragmentation, and number (twenty-four or more per animal), it does indicate that they were consumed with enthusiasm. They make up 50% of the Malkata ensemble. Sometimes ribs would be cut in half, as is common in modern butchery. This curtails their length and makes them more manageable. There are

Fig. 38

several examples of ribs covered by cut-marks that indicate filleting. Also present are cut-marks indicating division and articulation. One wonders how they were cooked — they are never shown in pots. Perhaps the Egyptians were the first aficionados of barbecued ribs?

Cuts that are clearly ribs are labelled *spḥt* on Neferuptah's offering-table (Fig. 27).

Foreleg: *ḫpš*

This is shown being removed whole from the animal after slaughter, and was probably the first limb to be removed from the carcass (see above and III.1.1). In its entirety it is comprised of the scapula, humerus, radius, ulna, carpals, metacarpals and phalanges (Fig. 28). Generally, in mummified and other funerary offerings (excepting **R15**, **S360**), the leg is cut off below the carpals as in the eight examples from KV35 and KV34 (Appendix II; Fig. 43), though sometimes the box in which it is placed is shaped like the entire leg (**S306**). There is not much meat on the metacarpals and phalanges. It is probable that the foreleg was removed in one go, but was then divided into its composite parts for consumption (see note 1, and above discussion for 'shoulder').

This is the joint most frequently shown being offered to the gods as well as to the deceased. It appears in several spells in the Book of the Dead (e.g. #28, #62, #106) and in the Coffin Texts (e.g. #279, #936).

Humerus: *ḫpš ḥri*, upper part of foreleg (?); *tpy n ḫpš* a part of the leg (Poesner-Kriéger 1976: 244), possibly meaning the upper part, i.e. humerus/thigh; *ḫnḏ* part of a leg (*WB* III: 314; Gardiner 1947a: 17-8).

A total of twenty-six individual humeri victual mummies with meat attached were found in New Kingdom burials (Appendix II; Figs. 42-

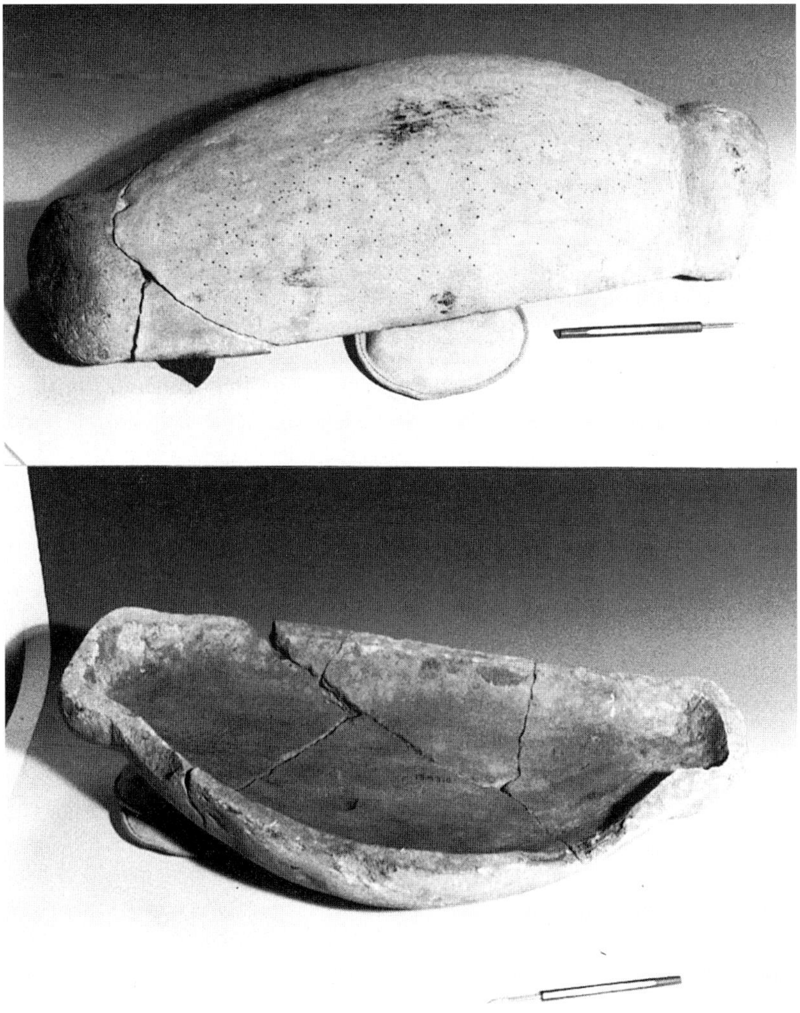

Fig. 39

43). Four humeri were found in the Middle Kingdom tombs at Dahshur, and five in the tombs of Nag-ed-Deir (Fig. 45). From Malkata four showed evidence of butchery, three of which indicated dismemberment at the join between the humerus and the radius and ulna (Fig. 30). It made up 7% of the findings from Malkata (Fig. 44). Humeri are very rich in meat. In modern Egypt a cut from any leg of the animal is called "*raj(g)al*", the word for leg.

Fig. 40

N2-7 (Appendix I) could represent the humerus or the tibia of a bovid (Fig. 39).

Radius & Ulna: *ḫry ḫpš, ḫnḏ* part of a leg (*WB* III: 314; Gardiner 1947a: 17-8).

These two bones are treated as a group due to their proximity and interdependence in the skeleton. Two such combinations appear in the funerary offerings of Nag-ed-Deir (Appendix II.ii; Fig. 45). Fourteen such joints were found amongst the victual mummies, with another found in conjunction with the metacarpals and carpals, and four independent radii (Figs. 42-43). On the whole, it seems that the Egyptians thought of the two bones as a unit when they jointed the animal. Further secondary jointing would have separated the two bones as is indicated by the butchery marks on one ulna and two radii from Malkata.

Hind Leg: *mnt*

The femur, tibia, tarsals, metatarsals and phalanges comprise the entire hind leg. In two- dimensional representations the hind leg is sometimes shown being removed in its entirety, although the foreleg is more commonly shown thus. No hind legs appear in the Dahshur offerings and only

Fig. 41

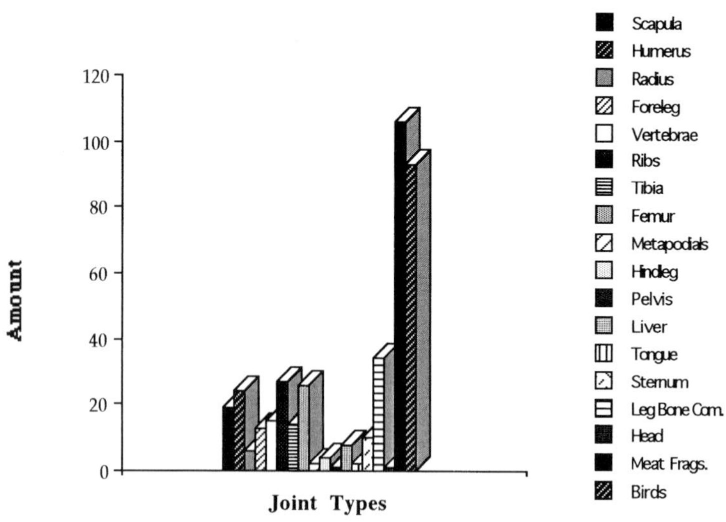

Number of Each Joint

Fig. 42

	KV34-35	KV62	KV43	KV36	KV46	Isitemkheb	QV46	MMA1021	TT358	KV60	TOTAL
Scapula	12	4	0	0	1	0	0	2	0	0	19
Humerus	16	4	0	1	1	0	0	1	0	1	24
Scapula & Humerus	1	0	0	0	1	0	0	0	0	0	2
Radius	4	2	0	0	0	0	0	0	0	0	6
Radius & Ulna	14	0	0	0	0	0	0	0	0	14	
Radius & Humerus	1	0	0	0	0	0	0	0	0	0	1
Leg bone	0	0	0	2	1	1	0	0	0	2	6
Foreleg (to carpals)	7	0	0	0	1	2	1	1	0	1	13
Tibia	8	4	0	0	1	0	1	0	0	0	14
Tibia to Metatar-sal	0	0	0	0	0	0	0	1	0	0	1
Tibia & Calcanem	1	0	0	0	0	0	0	0	0	0	1
Tibia & Tarsals	8	0	0	0	0	0	0	0	0	0	8
Femur	20	1	0	0	0	0	2	0	0	0	23
Femur & Patella	0	3	0	0	0	0	0	0	0	0	3
Metapo-dial	0	2	0	0	0	0	0	0	0	0	2
Hindleg (to tarsals)	0	0	0	1	0	2	0	1	0	0	4
Pelvis	0	0	0	0	0	1	0	0	0	0	1
Liver (portions)	5	0	0	1	0	1	0	1	0	0	8
Tongue	0	2	0	0	0	0	0	0	0	0	2
Head	0	0	0	0	0	1	0	0	0	0	1
Meat Fragment	34	14	0	0	0	4	2	26	21	5	106
Pigeon/ dove	0	6	1	2	1	0	0	9	0	0	19
Duck	3	1	2	1	2	4	5	13	1	0	32
Goose	3	4	3	2	3	0	0	20	0	1	36

Fig. 43 — Joint Types Found in Victual Mummies

one comes from Nag-ed-Deir (Appendix II.ii; Fig. 45). The New Kingdom is richer in hindlegs as offerings — four examples come from the victual mummies (Appendix II; Figs. 42-43), although most stop at the tarsals.

Femur: *mꜣst*; *mst*; *ḏnḥ*; *iwf n pḥwi* ham, i.e. femur (Barta 1963: 119); *iwꜥ* femur/humerus (Verhoeven 1984: 23); *iwꜥ* (*WB* I: 50).

All these words save *iwf n pḥwi* are thought to be used for the thigh, i.e. the upper part of the hind leg. The word *mny* means part of the hind leg, and could mean femur. Although two- and three-dimensional representations and offering lists would suggest that the foreleg was the

Area	E	J	K	L	P	TOTAL
Head Bones	196	90	3	17	4	310
Astragalus	4	7	0	1	0	12
Calcaneum	4	5	1	0	0	10
Carpals	5	17	2	0	5	29
Femur	16	7	7	3	5	38
Humerus	26	27	4	2	4	63
Metapodials	32	37	2	11	3	85
Patella	1	6	0	0	0	7
Pelvis	24	26	7	1	7	65
Phalanges	28	32	0	5	9	74
Radius	13	10	0	5	2	30
Ribs	126	233	13	34	44	450
Scapula	17	18	0	2	1	38
Sacrum	5	0	0	0	0	5
Tarsals	2	3	5	0	0	10
Sternum	5	0	1	0	0	6
Tibia	27	12	5	2	4	50
Ulna	7	7	1	1	1	17
Vertebrae (Indet.)	119	138	19	3	24	303
Atlas	5	0	0	0	0	5
Axis	3	0	0	0	0	3
Caudal Vert.	4	1	0	0	1	6
Cervical Vert.	9	6	1	2	0	18
Lumbar Vert.	18	3	2	2	1	26
Thoracic Vert.	43	26	1	5	3	78

Fig. 44 — Elements per Area in Makata
(Areas defined in Chapter X)

standard offering, twenty-four mummified femurs have been found, with two additional femurs in conjunction with the patella (Fig. 43). The foundation deposits in Hatshepsut's funerary temple at Deir el-Baheri contained femurs. Two femurs, in conjunction with patellae were found in Dahshur (Boessneck, J. and A. von den Driesch 1982: 120-1; 124; Appendix II.ii), while no independent femurs feature in the offerings at Nag-ed-Deir. In the New Kingdom site of Malkata femurs make up 4.2% of the number of bones found, two of which show knife cuts indicating dismemberment (Fig. 30). Joints resembling femurs are depicted hanging on butchers' lines or in a pot from the Old to the Middle Kingdoms. The New Kingdom representations of butchery scenes are fewer and less secular than those of preceding eras, consequently there are fewer illustrations of suspended femurs.

Grave	7172	7235 C	7266	7290	7296 A	7304	7421	7454 A	7461	7481	7484	7497	7519	7521 A	7525 A	7539	7583	7595 B	7597	7621	7624 A	Total
Gender	F	F	M	N	M	N	M	M	N	F	F	F?	N	N	F+	N	M	N	M	N	F	13
Goods	P	P	√	√	√	√	√	√	√	P	√	P	√	√	√	√	√	√	?	P	√	15?
Bos	√	0	0	0	0	0	?	0	0	0	0	0	3	0	1	0	0	0	0	0	0	5+?
Capra	0	0	0	√	0	0	?	0	0	0	√	1	0	1	0	0	1	7	10	1	1	14+?
Ovis	0	1	0	0	0	1	0	0	0	0	0	0	0	0	0	0	0	0	0	0	0	2
Gazella	0	0	0	0	?	0	0	12	0	0	0	0	0	0	0	10+	0	0	0	0	0	22+
C/O/G	0	0	√	0	√	0	0	√	√	√	0	0	√	√	0	0	0	0	0	0	0	7?
Scap	0	1	0	0	0	1	0	1	0	1	0	0	1	0	0	0	0	0	1	0	0	6
Humer	0	0	1	0	0	0	0	0	0	0	0	1	0	0	0	1	0	0	0	0	0	3
Radius	0	0	0	0	0	0	0	0	0	0	0	0	0	0	0	0	0	1	1	0	0	2
Ulna	0	0	0	0	0	0	0	0	0	0	0	0	0	0	0	0	0	1	1	0	0	2
Phalan	0	0	0	0	3	0	0	0	0	0	0	0	0	0	0	0	0	1	1	0	0	5
Leg-bn	0	0	√	0	0	0	0	√	√	√	√	0	0	√	0	0	0	0	0	0	0	6?
Femur	0	0	0	0	0	0	0	1	0	0	0	0	0	0	0	1	0	0	0	0	0	2
Tibia	0	0	0	0	0	0	0	1	0	0	0	0	0	0	0	1	0	0	0	0	0	2
Tar/Car	0	0	0	0	0	0	0	2	0	0	0	0	0	0	0	0	0	0	0	0	0	2
Metap	0	0	0	0	0	0	0	2	0	0	0	0	0	0	0	0	0	1	1	0	0	4
Astrag	0	0	0	0	0	0	0	1	0	0	0	0	0	0	0	0	0	0	0	0	0	1
Calc	0	0	0	0	0	0	0	1	0	0	0	0	0	0	0	0	0	0	0	0	0	1
Ribs	0	0	0	0	0	0	0	0	0	0	√	0	3	0	0	√	0	0	4	0	0	9?
Vert	0	0	0	0	35	0	0	4	0	0	0	0	0	0	0	7	0	0	0	0	0	46
Patella	0	0	0	0	0	0	0	0	0	0	0	0	0	0	0	0	0	1	0	0	0	1
Hd-bns	√	0	0	0	0	0	0	0	0	0	0	0	0	1	0	0	√	√	√	√	√	7?
Horn	1	0	0	1	0	0	2	0	0	0	0	0	0	0	1	0	1	2	0	1	1	10
Total	2	1	2?	1	38	1	2	14?	1?	2?	2?	1	4	2?	1	11?	2?	8?	10?	2?	2?	–

Fig. 45 Nag-ed-Deir (Identifiable Bones)

KEY

?	Possible	Scap	Scapula	Astrag	Astragalus	√ Present, number uncertain
M	Male	Humer	Humerus	Calc	Calcaneum	
F	Female	Phalan	Phalanges	Vert	Vertebrae	
N	Neuter	Leg-bn	Leg bones	Hd-bns	Head bones	
P	Plundered	Tar/Car	Tarsals/Carpals	C/O/G	Capra, Ovis, Gazella	

Tibia: shank/shin *sdḥ*; *mny* (?), *sd3ḥ*; *swt* shin (Hassan 1944: 117; Posener-Kriéger 1976: 245; Verhoeven 1984: 23); *swt* piece of meat (Barta 1963: 49), but could also be used to describe a tibia, especially as the determinative contains a bone; *ḫnḏ* part of a leg (*WB* III: 314; Gardiner 1947a: 17-8).

This portion of the animal was, like the femur, not stressed as an offering, although twenty-four tibia form a significant part of the New Kingdom victual mummies (Figs. 42-43). 5.5% of the Malkata findings are of tibia (Fig. 44). One tibia found in Malkata shows evidence of dismemberment, while two others show signs of filleting. There are no separate tibia in the offerings from Nag-ed-Deir or Dahshur.

On Neferuptah's offering-table (Fig. 27) two joints of meat are labelled *swt* which Kamal (1909: 10-3) translates incorrectly as shoulder. These two cuts, although differently depicted, are both leg joints.

Metatarsals/Metacarpals:

These joints are rarely found as offerings in Egypt. They only appear thrice amongst the victual mummies, and then only in conjunction with other joints to make up a leg. They do appear solo twice in the victual mummies of Tutankhamun (**S205** and **S206**). Only in Nag-ed-Deir (Appendix II, three examples; Fig. 45) and Balat (Appendix II.ii: **R40-43**) do any appear as independent offerings. Generally these pieces were cut off the animal and might have been discarded or distributed as food amongst priests or others. There is little meat on these bones and in modern times are amongst the cheapest and least important cuts. Examples of metapodials from Malkata's palace dump, E, show cutmarks indicating dismemberment.

Feet: *ḳbḥ* part of leg/foot (*WB* V: 26); *ḫnd* leg/foot (*WB* III: 313); *wḥmt* hoof (*WB* I: 340); *inst:* rear hoof/foot (Posener-Kriéger 1976: 242).

The foot is generally made up of phalanges — three or more, depending on the animal, and sometimes the metacarpals/metatarsals (see above). Only two examples of phalanges have been from Nag-ed-Deir (Fig. 45). Appendix II shows that two or three legs amongst the victual offerings contain phalanges, but most entire limbs stop short at the carpals/tarsals (Fig. 43). There is almost no meat on the foot and it is better used for soup and glue. Phalanges do appear in the archaeological record and make up 8.2% of the finds at Malkata. Despite the lack of meat on the phalanges, document from the temple archives of Neferirkare report that two *inst* a day were offered (Posener-Kriéger 1976: 242), so obviously they were consumed and perhaps also used for glue — although more than two *inst* a day would be necessary to produce any useful amount of glue.

Tail: *ḫbst*, *sd* (Posener-Kriéger 1976: 246).

These have not been found in funerary offerings or in foundation deposits. However, if *sd* does indeed mean tail, they have been distributed to the *ḫntiw-š* from the temple of Neferirkare (Posener-Kriéger 1976: 246), and feature in lists of meat in Amarna (Pendlebury 1951: 172). The tail is best used for soup (or stew) as it is poor in meat, the body of it being almost entirely made of bone and cartilage.

Viscera

Mummification gives us ample evidence that the Egyptians were aware of the different viscera and their locations in the body. The Egyptians were also used to consuming the viscera from animals. It is possible that some of the pieces of meat pictured hanging above the butcher in the tombs of Rameses III (KV) and Ipy (TT217) were elements of the viscera.

Liver: *mist*

Liver was consumed and offered as is attested in texts from Neferirkare's archives (Posener-Kriéger 1976: 243), the cut so labelled on Neferuptah's table (Fig. 27), and the eight pieces of liver that figure in the mummified meats (Figs. 43-44).

Spleen: *nnšm* (Gardiner 1947b: 245).

The name for this organ was known and appears in the onomastica, but it is doubtful that this was consumed. It is not generally consumed in modern Egypt, Europe, the Indian sub-continent, or the United States. The second writing for it appears on Neferuptah's offering-table.

Lung: *wfꜣ* (*WB* I: 306)

This might have been consumed, but it would be a low-prestige cut, much as it is in Egypt today.

Intestines: *imy-ẖt*, *mẖtw*

Intestines were probably eaten in a soup or fried. Dried *mẖtw* are also known from P. Anastasi IV: 14.3, where they are measured in handfuls (Caminos 1954: 205). They are nutritious, but are not regarded as of high status, or even very desirable cuts today. Ankhmahor's tomb at Sakkara shows the removal of a slaughtered ox's intestines.

Pancreas: *sḫn*

The word *sḫn*, thought to mean pancreas, is more convincingly identified by Dawson as kidney suet (Gardiner 1947: 253), a highly regarded form of fat. This definition is endorsed by Faulkner (1976: 241). Others have said this word means scapula (Posener-Kriéger 1976: 245; Kamal 1909: 10-3); the stomach (Verhoeven 1984: 234); or just a piece of meat (Zayed 1958: 132). The last writing listed above appears on a floppy joint on Neferuptah's offering-table (Fig. 27) that bears absolutely no resemblance to a scapula, although it might well represent a pancreas.

Stomach: *r-ib*, *mẖtw*

Tripe was eaten in ancient Egypt as it is today. Modern Egyptians regard it as a low-status food.

Tongue: *šsꜣw* (Faulkner 1976: 271); *ns* (Faulkner 1976: 139); *šsr* (Faulkner 1976: 272); *snk* (Faulkner 1976: 234)

According to Carter's notes (1932) one of the boxes containing meat from KV62 contained an ox tongue (Figs. 42-43; **S215**). Unless he mistook a piece of liver for a tongue, this is the only actual evidence for the consumption of tongue in ancient Egypt, although it is quite a feasible supposition that it was consumed in ancient times as it is today.

Heart: *ib* (Faulkner 1976: 14); *ḥꜣty* (Faulkner 1976: 162)

Four of Tutankhamun's meat boxes were labelled *ib*, but contained humeri (see below and **S413-4**, **S416**). Although the heart was shown being removed it does not appear on offering-tables or in offering-lists in tombs and temples. It is possible that the heart was examined for indications of the health of the animal, as was the blood, but it is doubtful that it was consumed. Perhaps this reluctance to consume the heart was due to the fact that the Egyptians believed, certainly for humans, that the heart was the seat of the soul/essence of an individual and necessary in the afterlife. Whether such niceties were extended to animal souls is unknown.

Pelvis

In the late New Kingdom, if not earlier, the pelvis was cut into portions, as shown by the mummified remains from the burial of Isitemkheb D (**S278**, **S280**, **S284**; Figs. 42-43). Fragments of pelvic bones have been recovered from Malkata (Fig. 44). The name of this joints is not known, though *pḥwyt* (*WB* I: 537), from *pḥwy*, hindquarters, might be related. The *WB* and Gardiner (1947) both translate this as rectum. There is one cut found in the butcher-shop model of Meketre (MMA20.3.10) which Winlock tentatively interprets as an unusual cut of sirloin which includes half of the pelvic girdle (Winlock 1955: 25). I would disagree with this identification as the joint looks more like a humerus or tibia alone, a proposition put forth by A. S. Gilbert in his article on the model (Gilbert 1988: 81). Butchery marks indicative of dismembering have been found at the edge of the acetabulum from area E in Malkata.

Pieces of Meat: *ḳnḳn*, joint (*WB* V: 56); *midꜣ* edible part of an animal (*WB* II: 45); *iwf* meat, *ḏbw* meat (Meeks 1980: 446); *ḥʿ* piece of meat (Barta 1963: 49, on Neferuptah's offering-table [Fig. 27); *iwf n pḥwy* ham, i.e. femur (Barta 1963: 119); *ḥwn* piece of meat (Barta 1963: 11; *WB* III: 55); *iwr* dish of meat — perhaps cooked?; *sms* piece of bovine meat (*WB* IV: 141); *stp.t* choice cuts (*WB* IV: 336); *wʿb.t* piece of meat (*WB* I: 284), *dp(t)* a joint of meat (Gardiner 1947: 243); *sb* an edible part of an animal (*WB* IV: 81); *dgyt* an unspecified piece of meat (Caminos 1954: 205).

Most of the words above are names for unspecified joints or pieces of meat. Chunks of meat feature in several spells from the Book of the Dead (e.g. BD 1, 58, 99, 122, 125). They could come from any part of the body — if they did refer to specific parts of the body, and no doubt the ancient Egyptians knew which words referred to which portions. Unfortunately it is impossible for us to determine this in most cases. The term *iwf n pḥwi*, which Barta (1963: 119) defines as the leg from the rump down or the ham (femur), might be better translated as meat from the rump, and would denote many of the "steaky" chunks of meat in MMA1021 and TT358 (see Appendix II; Figs. 42-43). The rump is one of the richest sources of boneless meat, and it is quite possible that many names refer to meat from there. The word *dp(t)* can mean loins as well as a joint, perhaps indicating a joint from the lower back (Gardiner 1947: 243), i.e. the pelvic area or the femur. A long triangular joint, labelled *dpt* is pictured on Neferuptah's table. Another word, *sb*, if it is related to the word *sbḳ*, calf (Faulkner 1976: 220) might also refer to a leg joint, or meat from the leg. A jar bearing this legend was found in the tomb of Hemaka at Sakkara, and is now in the Cairo Museum (number unavailable).

The term *stp.t*, "choice cuts", probably denotes a whole series of joints rather than any single piece of meat. The *ḫpš* is part of it, as (probably) are the *spr*, *iwʿ*, and *swt*, these being the most commonly offered joints in offering lists as well as in actuality. The Coffin Texts corroborate this by emphasizing the forequarters of the animal and add the *sẖn* and *ḥʿ* joints, along with liver and poultry to the selection of favoured parts (e.g. Spells 923, 924, 936, and 1021 [Faulkner 1978: 65-73).

No doubt the ovoid/circular boxes representing meat (**N12-17**) and the actual meat pieces of that shape (**S403, S409**; Figs. 42-43) are referred to by some of these words.

There is one unusual piece of meat that appears suspended on the meat line in both two- and three-dimensional representations — a triangular cut of meat. This is frequently the most common cut on the line. No such piece has been found amongst the victual mummies unless the rolled flank steak (**S307**, **S378**, **S384-5**) was initially cut into a triangle before being rolled. Winlock interpreted this cut first as the tenderloin or filet mignon, and subsequently as "other" pieces of the animal that were cut to this shape, and were obvious to the ancient Egyptians but not to us (Winlock 1955: 25). Gilbert reinterpreted the joint as the ilio-psoas muscle in the lower back. It is located near the junction of the thoracic and lumbar vertebrae where the pelvis and upper femur connect. However, there is one problem with this interpretation; there should only be two such muscle groups per animal. According to Gilbert the model's suspended joints show the results of the slaughter of only one animal; however, there are seventeen triangular cuts in the Meketre model, meaning that, if he is correct in his identification, eight and a half animals would have been slaughtered. He does have another suggestion for this inconsistency. He writes:

> Faunal analysis provides a plausible solution to this inconsistency. Except when it is cut transversely during the production of steaks, the tenderloin is left intact by todays' meat cutters, but in antiquity most muscles were apparently left intact. Even in the context of the modern, pre-industrial village, it is more common for butchers without power saws to cut around bones than through them. The preparation of steaks and chops is, by contrast, at variance

> with anatomy, and a far simpler means of dividing a carcass is to follow the natural paths of muscles and cut them only where they need cutting, at their skeletal attachments. More likely, the triangular cuts of the model were meant to suggest the generally fan-shaped to fusiform shape of most limb muscles as well as many on the torso. Muscle origins are often broad for anchoring at several biomechanically necessary localities, while their insertions on the bone to be mobilized are often restricted in size. Severed at both ends, these muscles would appear roughly triangular, and they would become long and narrow when relaxed or drawn out. If the bovine carcasses in Meketre's slaughterhouse were stripped by removing muscles individually, then the abundance of triangles is explained. The number of previously butchered animals yielding the hanging triangles can probably be reckoned safely as one (Gilbert 1988: 86-7).

Assuming that the representations are accurate, there are one or two problems with Gilbert's argument. One might question his premise that "in antiquity most muscles were apparently left intact" (Gilbert 1988: 86). Unfortunately, there is no reference for the basis of that statement. He goes on to give the example of modern pre-industrial villages. This argument should not be taken at face value as there are several pre-industrial or 'traditional' villages, including those found in contemporary Egypt, where butchers, once the initial dismemberment is accomplished, divide the animal with their cleavers and knives with no regard to natural muscle groupings or articulations. It is not necessary for these triangles to represent recognisable muscle groups — they could just be portions of the animal cut to this shape and sliced laterally, perhaps for drying (see Chapter VIII). If this is the case, some reliefs suggest that these cuts come from the foreleg, see below, and would allow for more than two per animal. Some butchers have suggested that the cut comes from the corner rump (biceps femoris), but this once again brings up the problem of two per animal (Fabish 1990).

Several tomb paintings showing these triangular cuts of meat hanging on lines also feature a butcher seated below, carving up a leg (Fig. 18, and TT79, TT112, TT104). These representations would suggest that the triangular cuts came from the thighs of the animal. This interpretation would solve the problem of the number of triangular cuts per animal as four thighs per animal, with four or more triangular cuts per leg would explain the large number of these cuts. As mentioned above, it is possible that these pieces were called *dpt*.

Further evidence regarding these peculiar cuts of meat comes from American Indians (Fig. 40). The Plains Indians are the only other culture the author has thus far encountered that produces these strange cuts of meat. Lamentably, the originating joint of these cuts is unknown.

Some of the boxes containing victual mummies from the tomb of Tutankhamun (KV62) bear hieratic labels which should provide the Egyptian names of the contents. Examination of the labels in connection with the contents of the boxes by Gardiner (1932), Carter (1932), and Cerny (1965: 17-8) has shown that the labels do not always correctly relate to the contents (see below). Out of forty-eight boxes twenty-eight are labelled, and it seems that few are properly labelled. Below is a list of the boxes with the labels they bear with discussions of the names in relation to their contents.

S183-6 62b, d, q, s Triangular boxes marked *ḏ3ḏ3*, all containing scapulae (Carter 1932, Gardiner 1932, Cerny 1965: 17). Clearly *ḏ3ḏ3* means head, a joint that has yet to appear in a box — the only example of such a victual mummy is **S268**. Could it be possible that the Egyptians meant 'from the front (of the animal)' in this case, a position for which the scapula qualifies? The consistency between the labels and the contents is remarkable.
S199-202 62h, p, v, qq All the boxes are of differing shapes: two oval, one lozenge, and one vaguely rectangular, all marked *ib*, for *ḥ3ty*, heart. They contain the humerus. No boxes containing identified hearts have been found so far, although Winlock misidentified some pieces of meat as hearts (**S413-4**, **S416**). The heart was certainly offered, but there is no mistaking a humerus for a heart. A somewhat far-fetched explanation for this labelling might be that *ḥ3ty* was being used here in the sense of being the forepart of an animal, with the understanding that the humerus is from the foreleg.
S203-4 62u, aa Each box is of a different shape: a squashed triangle, an oval, but both are marked *s3*, for back. They contain the radius. There is no explanation for the label which clearly means back, while the radius is the front part of the foreleg.
S187-8 62 a, dd Both are triangularish shaped boxes marked *i3t*, and contain lumbar vertebrae. In this instance, at least, it is clear that the labels correspond to the contents of the box.
S207-8, S210 62l, e, n Two rectangular and one triangular box, all marked *swt*, and containing femurs. There is one unmarked box **S209**, also containing a femur. Although *swt* is the accepted word for tibia (Cerny 1965: 17 n.2), confusion between leg joints, especially from the

same leg, i.e. the hind leg, is understandable. Perhaps the term was more a generic term used for leg joints rather than a term for a specific joint?
S211-14 62kk, oo, rr, t The first three are marked *ḫn*, and the last is marked *ḫ*, and all contain tibia. Černy (1965: 17) suggests that these labels are abbreviations for *ḫnḏ*, part of a leg (*WB* III: 314; Gardiner 1947a: 17-8), a definition that would apply to the tibia. All cases, save **S214** which is rectangular, are triangular.
S190-1, S195-6 62f, m, ff, hh These oval boxes are all marked *g*, and contain the first, second and third ribs. If *g* is being used as an abbreviation for *grg(y)t*, ribs, then these are correctly labelled. It is in interesting that only the boxes containing the first three ribs were so labelled.
S197-8 62ee, jj One oval and one triangular box, both marked *ḳ*, and containing the sternum. Cerny (1965: 18) suggests several words that *ḳ* might be abbreviations for: *ḳʿḥt* or *ḳbḥt*, part of the leg, *ḳnḳn* a joint, and most plausible, *ḳ3bt*, breast. Once again, the consistency of labelling and contents, despite the different shapes of box, should be noted.
S215-16 62o, i Two oval boxes marked *nnšm*, spleen, and said to contain the tongue of a bovid. It is possible that Carter mistook a liver for a tongue. This is possibly the case, as otherwise this collection of offerings would contain no liver (save perhaps goose liver, **S230**), a most unusual omission given its presence in almost all other similar assemblages. Furthermore, the appellation 'spleen' to the contents of the boxes containing the so-called tongue would make much more sense if the boxes contained liver, neighbour to spleen, rather than the tongue, a part of the anatomy far removed from both liver and spleen. This is the only cut of meat that could be called visceral (other than **S230**) in this collection of victual mummies.

Out of these twenty-eight boxes only eight (vertebrae and ribs) seem to contain what their labels indicate. It is possible that in the cases of the tibia, femurs, and sterna that the labels did also refer to the contents, and in opposition to Gardiner's (1932) belief, any confusion concerning the correspondence between the joint and the label is modern rather than ancient. Furthermore, if the suppositions concerning the scapulae and humeri being marked as coming from the forequarters are true, then these two cuts are also correctly, if not very specifically, marked. The considerable consistency between the cut of meat, the shape of box, and the label argue against any random filling of boxes. Gardiner calls the Egyptians "unconscionable muddlers" (1932), while Černy more gently suggests that the fact that so few boxes contain what is expected is due to the Egyptian's carelessness, and puts forth the idea that "the boxes had been marked in advance and that some were expected to contain parts of the body which finally were not included at all (incorrect, except in the case of the head — see Appendix II); such boxes therefore were systematically used for other joints", thus explaining the consistency between the labels and the joints (Černy 1965: 18). Perhaps the fault lies with the strict translations supplied by the modern scholar rather than with the ancient Egyptians. Unfortunately, this is not a question that can be answered here, and one can only enlarge on the variety of terms that the Egyptians used to name pieces of meat rather than discover the precise parameters of those terms.

VII

PROCESSING

After slaughtering an animal there is a choice between consuming it immediately or preserving it for later use. If consumption is immediate then the animal is killed, jointed in the case of mammals, and can be eaten either directly or after the joints/entire bird have been hung for a few hours. The author has found that joints of meat can remain medically safe to eat for up to eight hours in the Egyptian climate in summer if they are kept in a shaded area, and for twelve to sixteen hours in winter. The place they are stored makes a big difference as to how long and how well they keep. With poultry the length of standing time depends on the species. In modern Egypt geese and ducks are occasionally killed two or three hours before consumption, although this is uncommon; this is never done with chickens as they breed bacteria more quickly.

There are, of course, societal differences in the definition of good and bad meat. In Britain and other parts of Europe a 'gamey' taste is prized in wild animals, and hunted animals are frequently hung up until maggots appear and are then consumed after the removal of this extra protein source (Page 1991). Meat with such an infestation would be unacceptable for consumption in the Middle East and parts of Asia. This is perhaps an extreme example of the differences that occur from society to society in taste and hygiene and perhaps one that is also influenced by the colder climate in Europe which preserves meat for longer.

If the meat is not consumed immediately after slaughter it must be preserved in some way so that it can be consumed later. There were several ways of preserving meats available to the ancient Egyptians — drying, smoking, brining, a combination of any two of these methods, pemmicaning, or using fat, beer, or honey curing. Some of these methods could only be used on specific types of animal, i.e. only on birds, or poultry, or fish. In addition to yielding flesh for food fish also provide roe which can also be preserved.

Preservation of protein-rich food in economical and easily transportable quantities was very important as it contributed to the mobility of the both army and traders, on land and water, and provided a reserve source of meat protein in times of scarcity. It is unfortunate that so little physical evidence is available to shed light on the means of preservation most favoured by the ancient Egyptians: the only examples of preserved meats are the mummified offerings from New Kingdom Thebes, and excavators' references to pieces of meat and entire fish and birds found in the tombs of Deir el-Medina. Otherwise all information concerning the processing and preservation of meat comes from two- and three-dimensional representations, ration lists, jar labels and ethnographic comparisons.

There are two-dimensional representations of meat processing from all periods of Egyptian history. For poultry the majority of the examples date from the New Kingdom, and only two from the Middle Kingdom. Most of the early Old Kingdom examples show only the plucking of the birds, not evisceration and preservation. Sometimes poultry is shown suspended from a line 'settling/hanging' or perhaps drying, although that is less likely. Despite the paucity of this evidence, it is possible to form certain hypotheses concerning the preservation of birds from such sources in conjunction with the evidence from the mummified bird remains found in several Theban tombs (Appendix II; Figs. 42-43). Although fish were rarely represented in tombs in the process of being eaten, the number of fishing scenes and the many references to their consumption in ration and offering texts (Janssen 1975: 348-9; Helck 1961: 432-3, 469-840), as well as the presence of their bones throughout settlement sites (Malkata, shown in figures 44, 63-65, Memphis (Jeffreys 1989: 9), Amarna (Kemp 1984: 155), Kom el-Hisn (Wenke et al 1988: 17), East Karnak [Brewer 1986]) attest to their consumption, both in fresh and processed form.

Although jar labels might inform us of the jars' generic contents, they do not specify the type of processing they underwent prior to storage. The legend on jars containing meat (see Chapter X) reads: *iwf dr*. There is no exact translation of the word *dr*, so it has been translated as 'preserved', with the method of preservation being unknown (Hayes 19: 91-2; Pendlebury 1951: 169-70; Leahy 1978: 16). Thus, it cannot be used to determine any specific type of preservation, although "pressed" (Pendlebury 1951: 169-70) has been suggested. *dr* might imply pounding or beating of some sort as one of its meaning is subdue or repress (Faulkner 1976: 314-5). Another meaning of *dr* is to expel or remove (Faulkner 1976: 314-5), which, as Peet and Woolley suggest, might mean that the meat is removed from the bone before preservation (1923: 167). Certainly butchery marks on bones from Malkata confirm that meat was filleted and removed from the bone (Fig. 30; Binford 1981).

VII.1 Drying

Drying is the most common, quick[1], and straightforward way of preserving meat and is especially popular in warm and sunny countries throughout the world. This process involves cutting off pieces of meat from a (preferably) lean part of the animal[2] and then hanging them up in the sun to dry. Of course, there are many variations on this theme. Depending on the temperature, humidity, and thickness of the cut involved, sun-drying takes from one to three weeks and meat so preserved can theoretically last up to two years, although the last few months of this period are not always medically safe (Dahl and Hjort 1976: 169). Generally such meat is consumed within a year. The main problem with the process is that while meat is drying, the outer parts of the cut dry first, trapping some of the internal water, and leaving a moist inner section which can cause the meat to spoil partially (Heath 1990; Pyke 1970: 175; Dahl and Hjort 1976: 169). If any of this internal liquid is contaminated by bacteria the meat spoils completely and is inedible. This eventuality can be avoided by regulating the thickness of the meat to between one and two inches. Another potential drawback is that "sunlight has a destructive effect on riboflavin" (McLaren and Meguid 1988: 20), a type of vitamin B2, the lack of which can cause skin disease. This, however, should not have been a problem for the Egyptians as riboflavin can be found in legumes (McLaren and Meguid 1988: 20), a commoner food than meat which, in any case, they would also have eaten in its fresh state.

[1] Plains Indians dry entire steers. It takes a single worker only five to six hours to cut a fully grown animal into pieces for drying (Weltfish 1965: 217).

[2] Although the entire animal can be dried, the cut of meat favoured for drying varies from culture to culture. Plains Indians can dry an entire buffalo, including all the viscera (Weltfish 1965). Biltong in South Africa and Zimbabwe is made from the rump, and in other African countries from the flank; the Numiamut Eskimos use the scapulae, lumbar vertebrae, humeri, metacarpals, and phalanges (Binford 1978: 428-31, 100). The use of the metacarpal and phalanges is of interest as these elements seldom appear as part of the funerary offerings unless attached to the rest of the leg. However, the scene in the tomb of Djhutnofer (TT104; Fig. 18) shows a man carrying an ox leg (i.e. a joint containing the metapodial and phalanges) up to a room at the top of the building wherein sits a man who is cutting up joints of meat. Above him hangs a double "drying line" (?) with biltong shaped lengths of meat along with the triangular cuts; it is possible that this is a drying/processing room and the leg is one of the cuts that is preferred for drying by the Egyptians. It is quite possible that if the metapodials and phalanges were not cut off and dried, they were sold off or distributed to people as a cheap cut (see Chapter X). Also interesting is the use of the meat from the vertebrae; vertebrae figure in the mummified offerings (Appendix II; Figs. 31-32, 42-43).

Drying was (and still is) also the easiest way to preserve fish. All tomb representations of fish-processing can be interpreted as showing this process. Fish were first slit along the backbone, gutted, washed, and left open in the air and sun to dry (Fig. 46). Nowadays Egyptians still clean out their fish in this manner. Removing the intestines as soon as possible was the most important act of processing as fish spoils faster than mammalian/avian meat, especially out of water in the heat. Washing them out was, additionally, a vital prelude to their safe consumption, and was a further reason for the location of processing sites near water. Exposure to the sun not only removed the surplus water, but exposed the fish to the antiseptic agency of ultraviolet light. It has also been suggested that fish were cut along their backs so that the thickest section would dry more easily (Keimer 1939: 228). In accordance with a majority of representations, fish could have easily been dried on the ground, though in some cases they are shown suspended on a line as in the case of meat and poultry. The suspended method would have been optimal as air would circulate around the fish encouraging even drying. There is an unusual representation of suspended drying fish on a boat in Horemheb's tomb (Klebs 1934: 88). In some New Kingdom scenes drying fish have been decapitated. Scales were normally left on dried fish. Herodotus (II: 92) also mentions drying as the favoured method of preserving fish (Godley 1946: 379). In Pap. Anastasi IV (15, 7) there are reports of

Fig. 46

dried and preserved *Buri*, *šnʿ*, and other types of fish coming from the delta, and being distributed throughout Egypt and abroad[3] (Caminos 1954: 200, 210-1; Hayes 1951: 160). The Timna Temple and the associated smelting camps contained fish remains, including *Clariidae*, the Nile Catfish. "As Timna is situated in the desert, c. 30km north of Eilat, any fish consumed there had to be imported over long distances, presumably in a dried condition" (Rothenberg 1988: 244-5).

Not all the fish found in tombs that have been described as 'dried' (*šw* [*WB* V: 429]) have been split open in the manner shown in representations. The ones found in the tomb of Kha were gutted and split, although not opened, and examples from Deir el-Medina (JE315; JE44038) were not even gutted, possibly having been salted whole rather than dried (see VII.3). However, other fish from Deir el-Medina must have been gutted and dried as the ostraka state that they are dried, as do other literary sources (Griffith 1898: 97). The dried fish in the British Museum collection (EA36191) are gutted but not split open. This splitting and drying of fish seems to be unique to the ancient Egyptians, as Radcliffe notes that Greek and Copts "rarely split their fish before packing them in large earthen pots" (Radcliffe 1921: 355, n.4)

All representations of fish in the act of being consumed show closed fish, i.e. not spread open as they are when dried. This is not surprising for dried fish were generally cooked or re-hydrated prior to consumption, then as now, which caused them to lose their splayed-out form. There is one exception to this, in the tomb of Horemheb (Fig. 7), where people are shown consuming splayed-out dried fish.

For mammalian meats, variations on sun-drying include the addition of salt and spices, and pounding. Adding salt preserves the meat for longer. When the meat is cut into strips it is massaged with salt in a procedure similar to dry curing (discussed more fully under "Brining"). Through osmosis salt draws out all the liquid in the meat and causes it to dry more evenly and efficiently than just sun drying; the lack of water and other nutrients also makes the meat an inhospitable environment for micro-organisms, and the osmotic pressure within individual cells forces out bacteria. Salted and dried meat will definitely last for two years providing it is kept in a dry protected place.

[3] The location of the exact spot in the delta whence the fish originate is under discussion. Suggestions for Tjel's identity include Kom Medinet Ghurab and environs (Caminos 1954: 200, 210-1), Tell el-Habua (Kemp 1992), and Tell Abu Sêfah (Gardiner 1947a: 181).

Biltong is another variation of drying. The term "biltong" can refer to salted strips of meat, but it is more commonly associated with dried and spiced meat. Even now, biltong is made from the meat (see notes 1-2) of all types of domestic and wild animals. Spices do not just serve to mask undesirable flavours and smells (Niven, Jr. and Shesbro 1960: 310), but actually help the preservation process itself. Biltong is made by taking strips of meat (preferably fatless) up to three to five inches thick, coating it in a mixture of spices: salt, celery, garlic, cumin, coriander, and cinnamon (Susman 1992) (which were all available to the Egyptians [Manniche 1989]), and leaving it to dry. Honey can also be used on biltong (Susman 1992). One can make biltong with larger pieces of meat, but in general long thin pieces are used. They can be cut from any area of the animal, although the rump and sides are preferred (Susman 1992). In Egypt, it would take between a week to ten days to dry, depending on the weather. Biltong lasts for a year and more under dry conditions, although it is best consumed before the end of the first year (Susman 1992), and even in particularly damp and humid conditions biltong can last for up to a year, albeit it loses some of its flavour and becomes a trifle rancid. I have experimented on how long biltong lasts under a variety of different conditions: pieces of biltong were kept in the bathroom, bedroom, kitchen, and drying closet (control) of a flat in Britain. Each area had different humidities and temperatures although the whole appartment can be characterized as damp. After eight months the biltong in the bathroom, bedroom, kitchen, and drying-closet (bathroom and kitchen being the most humid environments) was examined. Apart from the drying-closet biltong, which was hard and very dry, the bedroom biltong was undoubtedly the best preserved. It was stiff, had no mould and was edible and tasty. No ill-effects were felt as a result of eating the biltong. The bathroom biltong was limp and grey as a mould had settled on it. The mould was wiped away and the biltong was edible but definitely slightly rancid tasting. No ill effects resulted from consuming the biltong. The kitchen biltong had a faint dusting of mould, but it was stiffer and tastier with none of the sourness that was characteristic of the damp bathroom biltong. Thus one can see that even in damp sunless conditions biltong can exist for eight months, if not more.

Other versions of the process of drying meat entail pounding before drying it to thin the meat into even slices. This also helps to expel a great deal of the water that is so damaging to the successful preservation of the meat. After the meat is pounded thin it is hung up to dry on a line. This method was common amongst the Plains Indians (Weltfish 1965:

212- 21). As a further measure salt can be rubbed into the meat before it is dried, although the thinness of the meat does not make this essential.

The last permutation of dried meat is that resulting from the boil-and-dry process. Chunks of meat are first boiled in water. Heat and boiling are very effective in killing micro-organisms (pathogenic and putrifactive agents), especially surface micro-organisms. The meat is then put in an airy and sunny place to dry. This method, currently used in Nigeria and China[4], and historically used by the Plains Indians (Weltfish 1965: 217-21), is most efficient (but not exclusively so) when small pieces of meat (c. 20 cm) are thus treated as they dry rapidly thereby giving fewer chances for microbial growth and infection. However, this method does extract the water-soluble protein and liquid fat from the meat, thereby lowering its nutritive value.

The Egyptians did not dry birds as they did fish and meat. The closest they came to drying birds is the case of the quail. According to Herodotus, quails, after migrating across the Mediterranean to Egypt, fell to the ground from exhaustion and were soon despatched. Some of them were hung up like joints of meat and eaten shortly thereafter (Godley 1946: 365). It is more likely that these birds were hung up to improve the taste of their meat as they were eaten raw[5] (Lloyd 1976: 335; Fig. 13). Indeed, all flesh improves in flavour if it is hung up temporarily while chemicals reach equilibrium, *rigor mortis* passes, and the meat softens. This holds true both for mammalian and avian meat, and butchers and poulterers worldwide practice the hanging of their product. This would explain the suspended joints and birds on lines over mammalian and avian processing sites in tomb representations.

There is only one tomb, that of Ipy (TT217), whose decoration suggests that birds were also dried (Fig. 16). The birds are hung up on lines and it has been suggested that they were "either cured whole or cut in slices, which, after being hung up to dry on lines stretched between two posts, are potted in salt" (Davies 1927: 62). It is by no means obvious that the strangely shaped objects on the line above the duck evisceration scene are slices of drying duck. Unfortunately, no parallel to the scene exists, and contemporary ethnographic information, as well as information

[4] In China chopped up pieces of meat are boiled with a variety of spices until the water evaporates. Then the meat is deemed ready to be hung in a well ventilated area and dried (Huang 1992).

[5] There is no way to know whether raw birds or even fish were consumed during the Pharaonic period. The only references to the consumption of raw meat comes from the Graeco-Roman period.

from butchers in the United States and England, provides no examples of drying fowl (see below for salting and drying). Fowl flesh (with the exception of ostrich, which makes what is reputed to be 'delicious' biltong) is unappetising when dry. I have experimented with this and the results are tough, inedible and very unpleasant. Rehydration or soaking the meat in a sauce might ameliorate the taste, but from my experience this is highly unlikely. Furthermore, the chances of disease and pests in dried fowl flesh are much higher than in mammal or even fish meat as fowl has a propensity to spoil quickly, is particularly prone to bacterial infestation after death, and does not dry thoroughly.

All the above methods of preservation were available to the Egyptians. Tomb decoration and models support drying as an option, showing pieces of meat and sometimes poultry strung up on a line, as well as gutted fish lying in the sun. In the case of beef or mutton, most scholars interpret these hanging pieces of meat as scenes of drying meat rather than a way to store freshly butchered meat until its disposal (e.g. Ghoneim 1977: 163-64). This might be the case with triangular pieces of meat common especially to Middle Kingdom tombs[6], but it is hard to believe that joints of meat containing tibiae and femurs are being dried in this manner[7]. It is quite possible that once the animal was butchered it was cut into pieces; some of the joints being for further processing or immediate consumption, while others were immediately cut to the shape necessary for processing, which in the case of drying, would start at once.

The long thin pieces of meat represented on talatat blocks from the Aten Temple in Karnak (Luxor Museum 233) and the paintings in the Theban tombs of Ipy (TT217), Amun Mose (TT254), Djutnofer (TT104; Fig. 18), and Rameses III (KV11) look very much like modern biltong[8]. These pieces are interspersed with other pieces of meat that have been cut from a butchered ox depicted nearby. It would seem reasonable to assume that the long thin strips are drying biltong (any intermediary phase showing the biltong being rubbed with spices is not depicted) while the larger pieces of meat are awaiting transport and consumption.

[6] For references to the shape of the cut see Chapter VI. Miller (1990) reported seeing photographs of Sioux drying meat c. 1900, and the cuts shown were triangular fillets of uniform thickness cut for even drying and thus avoiding spoilage.

[7] It is possible to dry meat with bone after boiling, although it is not effective for any length of time and would take up a lot of boiling time and fuel.

[8] Some modern variations of biltong-manufacture include immersing the meat and spices in brine for 24-48 hours before drying in the sun, although this method is not traditional (Susman 1992). It might be possible that scenes showing cauldrons full of joints of meat simmering over the fire are illustrating such a practice.

Furthermore, in the funerary offerings found in the tomb of Amenhotep II Lortet and Gaillard list long narrow pieces of meat that were salted and dried[9] (Appendix II: **S130-47**, amongst others). These could have been a form of unspiced biltong, although, as only some have been examined, and none have ever been chemically tested, it is possible that those impregnated with spices that were available to the Egyptians. However, the ones examined were odourless, although any scent indicative of spices could well have evaporated over time. In Deir el-Medina Bruyère describes finding dried meat in small silt vessels (Bruyère 1937: 160) but he does not elaborate on the shape or chemical composition of the meat.

Drying meat using salt is common in contemporary Egypt. Pieces (chunks or strips, much like those found in Amenophis II's tomb [Appendix II.i]) of meat have dry salt rubbed into them. These are tied with string and hung or put in a windy place for two to three days and are quite safe to eat for three to five months. In Siwa Oasis the chunks of meat are threaded on strings and allowed to dry (Fakhry 1973). Unlike biltong meat cured thus is not eaten raw, but the salt is washed off and the meat cooked as if it had been fresh. Dehydration is used, so this meat is called *Lahma Mujfuf* (Khattab 1991).

There is less evidence for pounded-and-dried meat. Some reliefs, mainly from Old Kingdom Memphite mastabas, show a man near a butchery sequence beating what appears to be a piece of meat with a stone (Forbes 1965: 193; Klebs 1934: 91). This has also been interpreted as a non-meat related activity, or, less likely, the hammering of meat to soften it before it is cooked (Klebs 1934: 91). Nearby there is often a line with hanging cuts of meat, but this is not as common as in the Middle Kingdom. However, the term "pressed meat" (Pendlebury 1951: 169-70) might refer to pounded-and-dried meat. This is an attractive interpretation as the size of the jars bearing this inscription (Chapter X), especially the mouths, are too narrow to permit the passage of a joint with a bone, but dried and somewhat flexible pieces of meat could have been put into the jars prior to storage and transportation without difficulty. There are at least two representations from the Old Kingdom and one representation from the Middle Kingdom that show meat being cut into thin pieces, pounded, and then hung up to dry (MMA08.201.1,

[9] Most of the listed pieces and examined examples in the Agricultural Museum, Dokki, are not as long as those in the reliefs. One should note that representation of long slender pieces of meat hanging from a line, with the sole exception of TT104, all date from the Amarna period and thereafter. Of course this might be due to the incomplete publication of the Theban necropolis as well as an accident of preservation.

UC14311, and Davies 1920: pl.8). Pounding and sun and air drying meat is a common way of preserving meat and is still used by Africans and Native Americans today.

The boiling and drying option might be considered a possibility, but the meat would have to be boiled for several hours (depending on the amount and the thickness of the meat) which would be inefficient in terms of fuel. The scenes that show large joints with bones in them being boiled (?) in a pot near the hanging meat line and butcher might argue for this method of preservation, but are not entirely convincing as they could represent cooking for consumption in the near future, rather than long term preservation (Fig. 41).

VII.2 Smoking

To smoke meat, it must first be cut up into strips (like biltong) or large chunks, put on a frame and then placed over a fire preferably made with some aromatic wood (e.g. hickory, cedar). If fish or poultry is to be smoked, then the entire animal is placed over the fire, preferably on a rack of some sort. To create smoke, the wood would either be slightly green or the fire would be covered by some damp organic material. This process is best performed in an enclosed space which has air to circulate the smoke on all sides of the meat, unless the meat frame is stretched directly over the fire. One must guard against humidity in this process — not a problem in ancient Egypt. If large joints of meat with bone are being smoked, then a smokehouse would be the only way of effectively smoking the joints. Three days of smoking per pound is the recommended length of time for this operation (Pearson and Tawber 1984: 286).

Smoking meat is a superior form of preservation than drying (Dahl and Hjort 1976: 169) as it conclusively kills any microbes that might be in and on the meat and simultaneously reduces the amount of water present in the meat, improving the quality of preservation (Pyke 1970: 174).

The evidence for smoking or drying-and-smoking (more common ethnographically) in ancient Egypt is minimal. Although some scholars believe that the primarily Middle Kingdom representations that show cuts of meat (and occasionally poultry) suspended on a line near a fire illustrate the smoking of meat for preservation, this is improbable. Firstly, some of the joints of meat contain bones and would therefore not be preserved by smoking as this would take several weeks to accomplish effectively[10]. Secondly, there is a pot on the fire which severely limits the amount and direction of smoke available for the task. Thirdly, the joints are hung at some distance from the fire, though this might be due to Egyptian representational convention concerning perspective and distance. As far as physical evidence goes, Bruyère thought that some of the meat recovered from the sepulchres of Deir el-Medina might have been smoked, but had no convincing evidence of this as the meat disintegrated shortly after excavation (Bruyère 1937: 107). If the pieces of meat found in the Theban necropolis (Appendix II) were smoked it would be very difficult to determine now as any smoky aroma would have entirely dissipated. Pertinently, Egypt was not rich in fuel, especially aromatic fuel, and it is therefore highly unlikely that out of all the technologies available for preservation, the Egyptians would favour smoking (Forbes 1965: 192).

Although technically possible, there is no evidence for the smoking of birds or fish in ancient Egypt. However, one might suggest that birds were pre-roasted as a method of preservation since cooking would kill any infesting bacteria and guard against further infestation. Tomb scenes and models of the Middle Kingdom almost always show a man sitting next to the workers, roasting a bird, which may have been part of the payment or the lunch of the workers. It is also possible, though less likely, that birds were roasted and then stored for later consumption — just as pre-roasted birds are found in modern grocery stores. Birds so prepared would last (if kept in a shaded place) up to forty-eight hours. If they were to be eaten after then they would have had to have been re-heated thoroughly to kill any bacteria that might subsequently have bred inside the bird during this period.

If fish were smoked this would have been done only in conjunction with drying as drying is necessary to rid the fish of unwanted water and oils. The smoke would have provided a measure of flavouring and served to kill any bacteria that might have settled (Proctor 1972: 148).

The smoking of any type of meat is not practised in Egypt today, although it is common in areas with a plentiful supply of fuel: Palestine and other parts of the Near East as well as some parts of Africa.

[10] If the meat were dried and smoked or salt-boiled and smoked this would be feasible (see 'Brining' section).

VII.3 Salting & Brining

This is the method that most scholars believe was most commonly practised in ancient Egypt. There are two ways in which salt can be used for curing: dry (salting) or wet (brine). Dry-curing would entail packing in salt and wet-curing would mean immersion in brine. Both dry and wet salt-curing are contemporary techniques (Montagne and Gottschalk 1961: 358), and could have been used on all types of animal.

Meat brining means putting meat as chunks, fillets, or even cuts containing bone in a solution of salt and water, with occasionally a few other ingredients, and sealing it. This treatment is similar to that adopted for the curing of ham, or making salt beef. It takes several weeks, especially for larger pieces of meat, and meat thus cured lasts for one or more years depending on the amount of salt in the solution. The variations that occur in the form of dry brining, brining and drying or partial drying (Pyke 1970: 175), brining-and-boiling and brining-and-smoking. Brining is most commonly used for preserving pork, perhaps as the salt improves the taste and is an effective agent against the harmful micro-organisms that find a good host in the pig.

The basic modern brining recipe is as follows, although the operation can be performed without both the nitrate and the nitrite:

20% NaCl salt
2% KNi saltpetre/nitrate
200p.p.m. NaNi nitrite (nitrite not necessary to cure)
1% sucrose (any sugar or honey will do)
Water (Ingram 1962: 272)

This is the preferred modern method of meat preservation as salt and saltpetre prevent the activity of normal putrefactive bacteria in meat; only organisms exceptionally tolerant of salt or those that can obtain oxygen by reducing the nitrate or nitrite will survive in a solution of brine — that is to say very few (Ingram 1962: 272). The only health-related problems with brining are that, if any disease (e.g. salmonella) is present in the meat or the brine, it will be carried through the medium of brine and spread throughout the meat if there is insufficient salt. Brining also affects the nutritive value (see Chapter X) and composition of the meat since "when the salt penetrates the meat, (the) water and substances dissolved in it leave the meat. Among those substances are soluble protein, glucose and glucose phosphates, and lactic acid, derived from breakdown of muscle glycogen during rigor mortis in the carcass; probably small amounts of other sugars (e.g. ribose) and amino acids and pigments" (Ingram 1962: 272).

Herodotus (II: 77) writes that the Egyptians pickled birds in brine and stored them in large pots (Godley 1946: 365; Lloyd 1976: 335), and then consumed them raw. In modern Egypt, salted meat (there is no record of salted birds) is always cooked before being eaten.

Tomb paintings show birds being strangled, plucked, eviscerated and then put into large narrow-necked amphorae (Fig. 12). The amphorae possibly contained salt, either in a solid or a liquid state. Although it is easy to see how birds could have been put into the amphorae when newly dead and malleable, how they were taken out when hardened and stiffened by salt is not so obvious. It is possible that the entire neck of the amphora was removed, thus giving easy access to the birds: according to Hope, several of the large amphorae found in Malkata were found with their whole necks removed (1978: 48-58).

A large amphora containing preserved birds such as those represented in paintings has been found in the tomb of Kha (Schiaparelli 1927: 159; Fig. 47). My examination of these birds showed that they had been plucked, beheaded, and eviscerated, and some had their feet and wingtips removed, although quite a few still had their feet (see above). They had been preserved in salt, but it is unknown whether this was in solid or liquid form. It is more probable that the latter, i.e. brine, was used, as a solid salt would still be present — unless the birds were removed from a separate preserving pot and then put into the amphora once they had been preserved. This seems fairly impractical as it would have entailed the use of several vessels and to remove the birds from salt and expose them to external elements would tempt bacteria and risk spoilage. So far, no complex chemical analysis have been carried out on these birds, although a basic test performed in Turin has shown that they were preserved in salt. Birds found in squat pottery vessels from Balat (Mastaba V) were also preserved in salt, most probably in liquid form, due to the reasons mentioned above and the softness of their bones (Chaix 1990; Appendix II: ii). Certainly in Graeco-Roman times birds were preserved in brine as Herodotus says they were pickled rather than preserved dry. Jar labels from Malkata and Amarna also attest to storage jars containing water-fowl, although the method of preservation employed is unknown (Hayes 1951: 96-70; Pendlebury 1951: 183).

Herodotus also mentions brining/pickling fish (Lloyd 1976: 80). This involves putting the gutted fish into a salt and water mixture in a large vessel, such as an amphora, sealing it securely and leaving it for ten to

Fig. 47

twelve days. This is considered by some nutritionists to be the safest way of preserving fish (Dumont 1977: 136). However, the higher the content of salt and water, the less protein can be derived from fish preserved in this way, due to being immersed in a solution (Proctor 1972: 146). Another problem with brined fish is that certain micro-organisms or even dermestid beetles reproduce and live in the liquid salt medium (Proctor 1972: 148). It is quite possible that the amphorae shown in the vicinity of the fish-processing scenes were used for both wet and dry salting.

All salted and dried fish would have had to have been washed and soaked prior to consumption. Generally they would have been cooked as well, although dried or dried-and-salted fish could have been consumed without cooking, although this would have been inadvisable due to the presence of bacteria and pest infestation.

Using brine in conjunction with other technologies is also popular as a method of preservation. One can combine partial drying (air drying) with pickling in brine, again a technique favoured for pork (Pyke 1970: 175), as well as poultry. This method has the advantage of being quicker than simple brining and is more efficient in eradicating bacteria. Ducks and pigeons have been found in Deir el-Medina that might have been salted and dried (Bruyère 1932: 156, 200-1). None has ever been examined for traces of salt so it is unknown if they were dried or salted or both. It is possible that an amphora from the tomb of Kha (TT8; Fig. 47) contains salted poultry.

Preserving ducks by a combination of salting and drying is common in China today. After the duck has been slaughtered it's entrails are removed and it is washed thoroughly, both inside and out, and hung up to dry for a few hours. Once it is dry (two to three hours later), it is put into a squat jar which is filled with soy sauce (the salt element), sugar, alcohol and spices. A heavy stone is placed on the duck to insure its total and continual immersion. It is left to soak thus for three to seven days, after which it is removed and hung up to dry in a well ventilated place. It can hang for ten days or more, and it can generally be used after fifteen or twenty days, although the hanging period can be extended to a year without ill effects (Huang 1992). This method of brining was available to the Egyptians if they used a salt solution.

Brining followed by smoking (and incidentally drying) is also practised, though this method is done more to achieve a smokey flavour, rather than from any necessity of preservation. This is done only with mammalian meat. Brine-cured meat is taken from its vessel and hung in a smoke house for three to five days or more so that it dries out acquiring the flavour of woodsmoke as it dries. Brining and boiling is yet another variation on brining. Instead of just immersing the meat in salt and nitrate solution and sealing it, this type of brining involves boiling the meat in the solution. This is very effective in killing off almost all types of bacteria that could be expected to infect the meat and also preserves the protein content more effectively than other methods (Pyke 1970: 175). This can also be combined with drying/smoking after boiling. This last method finds some favour with Egyptologists (Klebs 1922: 104).

Dry curing generally only uses salt, although in Europe the curing ingredients are usually,salt, sugar, and nitrate/nitrite when quadrupeds are being cured. These are added to the meat by massaging them on and rolling and packing the meat in the mixture. The moisture from the meat is drawn out to form a brine and the curing ingredients are transported into the meat by diffusion. This procedure takes a few weeks and is similar to mummification (see above in "Drying"). The joint/cut can be smoked thereafter for further protection or flavour (Pearson and Tawber 1984: 286).

Fowl could also have been dry-cured just using salt. Such a method of salting and drying fowl is also practised in China today. After the bird is cleaned, washed, and free of the washing water, salt (occasionally with added herbs and spices) is rubbed into the body cavity and onto the skin of the bird. It is put into a jar and pressed down with a stone for three to four days, after which it is hung for at least a week, after which it is ready to be eaten. This also lasts for up to a year (Huang 1992). In Egypt the birds can only be hung to be initially dried of the cleansing water for a very short period before becoming infested with bacteria and insects. In China this form of drying is done in the autumn and winter due to fear of spoilage (Huang 1992). The dry salting method might have been used in Egypt, and could have been the one used for Kha's poultry (Fig. 47). It is difficult to determine whether Kha's fowl-filled amphora contained liquid or solid salt. From the only inspection possible it would seem more likely that the birds had been preserved in a solution of salt as there were few visible remains of solid salt. Had they been packed in salt there would have been more residue.

An unusually prepared type of duck enters into this section. Several examples of spatch-cocked-type ducks have been found, primarily from Deir el-Medina. A few of these are in the Louvre (Louvre E14551)[11] and

[11] Louvre E14551 from Deir el-Medina from the time of Hatshepsut/Tuthmosis III is a plucked, beheaded, and spatch-cocked duck. It measures 14 cm from the tip of the tail to where the head is cut off, 12.2 cm. in width. It retains its feet, but the wing-tips are trimmed. Two other similar birds without museum numbers are supposed to be from the same area. The large one is beheaded with both feet intact and measures 16.0 cm × 12.0 cm and is damaged slightly; the smaller bird, prepared in exactly the same manner as the two above, measures 14.0 cm × 13.0 cm.

were examined by me. "Spatch-cocking" is a specific way of preparing a fowl for cooking by cutting down along its backbone after it has been killed and plucked, and opening it out (Hume and Downes 1963: 240). Generally the feet and wingtips are cut off and the bird is grilled. The birds found in Deir el-Medina are cut open, but unlike modern spatch-cocked birds are slit down the ventral side (ventral slits are made in evisceration, as shown in tomb reliefs [Table II]) and then flattened out. Deir el-Medina poultry does not show any definite signs of grilling in the form of burn marks or discolouration of the skin. Indeed, they look almost uniformly grey so one cannot tell if they have been cooked, although one might expect a blackness rather than greyness had this been the case. It is more likely that the birds were spatch-cocked and then left in the sun to dry, possibly after being massaged with salt. They could also have been smoked. In modern Egypt this style of preparation is unknown, and it is infrequent in the West.

For fish, dry salting would involve kenching — rubbing the fish with salt after it had been gutted and washed — and hanging out to dry in the sun and air. It is quite possible that the drying fish shown in tomb reliefs were also salted as the procedure is identical excepting the additional step of salting the fish before drying. Salt would definitely lengthen the 'shelf life' of fish as it expedites drying by drawing out water from the fish and killing bacteria because it has antiseptic properties, whether in a soluble or solid state (Proctor 1972: 148). It is possible to smoke fish after salt-drying them as well as after straight drying, although it is doubtful that the Egyptians did this.

Another possible method of dry salting would be layering fish (gutted or ungutted) with alternating layers of dry salt in a large vessel, like an amphora, sealing it tight and leaving it for a few days. This would explain the presence of amphorae in certain fish-gutting scenes. *Bulti* and other fish from Lower Egypt are preserved in jars, probably amphorae, in Malkata (Hayes 1951: 104, 160). They also form part of Kha's burial goods (Schiaparelli 1927: 159-60, fig. 142).

The modern *faseekh* is made in this manner. Fish are washed, gutted, and rubbed with salt and then packed in a pottery (or now a metal or a plastic) container in layers separated by dry salt. Traditionally the last layer of salt is covered by heavy stones to weigh down the salt and fish allowing the former to permeate the latter fully. The fish are allowed to stand thus for seven to ten days, after which they can be flipped and allowed to stand for another week or so. During this time the bodily fluids drain away and the salt solution penetrates the fish making it firm and fairly odiforous. In Upper Egypt this preparation is called *melouha* (Koraiem 1989).

As mentioned above, the most crucial ingredient for brining and some forms of drying is salt, as this agent kills bacteria by increasing the osmotic pressure in the cells when it draws out the water from the meat. Salt curing is a very simple way of preserving meats of all kinds. Salt was widely available in ancient Egypt, as it is in Egypt today. It is found in Western Egypt in the Wadi Natrun, Siwa, Kharga, Dakhla, Farafra, near Memphis, Barnugi, Thebes, el-Kab, and Lake Mareotis (*LÄ* V 371). It was also extracted in the delta from sea-water, inland from the marshes and even from the soil of the desert itself (Forbes 1965: 178), whilst local deposits were scattered throughout Egypt. Salt sacks, cakes and bricks have all been found in tombs (Bruyère 1937: 152) and are mentioned in offering-lists in temples (Papyrus Harris). They formed a part of the ships' supplies, and were common items of trade (Janssen 1961: 83). We know that the Egyptians were capable of salting as natron (a form of salt) and salt (NaCl) were frequently used in preservation of formerly living organisms, including people and fish (Lucas 1948: 304). It is quite possible for the Egyptians to have used dry as well as wet curing.

Joints of meat or entire birds wrapped in bandages (sometimes placed in coffinets whose interior is coated with pitch) have been found in New Kingdom Theban tombs (Appendix II; Figs. 42-43, 59). These so-called victual mummies were examined by the author not only to determine the animals and joints offered, but the method (if any) of preserving these offerings. Most of the victual mummies were wrapped in bandages, although some are (now) loose. Most are brown. It has been suggested that this browning was given to victual mummies by the application of very hot resin on the wrapping which might have cooked the exterior surface of the mummy (Lortet and Gaillard 1905: 248). Actually, the hot resin was more often applied to the coffinets containing the mummies in order to firmly seal and water-proof them. It is more likely that the resin only accidently adhered to the bandaged joints when they were placed inside the box and had no part in the process of actually preserving the meat.

The mummified joints were tested to see if they contained any applied chemicals/elements they might have been used in the preserving process. The first test consisted of tasting all the joints from the tomb of Amenemhat Q (see Appendix II) located in the California Academy of Sciences, many (about 85%) from the American Museum of Natural History, as well as several (approximately 90%) of the joints from the tomb of

Meryetamun[12], also located in the AMNH. These joints all tasted salty, while, in all but three or four instances, the bandages did not. Further tests consisted of microscopic examination of several of the joints. These examinations revealed white residues and crystals of what was probably salt. Only in two instances did the bandages show any sign of salt. One sample of meat was obtained from the AMNH (A95-3161, from the tomb of Amenemhat Q), and two small pieces of this (the interior fibres) were examined using a SEM that was hooked up to a LINK system and could analyse the inorganic elements in the meat sample[13] (Figs. 49-54). The magnification showed some crystals and elements on the sample which, although they could not be positively identified from their structure as seen on the screen, could be identified by the element analysis. The first sample revealed a globular 'colony' on one side. This was more closely examined and revealed a series of multi-faceted nodules (Fig. 49). These showed an extremely high occurrence of chlorine, and significant amounts of sodium (Fig. 50). Other areas of the sample were tested and showed some spiky growths on the fibres (Fig. 51) which also showed a high chlorine content (Fig. 52). The second sample (from the same joint) was then examined and showed the presence of both sodium and chlorine (Fig. 53), although in lesser quantities than in the first sample (Fig. 54). Other portions of both samples were tested, and though they did not always show crystals, they all had some sodium and chlorine present[14], indicating the presence of salt. It cannot incontrovertibly be proven that this salt was added to the meat for preservation rather than being a precipitate from the ground water; however, the sample comes from the interior of the joint, thus making it less likely that the salt was from the ground. Furthermore, most of the bandages taste-tested were not salty, while the meat they protected was, thereby implying that the salt on the meat comes from human intervention rather than from the ground.

[12] It should be noted that most of Meryetamun's victual-mummies were in a large box and had no traces of resin. See Appendix II for full discussion.

[13] The machine works as follows: the sample, coated with gold or a gold-palladium mixture is placed in a vacuum compartment and bombarded with electrons. These electrons agitate the sample and any inorganic elements or chemicals in it, which, in turn, emit electrons. These are interpreted in two ways. Firstly, the show up as an image on the screen which can be photographed (Figs. 49, 51, 53), and secondly, the type of electrons given off can be analysed and their originating elements identified (Figs. 50, 52, 54). I am grateful to D. Samuel and C. Putnis who made the SEM available to me, and to C. Putnis for her help with analysing the elements.

[14] Putnis (1992) suggests that the silica appearing on the graph is a replacement element due to the gradual fossilization of the meat.

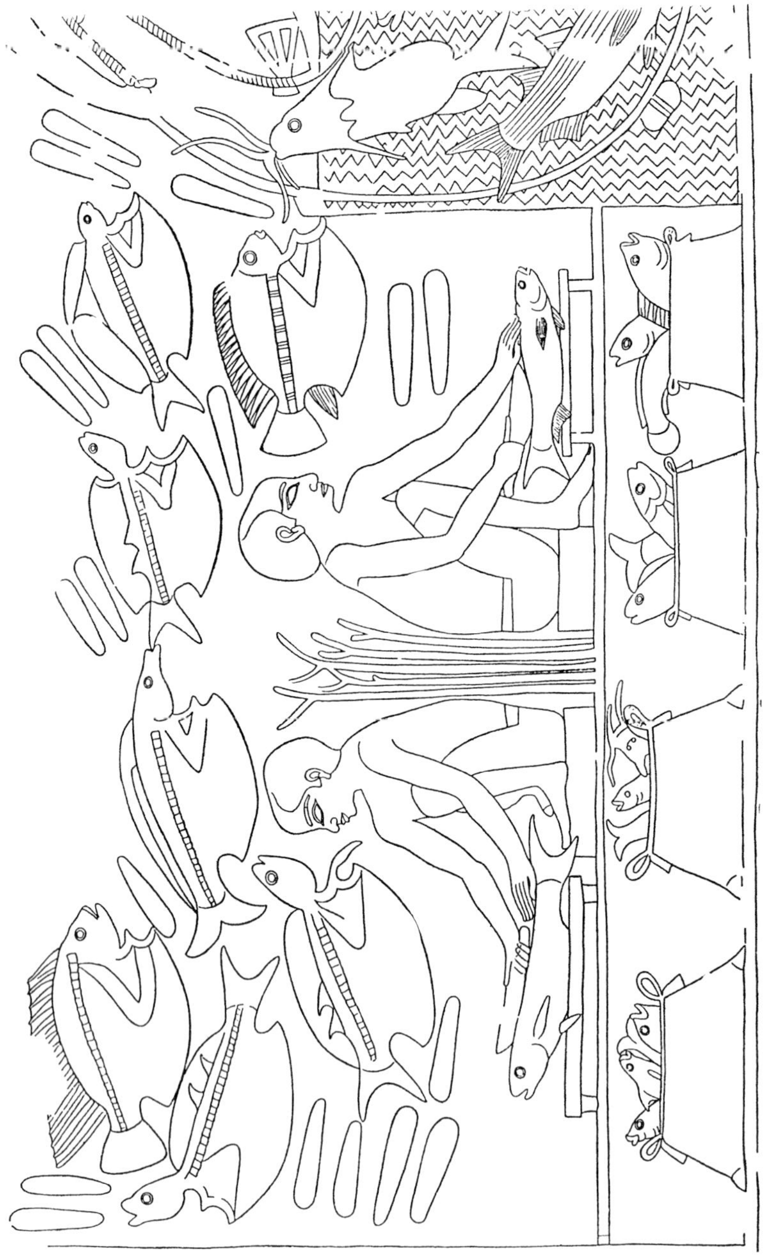

Fig. 48

A further interpretation of the Middle Kingdom paintings of butchers, cooking-pots and the suspended lines of meat is that they depict the process of boiling the meat in brine prior to subsequent drying. The only disadvantage to this theory is the length of time and amount of fuel it would take to brine-boil the joints shown in the drawing as most of them are 'on the bone': models do not clarify this process at all. Furthermore, in the Workmen's Village in Amarna salt crystals were found adhering to a deposit of broken pottery vessels found close to buildings tentatively identified as pig butchery sites. The crystals were subsequently identified as pure salt, suggesting that pigs were salted and cured here (Kemp 1986: 73-4). However, as our knowledge of the ground chemistry of this area is uncertain it is possible that the crystals derived from having been buried in the ground rather than from use, so there is no undisputed proof of bringing. However, given the ubiquitousness of salt and mummification in ancient Egypt, it seems likely that salt curing, whether brining or packing, was a favoured method of preserving meat. That it was a favoured method for fish preservation is definite as Egypt was famous for its salted fish from Pharaonic times onward. Lloyd writes that the "pickling industry was probably highly prosperous; for it even exported its products to Syria" (Lloyd 1976: 80), while Herodotus mentions its popularity in the fifth century BC. Furthermore, "Diodorus ... tells us that practically all fish were at once pickled (brined) or salted", a statement confirmed by Julius Pollux's mention of the Egyptian *tariche*, especially that from Canopus, being exported far and wide, certainly to Palestine, whither "the Egyptian fish came in baskets or barrels" (Radcliffe 1921: 355). Salted fish also feature in the Coptic period (Crum 1939: 44).

For fish, at least, it is likely that the Egyptians practiced both kinds of salting as well as plain drying. Whatever process was chosen was dependent on the prosperity of the individual performing it as obviously any process involving salt would have been more expensive than one without. The amount of salt necessary for each recipe would also have been dictated by economic feasibility, and the vessels necessary for each preparation would also have had to have been considered in terms of cost. Consequently, plain drying would have been by far the most accessible and the cheapest of available methods, while a *faseekh* type preparation would have been the most expensive. Between these extremes lay the option of brining.

Either method using salt would have been the safest way of preserving fish, as salt not only kills micro-organisms but also protects it from

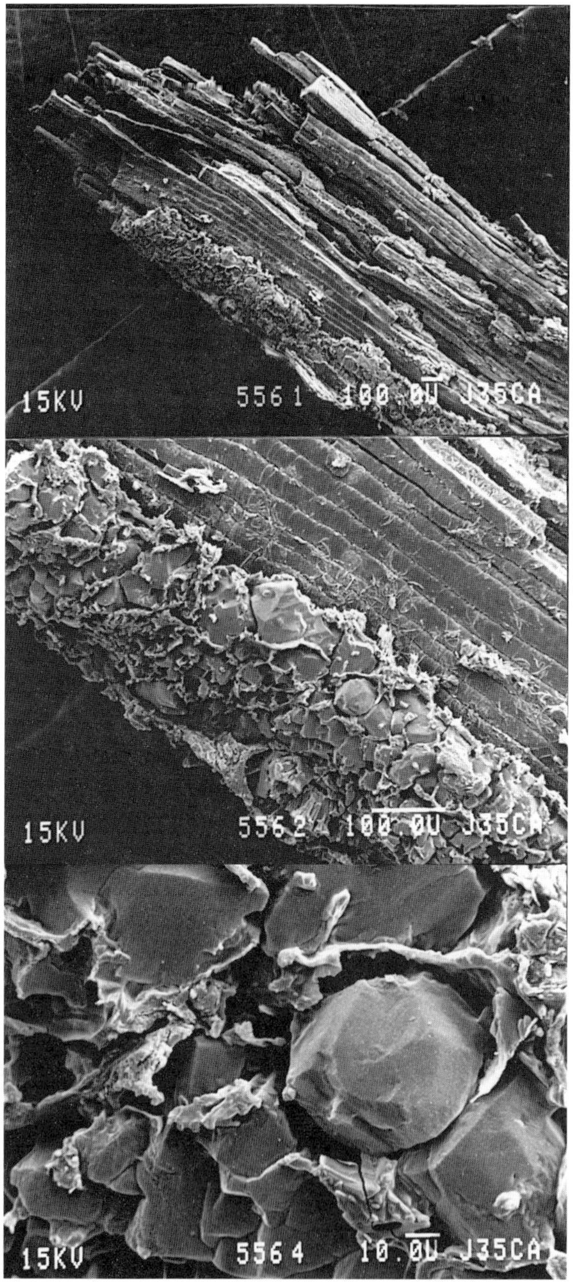
15KV
5561
100.0U J35CA
15KV
5562
100.0U J35CA
15KV
5564
10.0U J35CA

Fig.49

insect attack (Green 1967: 331-4). Kenching might have been a more effective way of safeguarding fish against pests and bacteria as more salt is applied to the fish, thereby rendering it more inhospitable to pests. The bacterial contamination in fish is not just determined by the method of preservation, but by the way in which it is caught (Shervan 1962: 170). If fish were caught in a net and dragged along the river floor through the mud, some would inevitably die in the process before reaching the surface. This would have provided greater opportunity for bacterial infection to occur. The line, harpoon or trap would have been a preferable alternative as fish could be thus removed from bacterially rich environments before death.

VII.4 Special methods

There are some methods of preserving meats that are unique to certain animals.

VII.4.1 Pemmican

Pemmican derives its name from a mixture made primarily by North American Indians. It consists of dried lean mammalian meat which is pounded and mixed with melted fat and berries (generally Sarvis berries, although grapes, plums, and wild cherries are also pressed into use) so that it makes a paste which is then formed into cakes. Pemmican lasts for up to a year, sometimes longer, if it is kept in a cool place where the fat will not become rancid. Although this has never been argued as a possible method of meat preservation used in ancient Egypt, it would have been well within the technical means of the Egyptians.

There is one method of meat preservation used in modern Egypt similar to pemmican cakes in its use of meat being mixed with fat. Chopped-up pieces of any meat (excepting chicken and fish), preferably boneless, and from any part of the animal[15], are placed over a fire in a wide-mouthed cooking pot (originally pottery) before fat and salt are added. This is cooked for three to five hours in the open air. After cooking the meat is

[15] In modern Greece a similar type of mixture called "Sigilina" is made out of pork. The Greeks prefer to use the fatty shoulder joint, and therefore do not add any salt. They also use no salt. The mixture is traditionally stored in a pottery vessel, and is scooped out and cooked, much like bacon. Sigilina lasts for six months, approximately (Notaris 1990).

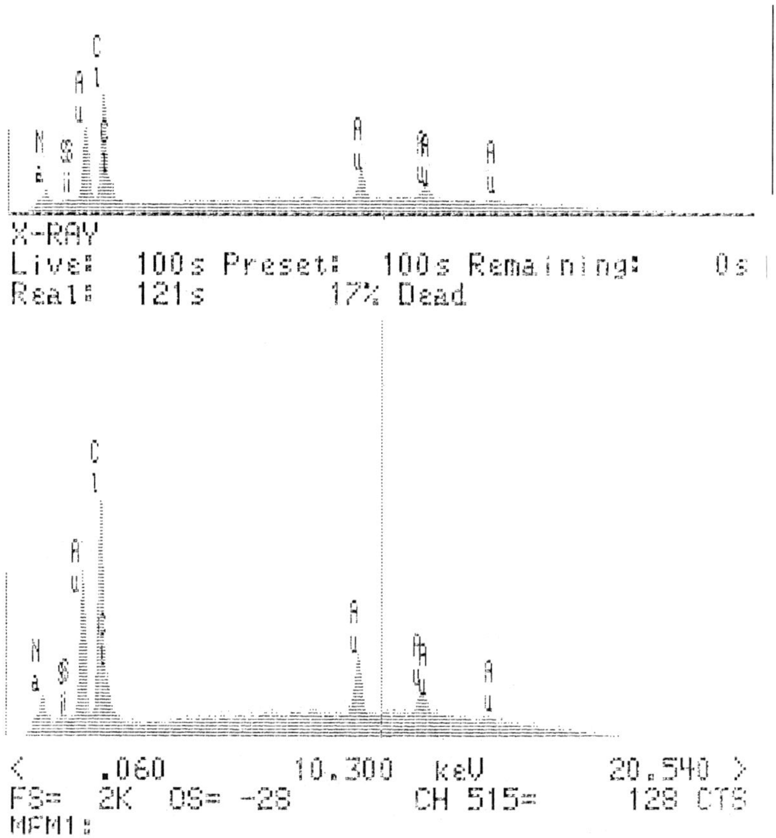

Fig. 50

allowed to cool before being covered, sealed tightly and stored, ideally, in a cool dry place (generally in a store-room or under the stairs). The meat can be ladled out and cooked further (generally with vegetables) whenever necessary. It lasts up to three to four months after the initial treatment, sometimes more, depending on season and place of storage. Such a mixture is also common in other North African countries such as Morocco. The name of this type of meat is: *Lahma Mahfooz*, meaning safe meat (Khattab 1991). There is no archaeological evidence to support this method of preservation. However, New Kingdom meat jars bearing the legend, *iwf dr* (pressed/boned meat?), could contain meat processed in this way, in addition to that of pounded meat (see above) and this should not be ignored as a possible means of preservation.

These, then, were the methods of preservation available to the ancient Egyptians. The question remains, as to which they favoured? Archaeological evidence suggests that drying and salting seems to have been the easiest and most popular technique. The pieces of meat from Theban tombs are salted, there are strips of biltong pictured on talatats and tombs, and the triangular pieces of meat suggest in their depiction that they are being dried, possibly after having been beaten. Salting is another strong contender, especially for pork (Kemp 1986: 74). Furthermore, the mummified meats were showed evidence of salt. There is, however, no reason why the ancient Egyptians should have restricted themselves to one type of preservation when so many were available[16]; different methods are good for different types of meat (e.g. brining for pork) and involve different costs in terms of time, energy and money. It is very unlikely that smoking was used in ancient Egypt as it would have taken too much of all these. It is quite probable that the Egyptians used various methods depending on the cut of meat, the animal the meat was from, and the materials and labour involved. Poorer families probably just dried any excess meat they might have, whilst wealthier ones would use dry-or wet-salt methods, sometimes in conjunction with drying.

IV.2.4 Fat

Goose fat is a common source of fat in ancient Egypt (Chapter VIII), and it is possible that both geese and ducks were preserved in goose (or other) fat. The birds would be plucked, eviscerated and put into a fat-filled amphora, much as is done in other parts of the world today (Montagne and Gottschalk 1961: 358). This is an alternative to the salt hypothesis for the amphora mentioned above, especially since 'confit' of goose or duck can last up to a year in a cool place (less in Egypt) (Strang 1991: 36-9; Pinch 1992).

IV.2.5 Honey

Another method of preservation that has been suggested in the case of poultry is honey (D'Auria et al 1988: 142). Honey has long been known

[16] Modern Egyptians use both the pemmican and the drying method, sometimes even in the same household. Dried meat is easier to carry and lighter, so it is favoured for long trips, while pemmicaned meat is used at home, especially in times of scarcity.

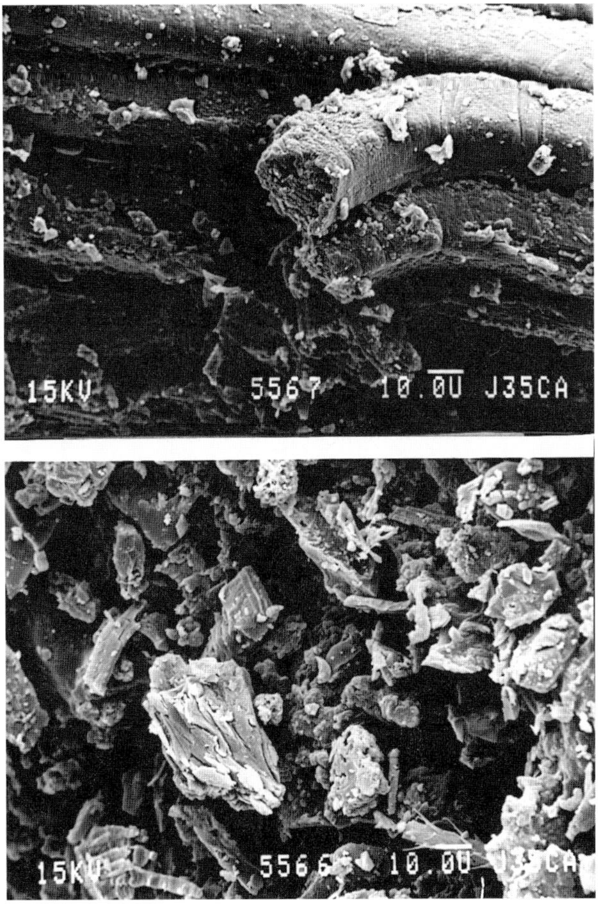

Fig. 51

as a preservative and is frequently used in curing food, especially hams. Honey is an excellent preservative as micro-organisms are unable to survive in an environment with such a high sugar content[17], although this will not deter moulds (Heath 1990). These are, however, less dangerous and can be washed/wiped off. It has even been suggested that honey was used to mummify the body of Alexander the Great. The idea that the Egyptians used honey for curing and preserving is based on the example of birds from the tomb of Queen Merytamun. These were recently studied by M. Leveque of the conservation department of the BMFA, who

[17] See above for the use of sugar in the Chinese brining method.

suggests that "the birds, which may have been cooked, were thinly wrapped in linen that had been saturated with an oily or resinous substance, perhaps even honey, intended to enhance both the preservation and the flavor of the food" (D'Auria et al 1988: 142). Personally I doubt that honey was used. It is true that it was available in Egypt and that the Egyptians were aware of its preservative and curative powers, as is shown in several of the medical texts, but there is absolutely no evidence that they used it with meats. Honey, the only sweetening agent in ancient Egypt, was a valuable commodity, making it unlikely that it would be used for the preservation of funerary offerings, the majority of which are preserved in salt[18]. Also it is questionable whether the ancient Egyptians would have mixed such extreme tastes. All their food was very simple and with little evidence of mixing of ingredients, especially meats. This holds true of contemporary Egyptian food which never mixes sweet and sour and salty. Unfortunately it was impossible to test any ancient poultry flesh to see if it showed any traces of sugar or salt; perhaps this test could be performed at a later date.

VII.4.4 Beer

Bruyère suggests that fish might have been preserved in beer (Bruyère 1937: 104). He bases the hypothesis on an item in Rekhmire's list of funerary offerings which he translates: "curches de poissons (?) confits dans la bière, 2 amphores" (Bruyère 1937: 104). There is no real textual or practical reason to believe that this was an Egyptian practice.

VII.4.5 Roe

Fish roe was one of the main by-products of fish-gutting. They make their first recorded appearance in the representations in the Memphite tombs of the Fifth Dynasty and later (Fig. 48). Any earlier representations that might have been painted are lost (Harpur 1987: 93). The roe, known in Egyptian as *swḥt* (*WB* IV: 73), is from the ovaries of the Mullet (*Mugil cephalus* and *M. batarekh*) and is still an important food items in Egypt (Darby et al. 1977: 372). Egypt has been famed for mullet roe

[18] This could be influenced by the salt in the soil where the meat was found, it cannot always be held to be a factor, as much of the meat excavated has been in wooden boxes.

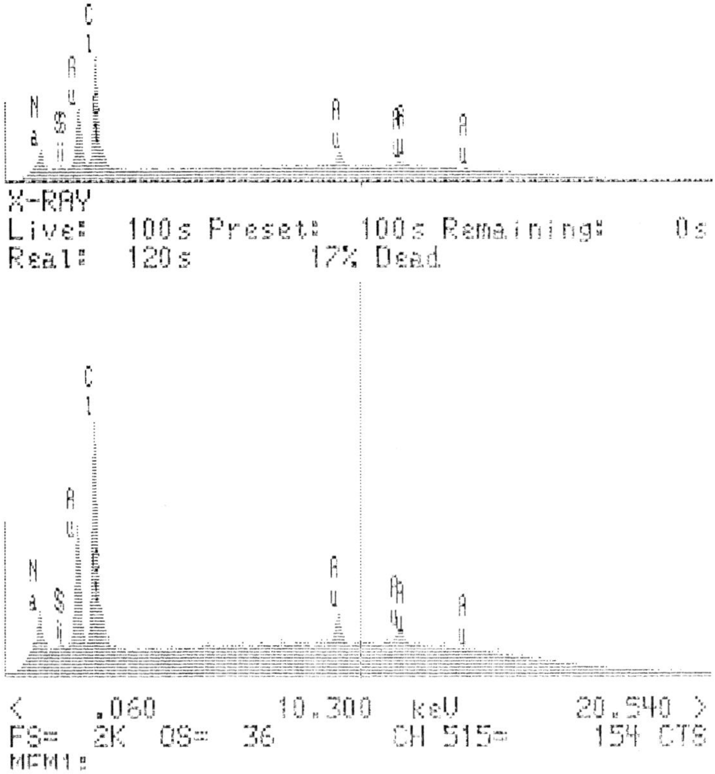

Fig. 52

(*battarah*) since 1657 (Vandier 1964), and it was widely exported throughout the Middle East and even parts of Europe in the second half of the nineteenth century (Keimer 1939: 234).

Battarah is made today by removing the ovaries, which are two oblong masses connected by a membrane, cutting this membrane, and washing the two separated masses with salt and water. This is a further argument for the adjacency of fish-processing sites to water. The ovaries can be pressed between planks to remove the salted water to speed up the drying procedure; this was commonly done in the nineteenth century when roe was exported, but it is doubtful that this latter part of the procedure was performed in ancient Egypt. Certainly roe-pressing with planks does not figure in any extant reliefs. The final step is drying the *battarah* in the sun and air (Keimer 1939: 217). A variation on this method of preparation is found in smoking the salted ovaries, although again there is no evidence for this practice in ancient Egypt (de Morant 1973: 68).

Fig. 53

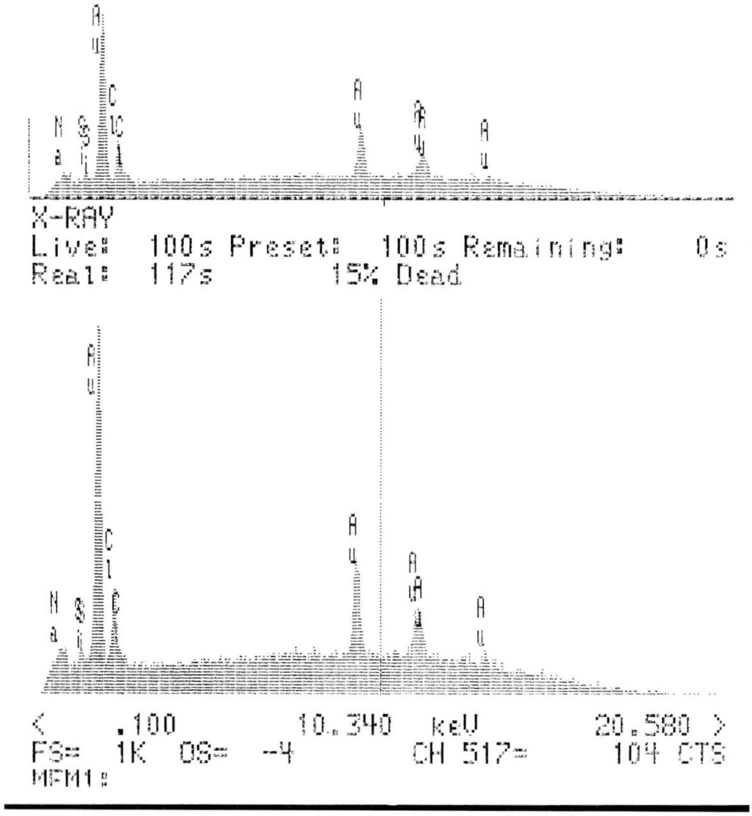
X-RAY
Live: 100s Preset: 100s Remaining: 0s
Real: 117s 15% Dead
< .100 10.340 keV 20.580 >
FS= 1K OS= -4 CH 517= 104 CTS
MFM1:

Fig. 54

VIII

FAT AND BLOOD

Two immediate by-products of meat that are valuable food sources and were available to the ancient Egyptians are fat and blood. There is an abundance of archaeological evidence to support the use of the former, while debate surrounds the use of the latter.

VIII.1 Fat

Fat was as greatly prized by the ancient Egyptians as it is by their descendants[1]. It is the second largest component of meat. The amount of fat in meat varies tremendously between types of animal and joint. Fat is a concentrated source of energy, and provides 10% to 40% of our total energy, and also contains vitamins A and D (McLaren and Meguid 1988: 33-4). In addition to its nutritional value and flavour, it can be used as an energy source for lamp-oil, for cooking, both as an additive and as an item of food in its own right, and for medicines and unguents. One of the reasons for the predilection for fat is that people tend to crave it because makes them feel fuller faster and longer as "fat delays the emptying of the stomach" (Passmore and Eastwood 1986: 205). Goose fat (VII.4.2) was commonly used in medicines, as were such diverse fats as ox (Ebers XXXIX, LIII, LIV, LXIII, LXVII, LXIX, LXXI, LXXVII, LXXX [Ebbell 1937]), hippopotamus (Ebers XXVI [Ebbell 1937]), and antelope (Ebers XXXI, LII, LXXX [Ebbell 1937]).

[1] Today sheep fat, *leah*, is more available than that of beef as sheep are more frequently slaughtered and the fat is more accessible than that of a bovid. It is highly prized and even sold independently of meat. This is used as butter, or pieces are fried and put in with vegetables. It is an honour and a mark of respect and hospitality to be given pieces of fat if one has been invited to a meal, and to offer fat is an indication of wealth. This is so with Copt and Muslim alike. There is a belief that fat bestows strength and virility so during wedding feasts lots of *leah* and sheep meat is given to the groom for these virtues and prosperity (Khattab 1991; observations and interviews in Middle Egypt 1991). This predilection for fat is rooted in the fact that fat takes longer to digest than any other substance, thus giving the impression of satiety and fulfillment for longer periods of time (Passmore & Eastwood 1986: 205; Farb & Armelagos 1980: 33-4).

The Egyptians displayed their preference for fat starting with their livestock, a tendency is especially obvious in the New Kingdom when obese cows are shown being led to slaughter, some unable to walk independently due to their extreme weight, and therefore being pulled on carts (Redford 1988: pl. 31; Fig. 11). Amarna reliefs show these animals being force-fed (MMA21.2.25, Figs. 1-3; Davies 1903: pl. XXIX) and the North Palace contains a stable similar to the ones depicted (Newton 1924: 295) with tethering stones fixed in the ground next to feeding troughs to facilitate force-feeding and fattening. The models of Meketre contain a stable where animals are being fattened and hand-fed (Winlock 1955: 22), and paintings in Tomb 3, at Beni Hasan, show oxen, goats, oryx and birds all being force-fed (Newberry 1893: pl. 30). Meat animals are already prone to fat as they are castrates, an operation that encourages fat deposition (Gilbert 1988: 77). This tendency is further encouraged by force-feeding the animals and depriving them of exercise, thereby producing animals whose weight is made up of fat rather than muscle (meat).

Fat is obtained from an animal after butchery. All the fat, the majority of which is found around the kidneys[2], or, in the case of sheep, in the rump, is cut off (Pyke 1970: 17), placed in a large pot (pottery or bronze), and is boiled. The fat is periodically stirred and the excess water resulting is ladled off. After some hours of boiling (dependent on the percentage fat content) the hot liquid is allowed to cool and is pressed by a paddle, generally of wood, to remove the last of the excess water. It is then considered "rendered" usable fat, and is poured into jars in a liquid state and stored (Hayes 1951: 91). The ancient Egyptians used pottery vessels as well as bovine horns for the storage of fat (Bruyère 1937: 152, 199-200). These vessels varied in size, ranging from small pots to large amphorae (Bruyère 1937: 152, 199-200; Hope 1978; see VII.1), but most had in common small mouths that could be easily sealed. Although many fat containing vessels were neither glazed nor burnished (the Malkata amphorae were, for the most part, burnished [Hope 1989: 89-90]), both these techniques would have helped to keep the fat from seeping through the pottery and spoiling. Bovine horns were used to store fat since they are not porous and can therefore better protect it against rancidity, and have the added advantage of being able to hold small, easily transportable amounts of this expensive commodity.

[2] The word *gnn*, suet, is thought to describe this item (*WB* V: 176). Three jars containing this have been found in Amarna (Pendlebury 1951: 169-74).

According to some scholars, notably Naville (1930: 9; Kitchen 1974: 195), fat-rendering has been depicted in wall paintings. One such example is in Seti I's temple at Abydos, (Calverly's Room 17, a room off the butcher's yard, Figs. 23, 55). The depiction shows a priest overseeing two men (also priests?) cutting up pieces of fat on a table, using a leaf-shaped knife; this fat is destined for a stove (?) which seems to consist of a pot for boiling over a place for a fire (?). Unfortunately, this portion of the scene is so badly damaged one cannot determine the construction of the boiling apparatus and the text is too limited to be of help. All it says, is effectively "fat being chopped up". The next vignette in the sequence shows a cloth or net containing the boiled fat being wrung in a press with a large vat below to catch the liquid.

A similar press, but of a different colour, is located on the north side of the room. This is thought to be a wine press (Naville 1930: 9; pl. III and IV). This depiction, thought by Naville to depict fat-rendering, does not however exactly tally with the way in which fat is usually rendered. For one thing, the fire necessary for boiling is not shown, though this might be due to the damaged state of the drawing. The removal of the fat is also not shown; we only have the rather abbreviated and damaged legend that suggests, due to the presence of the word *ʿḏ*, that the activity depicted is related to fat. Finally, the press used for rendering is odd for two reasons; the first is that one can press and remove all water from the fat in the boiling pot without recourse to an additional press. Such presses are generally associated with the making of wine or perfume (Newberry 1894: pl. 31.7; Newberry 1893a: pl. 12). Even Naville suggests wine-making for the other press on the north wall. Secondly, if the press is a way of ridding the fat of water, why go to all this trouble if Naville is correct in his assertion that the fat was used for lamp fuel (Naville 1930: 9)? Thus, the identification of this scene as fat-rendering is not secure, principally on account of the press, for which no archaeological, practical or ethnographic parallels exist.

A badly damaged scene in KV11, Room C, also seems to represent fat-rendering[3] as a part of the processing of an ox. A large metal cauldron with a row of jars above it, is situated next to a scene of drying meat, and above a depiction of what might be a press similar to the one featured in Abydos. There also seems to be an accompanying activity of chopping-up (?) the fat prior to rendering. Again, the identification of this as a fat-rendering scene is not secure, although it is a possibility. The depiction

[3] I am grateful to Dr. M. Marciniak for the use of his drawings of the tomb.

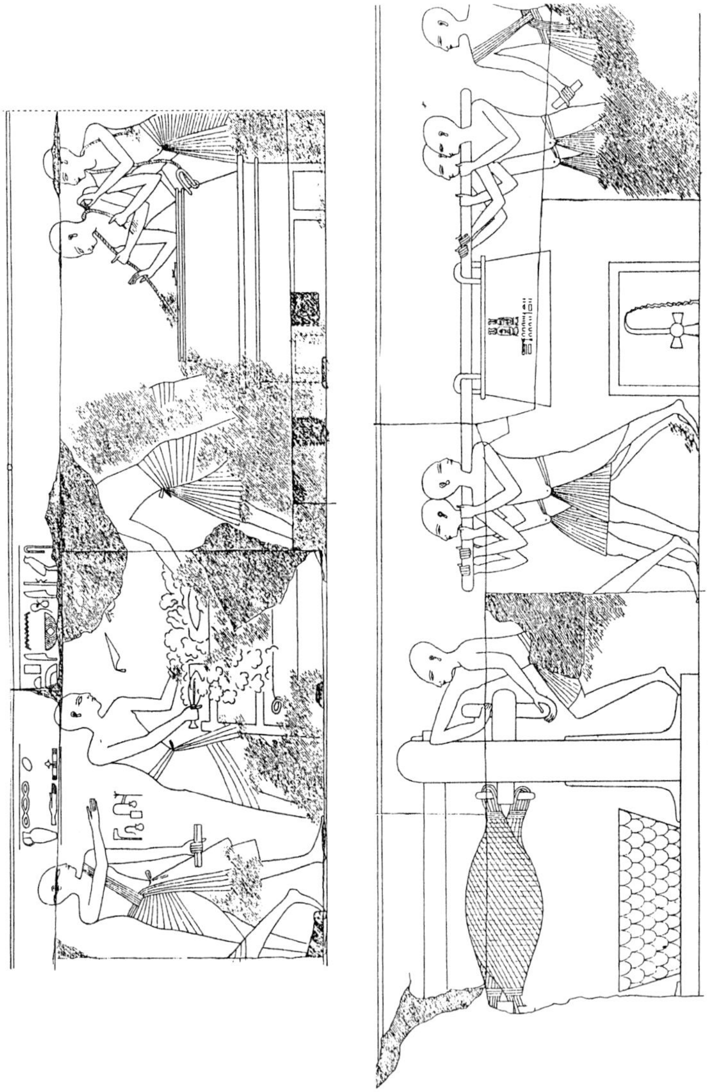

Fig. 55

of other methods for rendering fat can be found in models of the Middle Kingdom as well as in wall paintings subject to reinterpretation.

The Middle Kingdom models, notably that of Meketre, frequently show a pot on a three-legged stand over a fire. The pot and stand are all painted red, while the contents are white and red. This has traditionally been interpreted as the making of blood pudding (see below) (Winlock 1955: 23; Gilbert 1988). I would like to put forward an alternative interpretation of the scene[4], namely that is shows the rendering of fat taken from a newly slaughtered animal in the manner discussed above. All that is necessary for the process is a fire, a pot, some water, and the fat: all available in the butcher's yard. The only problem arising from this interpretation is the colour of the pot's contents. The contents of the vessel are whitish with dabs of red which could be fat although there is no illustration of the removal of fat in the model. The white could also be a froth or scum from boiling the blood. One cannot securely identify the contents of the vessel on the basis of the paint on the model. The red in the contents does support the interpretation of blood rather than fat which would be white or yellow. It is possible that fat and some blood were combined to make a version of blood pudding, although there is no known modern counterpart for this recipe. There is an instance, in the tomb of Ti, when both blood and fat are together removed from the scene of butchery (Épron and Daumas 1939: pl. 50). Nonetheless, the fat-rendering interpretation should not be ignored as it would have to be done somewhere, and regularly, too, as the number of jars bearing the legend "*ʿḏ*" and "*sr*" attest, and ox fat has been found in vessels (Banks and Hilditch 1933: 269).

Fat was not just taken from bovids[5]; it was removed from both sheep, who have a great deal of it (as jars in Malkata attest) (Hayes 1951: 159; Leahy 1978: 12), and geese, as there are jars in Amarna whose contents are labelled as the fat of *srt* geese (Pendlebury 1951: 169-75), as well as literary references to its existence, including the fat of *r* geese in the list (P.Anastasi IV.14.3ff [Caminos 1954: 200-212]). Anyone who has cooked a goose knows how fatty they are; in fact, half of a goose's body weight is made up of fat (Pyke 1970). Goose fat would be rendered in the same way as that of a mammal, but many more geese would be needed to make up a similar amount of fat as that which comes from a bovid. Goose fat would certainly last for up to a year in a cool climate, less in Egypt (Strang 1991: 35-9).

[4] This suggestion has been mentioned briefly in H. Fischer (1976) note 14, p. 99.

[5] Fat of special types of cattle is mentioned: the fat of meshwesh bulls (Hayes 1951: 82).

To the author's knowledge only two references to pig fat exist, the first coming from a jar docket from the Workmen's Village at Amarna (Kemp 1985: 66-7), and the second coming from a hieratic jar label from Deir el-Medina (Koenig in Kemp 1985: 67). In the case of the former, the lard was manufactured in the Delta and brought to Amarna. It is possible that lard was also rendered at Amarna as pigs were abundant there. Lard is rendered in the same way as any other fat and is probably more plentiful than suet. It is possible that it was made in Amarna, but that, as yet, no known example has been found. Perhaps analysis of pottery vessels from the Workmen's Village at Amarna will show some evidence of lard. Pig-fat being used for softening the joints is mentioned in the Ebers papyrus (LXXXII [Ebbell 1937: 96]).

Fat rendering and overseeing it as well as matters related to it must have been a task with some status attached to it as is attested by the title "Overseer of the Rendering House" (see Chapter V).

VIII.2 Blood

Blood is a rich source of protein and vitamins, especially iron. Pastoralists and nomads throughout Africa depend on it for protein not only in times of scarcity, but on a regular basis. They consume it straight from the live animal by making a small cut in the animal's neck and sucking it, or by bleeding the animal more extensively, or by using the blood from a slaughtered animal and boiling it until "it is fairly consistent and can be used as a meat flavouring with porridge; or the men let it stand till it coagulates into a solid block, and, after roasting it in the embers of a fire, cut it up and eat it" (Evans-Pritchard 1940: 28). One can extract 16 to 17 kg. of fresh blood from a steer when it is hanging (Gilbert 1988: 80)[6]. Tomb decoration and butchery models suggest that the Egyptians also used the blood of the animal in a sort of blood pudding.

Quite a few Old and Middle Kingdom butchery scenes and models show a man catching the blood from the throat of a slaughtered ox (Fig. 56; Blackman 1914: pl. 11; Blackman 1924: 27; Montet 1925: 165;

[6] 8-10% of an ox's body weight is made up of blood. Modern cows have an average weight of 500kg, and therefore contain between 40-50kg of blood. However it is impossible to collect all this blood due to the practical difficulties of manoeuvering an animal to collect its blood, the fact that not all the blood flows out — even the sharpness of the knife used affects the amount of blood flow. Thus, a steer only yields 16-17kg of blood (Jackson 1992).

Fig. 56

Winlock 1955: 23) and then carrying it away. The main room, C, in tomb D2 in Meir contains a tethering stone and a basin (35cm in diameter) cut into the living rock which Blackman has identified as being used to catch the blood of the animal (Blackman 1924a: 27, pl. 25; Fig. 24). This is highly unlikely as the vessel is hewn out of the rock and cannot be manoeuvred to catch the blood of an animal, furthermore, how would the blood be removed to be cooked? It is far more likely that the basin was actually a food-trough for the offered animal, and that the animal's blood was not collected. If, indeed, it was collected, it is far more likely that this was done in a pottery vessel. Some models also show a pot containing what might be blood (an observation based on its colour) over a fire. As discussed above, this has been identified as the cooking of blood pudding (Winlock 1955: 23).

As mentioned above, a modern ox yields about 16-17kg. of blood when suspended. However, no Egyptian oxen are suspended in representations, which would make the collection of the blood much less effective due to the lack of gravity, the awkwardness of the recumbent beast's pose, and the shape and size of the collecting vessel. In fact, the size of the vessel in many of the illustrations suggests that very little blood was actually caught (Blackman 1914: pl. 11). Could it be that the blood in the bowl was used for a ritual purpose rather than consumption? Winifred Blackman suggested this after witnessing an identical scene in Egypt early this century when the ox's blood that was caught upon slaughter was smeared on a doorjamb on behalf of the child for whom the animal was slaughtered, the idea being that the sacrifice of the animal's life ensures the life of the child (Blackman 1927: 289; 82). Several Egyptian reliefs show a priest examining blood in order to determine the purity and health and therefore acceptability of the animal (Fig. 9; Balcz 1933: 19; Leibovitch 1953: 59). That blood could be used for some ritual other than purity-testing is a subject for speculation. If the blood were used for rituals or purity-testing, rather than for food, the cooking pots on the side might indeed have been for the rendering of fat. Ethnographic parallels from contemporary Egypt are absent due to the Islamic ban on the consumption of blood which might have even have influenced the Copts as they, too, do not consume blood or blood puddings. It is possible that the ancient Egyptians, like the Hebrews who were long resident in Egypt, did not consume blood as it was considered unclean — it certainly is not mentioned in offering-lists (Fischer 1976: 99; Leibovitch 1953: 59-60). It is, however, mentioned in recipes in medical papyri: Hearst Papyrus (II.4, XII.10, IX.17, IX.8 [Reisner 1905]), Ebers Papyrus (LXIII, LXV [Ebbell 1937]). Thus, it would seem that blood was used in medicine, although it is unclear whether or not it was eaten. No conclusive evidence exists to determine whether or not Egyptians consumed blood pudding or how they rendered their fat. The number of jars of fat and references to the finished product confirm that they did the latter, although there is doubt about the manufacture and consumption of the former.

IX

STORAGE AND TRANSPORTATION

There are two elements to storage — the containers in which the processed and preserved items were stored, and the structure in which these were housed. These both influence the length of time a product can be safely stored, but other factors, such as the item's "composition, resistance to attack by micro-organisms, (and) the external conditions of temperature and the gases in the environment" (Desrosier 1982: 47), also play a large part in the success of the storage operation. The vessels in which the products are stored are most responsible for repelling any micro-organisms or any pests, while the building in which they are housed is responsible for environmental regulation.

IX.1 Storage vessels

Most processed meat, with the possible exception of biltong and splayed-dried fish (Davies 1903: pl. XXXI), would have been stored in pottery vessels. Fats would also have been stored in such vessels (see VIII.1). There is only a small number of vessels that actually provide physical evidence of meat storage. The majority of vessels containing meat found in a funerary context were plates, bowls, or "flowerpot vessels" that displayed food and acted as receptacles for a meal/offering rather than as long-term storage containers (Bruyère 1937: 97; de Morgan 1903: 55; Arnold 1982: 53-4). With extant evidence from funerary and settlement sites, as well as wall decoration, two criteria can be used to define a successful storage vessel: it should be a shouldered container capable of being closed tightly (with the possibility of creating a vacuum for better preservation), and should be made of a less porous clay, such as marl, rather than of silt to prevent seepage. Jars of various shapes fulfilling this criteria and containing evidence of having stored meat products have been found in archaeological contexts. The two most significant types are the amphorae and the so-called meat jars (Fig. 57A-B). These vessels contained fish, poultry, meat, suet, and fat (Smith 1976: 181; Hayes 1951: 157-62; Pendlebury 1951: 169-75).

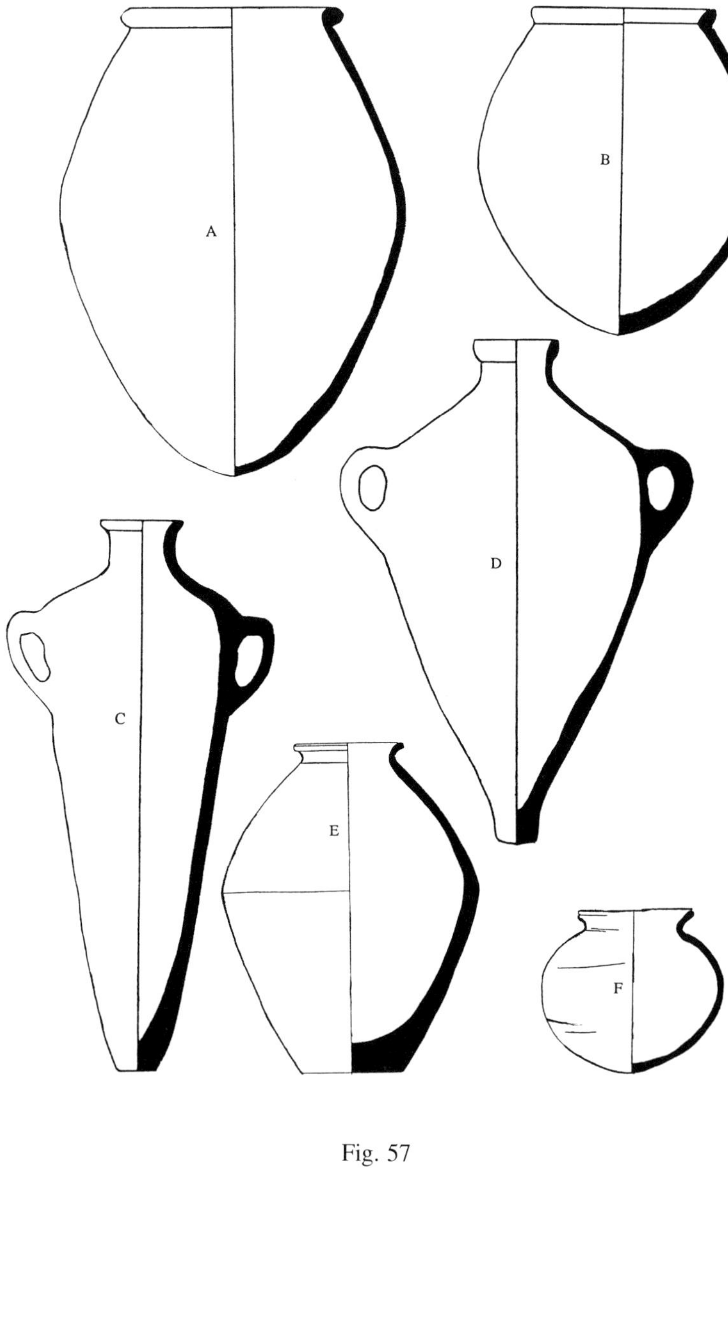

Fig. 57

The term amphora (Fig. 57C-D) here denotes a necked vessel generally with an out-turning lip and with a shoulder, either rounded or angled, and a body that culminates in a rounded or sharp point. They can have handles or not. Amphorae can be made of silt or marl, although the majority of large storage vessels were made of marl and marl-silt mixtures; "this fabric is fairly dense and when coated and/or burnished it becomes quite impermeable. The use of either a surface coating and/or burnishing for amphorae, irrespective of the fabric used, was obviously deemed advantageous by the Egyptian potters. Amphorae with plain surfaces are extremely rare" (Hope 1989: 89-90). Generally they are cream-slipped; any other decoration is extremely uncommon (Hope 1989: 17). Amphorae "were undoubtedly inspired by the so-called 'Canaanite Jar' ..., and copies of the prototype would seem to have begun during the XIIIth dynasty" (Hope 1989: 87).

In tomb scenes amphorae are pictured being filled with processed poultry, and indeed one such amphora made of marl clay was recovered from the tomb of Kha (Schiaparelli 1906: 159). Amphorae were not only used for poultry, but, as jar sealings and labels prove (Leahy 1978; Hayes 1951; Frankfort and Pendlebury 1933; Pendlebury 1951; Peet and Woolley 1923), they "were used primarily for the transport and both temporary and long term storage of (many) food stuffs, various beverages, incense and oil within domestic, religious and funerary contexts" (Hope 1989: 98).

Amphorae are especially well adapted to food-storage due to the shape of their necks — this is especially so if a warm substance is put in before sealing the jar. For instance, if hot meat/fat/poultry is put into a jar and the jar is filled to below the neck and sealed with an airtight (or close to airtight) stopper[1], the meat cools leaving a vacuum thereby creating an anaerobic environment which aids preservation. In modern wine vessels the top is a vacuum for this very reason (Heath 1990). Another practical reason for the particular suitability of amphorae for food storage is that the marl or mixed marl clay amphorae keep their contents cool and protected as they initially permit evaporation which leads to a stabilisation of the internal climate of the vessel, which in turn contributes to better long-term preservation (Hope 1989: 109). Due to the limited amount of ceramic work done on sites such as Malkata, where only the labels have

[1] Most of these vessels were sealed by placing a stopper of reed, pottery or mud (Hope 1977: 14) on the mouth of the vessel, followed by a flat-topped, moulded stopping in "the shape of an inverted bucket" (Peet & Woolley 1923: 161).

been examined, it is difficult to tell how many amphorae contained meat, fat, poultry or fish, and how many were stored in other storage containers — e.g. meat jars and their variants. It is known, however, that some wine amphorae were reused as meat jars (Petrie 1894: 33).

The term "meat jars" (Fig.IX.1A-B) is a term used to describe "store-jars with wide mouth and roll rim, slightly carinated body and pointed base" (Peet and Woolley 1923: 138) or "globular jars with carinated body, slightly pointed base, narrow outcurved neck" (Peet and Woolley 1923: 139; Kemp 1984: 136). There are many variations on these basic forms (Frankfort and Pendlebury 1933: 112, Type XII). The name appears to have been assigned to these jars as some of them bore labels (*iwf dr*) indicating that they contained meat, although several contained other products. These jars, like many of the amphorae, were made of marl clay which was quite dense.

The early excavators of Amarna believed, mainly due to the legend *iwf dr*, that these jars contained meat that had been removed from the bone (see Chapters VI and VII) and preserved in some way: pounded, pickled, salted (Pendlebury 1951: 170), an idea adopted by Hayes in his work on the Malkata material where he used the Amarna pottery typology as a temporary guideline (Hayes 1951: 96). Subsequent excavators also believe that meat jars were used for storing or pickling as they seem too small to accommodate a whole intact ox limb (Kemp 1984: 160 n. 9). These jars' maximum measurements are: 66 cms deep, and 30 cms wide (Peet and Woolley 1923: pls. XLVIII, XLIX, LIII; Fig. 57A-B). According to Hecker, "it would have been necessary not only to dis-articulate or dismember the major limb elements one from the other, but also perhaps to break them into smaller more usable chunks. This was most certainly the case if this meat, while still on the bone, was cooked in the much smaller and shallower cooking jars [Types 502 and 542]" (Kemp 1984: 160 n. 9). Hecker is correct in that the animal would have to be dismembered into major limb elements if they had to fit in the jars — these divisions are shown in the mummified meats (Appendix II) and in the Egyptian manner of jointing (Chapter VI). However, this is not a favourable method of preservation (see Chapter VII). It is far more likely, as Pendlebury and others have stated, that the meat was removed from the bone, preserved in some manner, and then stored in these jars. Bone is a waste of space, and by these means the marrow could have been consumed immediately. In a way, given the modern parallel of pieces of meat cooked in fat and salt being stored in a pottery vessel, one might favour this as a primary method of preservation; brining is also a method that lends itself to pottery vessels of this shape. The fact that frequently the hieratic labels on these jars gave the month, and sometimes day date of manufacture for meat (and fat)[2] commodities (Hayes 1951: 82; Pendlebury 1951: 169-72) would support this suggestion as this form of preservation has the least longevity of all the types mentioned in Chapter VIII. Out of the other methods mentioned salted or brined meat could also be stored in these jars. Biltong could be stored in jars, especially the small chunky sort, e.g. **S130**, but it would be unnecessary for longer strips as it could be put in baskets[3] or kept hung up — both more efficient ways of storing such strips as some scenes in Amarna tombs suggest (see below). Until the interiors of labelled pottery vessels, or, indeed, vessels in general, are chemically examined, it is impossible to ascertain the methods of preservation used on their contents.

Similarly shaped storage jars also of marl clay, albeit more globular, containing the remnants of preserved waterfowl and joints of meat were found by Valloggia in Mastaba V in Balat (Valloggia 1986: 208). The exact reconstruction of these jars is unclear as they were found in a fragmented state (Valloggia 1986: 208). The animals were found in two types of jars, the larger, more oblong jars of type inv. 1106, 1111 (Fig. 57E), or the rounder "cruches" (Fig. 57F), inv. 1462 (Valloggia 1986: 150-7; Valloggia 1991). These vessels would also have provided a preservative vacuum when sealed.

Although it is impossible to say if particular jar types were dedicated to meat and its by products, it is possible to get an idea of the basic shapes the Egyptians might have favoured for their storage. In modern Egypt the villagers store their *faseekh*, *Lahma Mahfooz*, oil, milk and water in unglazed pots generally of marl clay (Ballas) whose shapes are similar to the amphorae and meat jars used in ancient times. As discussed above, these shapes lend themselves, when sealed, to creating a vacuum and an anaerobic environment that repels microorganisms as well as the insects that can breed in such environments. This does, however, leave the problem of anaerobic bacteria (e.g. those responsible for botulism) that could multiply in this environment (Heath 1990). If the stored food were eaten without cooking, then it is probable that sickness, and in severe

[2] Malkata jars containing fat were reused more than any other type of jar — probably as the commodity was needed to be relatively fresh when used, so they consumed the fat quickly and reused the jars. It was probably fresh for no more than 6 months (Hayes 1951: 86).

[3] A discussion of the shapes, materials, and method of construction of the types of baskets that biltong could be stored in is beyond the scope of the present work.

cases death, would follow. If the foods were boiled or heated in any way, then most bacteria would expire, so rendering it safe for consumption.

IX.2 Storage buildings

After the food had been put into containers, the second stage of the storage operation was to house these containers in buildings. Information concerning storage areas comes from tomb paintings and the archaeological record, with additional data about the Amarna palaces and temples from the talatat blocks. Evidence from settlements shows that food products would have been stored in magazines, smaller storerooms, kitchens, cellars, or on roofs[4], depending on the size of the residence. A temple's holdings would be stored in a magazine.

In addition to the small storerooms that are part of a main house structure, and are frequently attached to the kitchens, not affording much extra protection for their contents (Bruyère 1939: 50; Peet and Woolley 1923: 49), additional storage areas in the form of basements or cellars have been found under the kitchens or/and the central chambers of several houses in Deir el-Medina (Bruyère 1939: 13, 29, 50, 269, 271; 1952: 35). Not all houses have cellars, perhaps due to the nature of the sub-soil (Bruyère 1939: 260). These cellars are reached by a flight of limestone stairs and are made of mud brick with some stone supports. The ceilings are of vaulted mud brick, and the floors of packed mud (Bruyère 1939: 29). They are approximately two to four metres square (Bruyère 1939: 66). Many of these basements contained a large amount of pottery, including amphorae and storage jars of different shapes and sizes (Bruyère 1939: 271).

Cellars are also a feature of houses in Sesebi. They have "small square mouths of dressed stone, usually with a stone covering slab. The (mudbrick) interior is small as a rule, beehive-shaped, and white-washed" (Blackman 1937: 150). Some Kahun houses also contained vaulted mudbrick cellars, "40 inches wide and about 60 inches long" (Petrie 1890: 24).

[4] The roof might have been used for storage in ancient Egypt as it is in modern Egypt. However, then, as now, it is more probable that cereal products and items of a more vegetal nature would be stored there rather than meat products. The latter would receive insufficient protection against heat and insects, and would spoil more quickly on the roof than anywhere else. Biltong is an exception to this as it can be stored in a light windy area — conditions similar to its time of manufacture.

Similar sized cellars were also found in Amarna houses. Like others they were generally rectangular, frequently with vaulted ceilings, and lined with plastered mud brick (Frankfort and Pendlebury 1933: 121; Peet and Woolley: 1923: 33; Pendlebury 1951: 22, 116, 122). They measured approximately 1m by 3m, on average (Frankfort and Pendlebury 1933: 52-3). A few were oval (Frankfort and Pendlebury 1933: 57) and one had a beehive vault (Frankfort and Pendlebury 1933: 73).

The archaeological evidence suggests that, whenever the soil and economics deemed it viable to construct cellars, they were used to store most food items. The vaulted ceilings common to most of the cellars would enable the air to circulate (depending on ventilation). Most important, the double insulation of the mudbrick and the surrounding soil would keep the interior diurnal temperature stable and cool, diminishing any threat of humidity. The double protection afforded by the roof and floor of the house above and the ground around would further protect the cellars' contents from temperature fluctuation and water and the resulting spoilage. The potential for insect infestation would not be decreased by this location, although it is possible that incense or herbs could have been used as deterrents, if not repellents. Rat traps (e.g. UC16773) would have been used to deter rodents. Well-sealed jars stored in these conditions would preserve the maximum shelf life of their contents, especially in the case of fat. Modern Egyptian houses rarely have cellars. Instead the cool protected space under the stairs is favoured for storage.

In houses, palaces, and temples the most favoured type of storage structure is the magazine. In Amarna the courtyards of some of the larger estates contained rows of long narrow roofed rooms of mud brick identified by excavators as storage magazines (Peet and Woolley 1923: 49). Communal storage might have been a possibility here. At Uronarti and Shalfak blocks of houses had attached mud-brick magazines (Reisner and Wheeler 1967: 10, 116). These domestic magazines are smaller-scaled versions of the magazines attached to temples, and in the case of Amarna, palaces.

Both palace and temple magazines were of mud brick. Some of these long narrow tunnels contained brick shelf supports which, in some instances, were whitewashed, and might have been used for stacking storage jars, while others had columns for roof supports, and a few contained bins, possibly for grain storage (Pendlebury 1951: 172). Magazines in the area identified as the police quarters are also of a similar construction (Pendlebury 1951: 131). Representations in tombs and on talatats show temple and palace magazines filled with amphorae containing food products, possibly including fish, *iwf dr,* and fowl, dried fish not in containers, and possibly bundles of biltong (Davies 1903: pl. XXXI).

The magazines of the Ramesseum, Uronarti (Reisner and Wheeler 1967: 13), Medinet Habu (Hölscher 1931: 59; 1941: 65), and Abydos are similar in construction to those at Amarna, although the excavators do not mention shelves or columns. The roofs of these magazines were vaulted (Ghazouli 1964: 144; Hölscher 1932: 64), as might have been the roofs of the Amarna magazines. In some the floor was made of mud tiles (Ghazouli 1964: 144). The thick brick vaults of the roof ensured that "the temperature inside these magazines remained equable throughout the year, thus ensuring the maximum preservation of the commodities stored in them" (Ghazouli 1964: 145). Dependent on ventilation, the vaulted roof would allow for circulation of air, and the thickness of the brick would protect the contents from the elements. Although many of these magazines would have served exclusively as granaries, others could also have been used to hang curing or cured meat. Hölscher believed that the rear part of Storeroom N in Medinet Habu "probably served to store goods ... (such as) jugs of wine or oil, honey or preserved meats" (Hölscher 1941: 65).

In Amarna and Malkata the word *šnʿ* (Gardiner 1947b: 206; Hayes 1951: 92; Pendlebury 1951: 172) has been thought to mean ergastulum or storehouse for food[5], and, indeed, in Amarna, jar labels with this legend have been found near magazines, thereby suggesting that these magazines were known by the name *šnʿ* (Pendlebury 1951: 106; 172; 211). The word *ʿt iwf*, or "meat chamber" is known from P. Bulaq 18's "section of the provisioning-quarters devoted to the preparation and storage of meat" (Quirke 1990: 35). This room, located in the palace enclosure, could possibly be construed as a chamber where meat was stored, although it might have been an area where meat was prepared for immediate consumption or long term storage.

The building materials used for all these structures also helped to protect their contents. Most of the walls were of mud brick consisting of mud, chaff, and some hard temper, e.g. bits of sandstone, limestone, chalk or clay. The mortar is made of gypsum, chaff, or clay; the plaster consists of mud and chaff with the odd limestone inclusion. Roofs were made of mud plaster and organic materials: palm/acacia logs, reeds, palm-leaves, palm-fibre rope, and grass (Lacovara 1990: 6, 8; Peet and

[5] The word might also mean a place of manufacture (Pendlebury 1951: 106; 172; 211). It is quite possible that manufacture and storage close together, as might have been the case with S.40.1 and the neighbouring magazines (see Chapters IV and VII).

Woolley 1923: 42). Floors generally were of beaten earth or mud tiles (Ghazouli 1964: 144).

On the whole, mud, the main component of these structures, is not a very strong conductor of heat. Fathy writes, "sun-dried earth brick is … one of the poorest conductors of heat. Partly because of its very low natural conductivity (0.22 calories/minute/cm2/unit thickness for bricks with 20% fine sand, 0.32 calories/minute/cm2/unit thickness for bricks with 80% coarse sand, as against 0.48 for baked bricks), and partly because mud is weak and necessitated thick walls, the mud-brick houses of Upper Egypt do remain remarkably cool for the major part of the day…yet mud-brick retains heat for a long time" (Fathy 1968: 63). Thus, the interior of any storage spaces constructed of sun-dried mud brick would be quite cool. Clay floors would only add to this. In modern Egyptian mud-brick houses the clay floors are always cool to the touch. Nowadays Egyptians sometimes sprinkle their floors with water to make the rooms even cooler by evaporation. This could have been used in ancient times, although care had to have been taken that there was not too much water, and none of it touched meat that was not protected in some way as otherwise it would have grown mouldy.

IX.3 Transportation

Meat products, like other commodities, were traded and transported throughout Egypt and abroad, whether for commerce or as sustenance for armies or traders (Janssen 1961; Wenamun's voyage). The primary mode of transport was on barges, via the Nile (e.g. TT40, TT57, TT217, D60), or on ships plying the Mediterranean or Red Sea (Borchardt 1913: pl. 11, 12; Lichtheim 1976: 227, Wenamun [2, 40ff]). Both live animals and processed foods would have been thus transported (Fig. 58). Land transportation was also used, especially over shorter distances or on any mining and/or quarrying expeditions within Egypt. Items were commonly transported by people and on the backs of donkeys, or, from the New Kingdom onward, in (ox-drawn) carts. If live animals were not transported via water they would have been transported on the hoof. The mode of transportation, and consequently, the journey-time would have influenced the type of containers in which processed meat was stored.

Owing to the course of the Nile, which ran the entire length of Egypt, water transport was most favoured. Heavy items such as obelisks, statues and other large stone wares, were transported by water, especially in the

Fig. 58

season of innundation when the height of the water and the strength of the current would make water transport the quickest and most efficient means of conveyance available. Grain (Gardiner 1941), vegetables, wines, beers, livestock, and meat products are shown being transported via water (Hornell 1943: 77; Naville 1898: pl. LXXIV; Davies and Gardiner: pl. 12).

In ancient Egypt, as in modern Egypt, barges were the favoured type of boat for transportation of goods. These were large flat-bottomed wooden boats with capacious decks for carrying a large number of items, and sturdy enough for carrying heavy items (e.g. obelisks, oxen, trees). The flat bottoms prevented them from sticking in the Nile mud, and, in case of this event, the flat bottoms kept the boat stable while the sailors dislodged it from the mud. Many barges had the optional feature of cabins, either of solid wood/matting, or of an open framework of wood, for storage. Some even had stalls for livestock (Davies and Gardiner: pl.12; Faulkner 1940: 3). The Causeway of Unas (Hassan 1955: 137; fig. 1) and Hatshepsut's Punt reliefs, among others, show all manner of objects being transported by boats and barges, both with and without cabins and stalls.

If Egyptian reports are to be believed, wooden boats of 51.83m (100 cubits), were built during the reign of Snefru; smaller boats of 18.2m (40 cubits) and 120 feet (80 cubits) were also built during this reign (Hornell 1943). Papyrus Harris mentions the construction of special boats made of acacia for transporting oxen (Birch 1876: pl. 112b, l.11). Grain barges, very similar to the barges used for transport of other food products, have been studied with the following results. Ramesside grain boats were approximately 18.75m in length, and 7.25m in width, and were like very large Sudanese *nuggars* which are used even today. These boats had a large carrying capacity (Hornell 1943: 77), and could easily accommodate oxen or a large number of storage vessels containing meat (and other) products.

As mentioned above, the best time for transportation was the innundation. At this time the river was fastest and strongest and capable of transporting heavily loaded barges. It was, too, a good time for transportation of meat products not only because of the height of the river, but also as the harvest would be long over and people could work on the boats, and, more importantly, this season would be better for transporting food-stuffs rather than trying to store them in moist conditions which might adversely effect their quality. Furthermore, it is possible that the innundation was a favoured time for slaughtering animals as there is less grazing land available for the cattle. Also, cattle caught in continually damp areas tend to get hoof rot, a painful disease which effects the overall health of the animal and the quality of its meat. Therefore innundation would be a practical season in which to cull the herds[6].

Boats other than barges would have been used on other bodies of water. Galleys were used on the Red Sea and the Eastern Mediterranean as there was not much wind and the sea was more choppy than the Nile.

[6] Unfortunately, I have been unable to work out any calendrical correlation of a concentration of cattle offerings on feast days with any specific season, although the culling of the herds might occur in the innundation whilst being independent of any religious significance.

The dimensions of these galleys were approximately 62.8m in length, and 21.0m in width (Gibson 1948: 24), providing ample room for the transport ot trade goods, whether of perishable or non-perishable commodities.

The amount of time and the length of a journey would influence the type of vessels in which meat products were stored, as well as affecting the condition of the food items. The length of time for a river journey would be influenced by the season, the direction of travel, the height of the water, the wind, the strength of the current, whether they travelled at night, the number of stops, and the number of sailors. Despite these factors, and the somewhat contradictory and incomplete evidence from ancient Egypt, it is possible to calculate average speeds and lengths of time for journeys made by the Egyptians. According to the Adoption Stela of Nitocris, it was possible to travel from Sais to Thebes in sixteen to seventeen days (Caminos 1964: 79); this is possible if one travelled about 57km a day; according to Herodotus the journey between Sais and Elephantine would take twenty days, with 58km a day being covered (Lloyd 1976: 35). This is contradicted by another document that gives seventeen days for a journey between Memphis to Hatnub, albeit during the time of a low Nile (Fischer 1975: 34) which would make for slower progress. Another source claims that the voyage between Memphis and Thebes would have taken twelve and thirteen days (Peet 1931: 486), with 59km a day being covered, while Herodotus says nine days are necessary for the voyage between Heliopolis and Thebes, although this calculation does not fit with most of the other journey times he gives (Herodotus II: 4-9; Lloyd 1976: 34-6). Later sources provide a clearer idea of journey times. An Englishman, E. Brown, travelling in 1673 estimated that he covered approximately 65 km. a day on a barge going downstream, although this number fluctuated depending on water-level and wind (Sauneron 1974: 92). A similar figure of 60 km a day is also suggested by R. Krauss (1984: 222), while Lloyd suggests 56km a day on the basis of different numbers culled from Baedeker's *dahabiya* speeds as well as Herodotus and other ancient and modern sources (Lloyd 1976: 34-6). Given the evidence, one can say that the Egyptians covered 55-60km a day, making the voyage between Memphis and Thebes possible between eighteen and twenty days.

As discussed above, the 'shelf-life' of preserved and pickled goods, as long as they remain damp free, varies from three months to two years, depending on the mode of preservation. Also, water transport is generally quicker and definitely cooler than land, thereby providing optimum conditions for storage of preserved foods. Thus, fish, and if jar labels are to be believed, meat, was transported by water in the Nile Valley as well as to further locations such as Syria.

Any unprocessed meat that had to be transported for any distance would have to be live, as is shown in some tomb reliefs. Fresh joints, however, could not be transported thus, although they could be present on board for swift consumption by sailors (Janssen 1961: 18). It has also been hypothesized that the crews of Egyptian boats "brought with them products of their native region, perhaps a part of their wages received in kind before the departure, to exchange them for different kinds of food in the local markets. Thus they would not have been condemned to live on their daily bread ration alone" (Janssen 1961: 8).

Land carriage was also used for conveyance and trade within Egypt, although it would be a more arduous means of transportation than water, and consequently less frequently employed. Unlike the Romans the Egyptians did not leave behind any roads — in fact even the Romans, obsessive road builders though they were, preferred to use the Nile for transport of goods and men. Land transport was probably not favoured in the valley, although it was the only way of conveying goods and men on quarrying and military expeditions in the desert or other land locked areas. The conveyances used by the Egyptians on land were limited. Until the end of the Middle Kingdom, and even during the New Kingdom, donkey caravans and porters would be the only 'vehicles' available (e.g. as in the tomb of Mahu [Davies 1906]). The only other way of transporting large objects (e.g. stones, coffins) was by sleds. Furthermore, these would be fairly inefficient for transporting vessels containing supplies and would considerably slow down an army. In the New Kingdom, the time from which the majority of our expedition information dates, horses and carts were used for transportation. Horse- or even donkey-drawn carts, such as are ubiquitous in modern Egypt, might have existed in Egypt, but they have left no trace of their existence either in the pictorial or in the archaeological record. The bullock cart has, however, appeared in the pictorial (Tylor and Clarke 1896: pl. 2) and the literary record (*Urk.* IV: 1232), as well as in model form from the coffin of Queen Ahhotep (CG52666, CG52668). For example, a stela from Wadi Hammamat, dating from the time of Ramesses IV claims that ten carts, with six yoke of oxen to (each) cart pulled supplies from Egypt through the Wadi Hammamat (Kitchen 1969: 13; l. 19-20). The exact nature of the supplies, *ḫr.t*, is unclear as it could be food or military/camp supplies. The issue is further confused as the stela continues by noting the presence of porters carrying bread, meat and cakes. Is this in addition to food supplies carried by the carts, or are the carts transporting non-edible items and the people the food? (Kitchen 1969: 13, 1.19-20; Breasted 1906: 196). Presumably the rest of the food would be acquired en route by the army, presuming it was travelling in settled or even fertile land. Regardless of what occurred on this particular occasion, bullock carts were a possible means of transporting meat products. Of course these conditions are not as good for storage as are those of water transport. However, if anything was being transported it would have been of a drier and saltier nature, so biltong would travel thus with more ease than would pemmican.

Animals that were to be consumed along the way would be driven, though the transport of their fodder, along with that of any pack animals, would be another burden that would hinder the speed of any expedition. In fact, food for the animals would be more cumbersome and harder to obtain than that of humans in the more remote desert areas.

Sand slowed progress even further. The journey from Qena to Qoseir has been known to take four to five days by camel via Wadi Hammamat, while the trip between Abou-ed and Berenice took eleven days by camel (Lloyd 1976: 36). The Egyptians did not have camels, consequently their beasts of burden were either donkeys or cattle. Neither of these beasts is noted for fleetness of foot, although both are steady and sturdy workers, and require little fodder (Foaden and Fletcher 1910: 802-4). One can safely assume that the journey through the Wadi Hammamat would take donkeys a day (if not two) more than it would a camel. Ancient sources are apparently silent on the subject of the length of time taken by land travel. Edel, in his work on the location of Yam (1955: 65-8), surmised that a donkey train of one hundred donkeys would travel 15km a day.

On overland routes food is exposed to heat and sun for longer periods of time and is disturbed much more than on board a ship, thus giving chemicals and bacterii a greater chance to intermingle and flourish. This adversely effected the food. Despite the longer 'shelf life' of most of the processed food, foods preserved in fat, e.g. meat or even poultry, would not travel well by land as the fat would become rancid with exposure to heat and motion. Brine-preserved foods would also suffer if thus transported for long since, if the vessels are not sealed properly, they stand a greater chance of being infected. Additionally, brine- and fat-preserved foods are less sensible choices for food on such expeditions as they are heavier to carry. Biltong or salted fish would be the ideal types of food for land expeditions as they are light, nutritious, and easy to pack and store — in fact, the former can be transported in baskets or reed bags rather than in pottery vessels. Thus land transportation would be used for expeditions rather than for trade as the ill-effect of land transport on the commodities would adversely effect their commercial value.

X

USING THE EVIDENCE: TOWARD A SOCIOLOGY OF MEAT

In ancient Egypt there was always some form of meat, be it poultry, fowl, game, sheep/goat, or beef that was available to the entire population. Beef would have been the most prestigious animal as, due to the size of cattle as well as their appetites, it was the most expensive meat to produce. However, the frequency of meat consumption and the type of meat consumed differed considerably between classes. Although the present state of knowledge concerning meat consumption is limited, one of the main goals of this study is to move to a full understanding of its distribution and consumption, that is, a sociology, of meat. The evidence currently available is quite limited and varies considerably in kind and quantity throughout Egyptian history. The analysis of this evidence is thus suited only to specific case studies which, it is hoped, are fragments of a larger picture and will, in time, help one to understand fully the practices of the time. As discussed in other parts of this work, the sources of evidence that can be drawn upon are, principally, the artistic and textual records that also shed light on so many other aspects of Egyptian life and culture, and the more unusual sources of evidence specific to this subject: actual bones and mummified joints consolidated through experimental and ethnographic work.

The limitations of the evidence concerning Egyptian society prevent a clear view of its social organization and the distribution of wealth. The idealised Egyptian view, as reconstructed by Egyptologists, sets the Pharaoh on top, below whom are nobles who serve at court, in the temples, and in the military. Below these are lesser nobles, merchants and traders, followed by peasants. However, as it would have been possible to be a wealthy peasant or a poor scribe, it does not necessarily follow that meat, being a luxury item[1], was only consumed frequently by the nobility,

[1] Meat (mainly beef) is considered a luxury due to its scarcity, the cost of rearing the beasts, and the difficulty in raising them in an arid country such as Egypt. It is not mentioned often in lists of provisions, and when it is, it is generally being distributed among the nobility, or the majority of it is; this is illustrated by account papyri, notably P. Bulaq 18 (Spalinger 1985: 219).

	KV34-35	KV62	KV43	KV36	KV46	Isitem-kheb	QV46	MMA1021	TT358	KV60	QV51	TOTAL
Gender	M	M	M	M	M/F	F	M	M	F	F	F	7M/ 5F
Ruler	Yes	Yes	Yes	No	No	No	No	No	No	No ?	No	3
Cattle	11-13	2	0	1+	2	1	1	1	1	1	0	23
Poultry	6	11	6	5	6	4	5	42	1	3	4	93
Gazelle	0	0	0	0	0	1?	0	0	0	0	0	1
Sheep/ Goat	0	0	0	0	0	1	0	0	0	0	0	1
Antelope	0	0	0	?	1?	0	0	0	0	0	0	1

Fig. 59 Minimum Number of Animals per New Kingdom Burial

Tomb	Mentu-hotep II	Nubhetep-tikhered	Ita	Khnumet	Asha-yet	Tomb 7	Tomb 10	Senebtisi	Wah
Royal	√	√	√	√	√	?	√	X	X
Cattle	1	√	1	√	√	2-4	2	1	1
Antelope	0	0	0	0	0	0	1	0	0
Poultry	2	√	√	0	?	0	0	2	0

Fig. 60 Middle Kingdom Offerings (Total Number of Animals)

Tomb	Mentu-hotep II	Nubhetep-tikhered	Ita	Khnumet	Asha-yet	Tomb 7	Tomb 10	Senebtisi	Wah
Head	1	0	0	0	0	0	0	0	0
Legs	3	?	0	?	?	0	0	0	1
Humerus	0	0	2	1	0	2	1	2	0
Scapula	0	0	0	0	0	2	1	1	0
Ribs	0	0	√	0	0	8	8	0	0
Femur	0	0	0	0	0	2	2	0	0
Patella	0	0	0	0	0	2	2	0	0

Fig. 61 *Bos* Bones in Middle Kingdom Tombs

√ = Present

but rather by those who were in the Pharaoh's inner circle, or who could afford it, or, indeed, by those who worked in the butchery business.

Almost all available sources of information about the ancient Egyptian diet pertain only to the upper echelons of society: the Pharaoh and the nobles of his court. The artistic record almost exclusively revolves around illustrating different aspects of the lives of the privileged. The food offerings found in burials, and indeed the burials themselves, were for the benefit of a wealthy élite (that is, at least, the majority of those that have been retrieved and studied)[2]. These food-offerings complicate the issue further as the question arises whether these were actual joints

[2] For an example of a lower status mummy see I:5 in Cockburn 1980.

that were consumed by the living, or were special funerary cuts devised for the dead? In fact, the butchery marks found on faunal material from settlement sites, as well as drawings in tombs and temples (see Chapter VI), point to the fact that they were one and the same. On the whole it seems feasible that the joints that were consumed would be identical to the joints offered, as there are a limited number of ways of butchering an animal, and a society generally adopts a single method that is used in all butchery, sacred or secular. Whether in life the tomb-owners would actually have consumed all the cuts provided to them in death is debatable. There is always some dispute as to whether or not funerary offerings consist of items produced specially for the tomb, rather than items that were in daily use. The meat-offerings from the New Kingdom are almost all from royal graves and of fine quality (see Figs. 43, 59). Perhaps these rather generous cuts were joints that the deceased would have liked to have consumed in life but did not; however, given the other evidence from texts and faunal assemblages, it is more likely that these joints were actually consumed in life by the deceased, and therefore included as part of the funerary offerings. What is interesting to consider, although there is no clear answer, is whether these joints of meat were offerings for the afterlife made to the deceased at the time of the funeral or whether they formed the deceased's share in the funeral banquet. It is more likely that they served the former function, although it is appealing to imagine the deceased partaking in the final feast with his/her mourners.

Information about the diet of the 'peasants', or ancient *fellahin*, comes from a few vignettes found in tomb paintings, extrapolation from what is known about the diet of other sections of the population, ethnography/ethnohistory, and common sense. It is only recently that faunal assemblages from settlement sites have provided information about workers and peasants as well as the nobility, albeit in a limited way. This is due in part to the paucity of such sites examined, the dearth of published information available, and in cases like Amarna, the fact that the classes residing in the Workmen's Village might have had a special, perhaps even privileged status similar to that of the workers of Deir el-Medina.

Following the conventional interpretation, the Pharaoh was at the head of the socio-political hierarchy throughout the span of Egyptian history. He was god, priest, and king, and would naturally have had the largest amount and variety of meat available to him. His lands would have provided him with wild and domesticated mammals and poultry as well as all available fish, perhaps including marine varieties. From the Old and Middle Kingdoms there is virtually no direct evidence concerning the Pharaohs' diet, while the artistic evidence from these periods is also not very revealing about their eating habits. The Pharaohs are merely depicted in front of offering-tables which are groaning with food, or standing whilst receiving tribute of food animals.

If the *ḥtp di nisw* formula had its roots in the real practice of early periods of Egyptian history it is possible that the Pharaoh had a great deal of control over cattle and other foods and goods. Hassan records that in the rock cut tomb of Debehen [LG90] in Giza there is a butchery scene whose legend reads: "Slaughtering the oxen for burial … *as a boon which the king gives* (*ḥtp di nisw*)" (Hassan 1943: 179). Whether this was merely a ritual formula or one which had its origins in a reality which changed over time is not certain, but the phrase suggests that at some point permission to own and slaughter cattle was required from the Pharaoh, although it is also possible that this boon was an 'extra' royal mark of favour. To some extent jar labels from the New Kingdom similarly confuse the issue of ownership of cattle as the processed meat often came from the holdings of nobles as well as the king, but it all ended up being owned by the king who probably used it to feed some of his court (see below). It is unclear whether or not the live cattle belonged to the nobles outright or were leased/gifted from the Pharaoh/temples (Kemp 1991: 191). Both seem to be possible (Castle 1992: 243-57). It is possible that the king might reward certain people with gifts of beef but no live cattle, wealth or lands. This last was included in rewards for military service, e.g. Ahmose son of Abana. However, this is not the place to discuss who owned the cattle, but rather, who ate it — the two are not necessarily the same.

The Middle Kingdom provides more direct evidence concerning the Pharaoh's diet in the form of food-offerings from the burials of members of the royal family (Appendix II: ii; Fig. 60). One example of a royal burial containing meat offerings, that of Mentuhotep II (**R4-7**), contains beef and poultry, although none of it is mummified, in contrast to the New Kingdom food-offerings. The Middle Kingdom burials containing meat offerings consist of beef, poultry, and perhaps antelope (Fig. 61), although the quantities are modest in comparison with the victual mummies of the New Kingdom. There are no known examples of whole mammals being offered; if any were, their remains have yet to be recovered.

The bulk of the information concerning meat consumption for all strata of Egyptian society comes from the New Kingdom where, in addition to the artistic evidence, there is physical evidence from faunal assemblages, as well as from victual mummies and the Pharaoh's bodies.

Physical Evidence: Victual Mummies

The majority of victual mummies (Appendix II, Figs. 36-37) come from the tombs of members of the royal family (8). The mummies are currently housed in various museums across the globe, notably the AMNH, the MMA, the CAS, the Egyptian Museum in Turin, the BMFA, the Cairo Museum, and the Dokki Agricultural Museum. As was mentioned in Chapter VI, these mummies consist of joints of meat or of entire birds that are wrapped in bandages. Most of these bandaged meats were placed in individual sycamore-wood 'coffins' shaped to the form and dimensions of the meat, the lids of which were joined to the body by dowels and holes. These boxes were plastered and whitewashed both inside and out, while the interior of several of these boxes was coated with pitch. The pitch which might have been used to seal the boxes and make them resistant to moisture, thus preventing the meat from deteriorating and smelling. It is possible that a roasted appearance (browning) was given to victual mummies by the application of very hot pitch on the wrapping which no doubt cooked the exterior surface of the mummy (Lortet and Gaillard 1905: 248). The method of making the 'victual mummies' is unclear, although it has been suggested that the meat was preserved by immersion in resinous natron (Lortet and Gaillard 1905: 307). Many of these victual mummies were examined by the author, the results of this study appear in the **S** catalogue in Appendix II. These mummies are generally entire birds or joints of beef, the most costly type of animal to be consumed. Only in rare instances are ovicaprids offered (Fig. 59). In many cases entire animals, rather than just joints, are offered (TT358, KV34-35, KV62, KV46, MMA1021, Isetemkheb D), which would mean that none of the meat would go in payment to priests, funerary workers, or guests at the funerary banquet — rather a waste of a valuable food source. Frequently more than one animal is buried in one tomb (Figs. 43, 59). These food-offerings are included amongst the funeral goods as they will not only provide the deceased with the coveted "choice cuts", but also are a show of wealth and power, much like the other luxury goods included in the burial. This is particularly so when several animals are offered. The animals found in tombs are also generally under three and a half years of age as shown by the unfused epiphyses of many of the joints (Appendix II). Younger animals have less connective tissue and more muscle, and are therefore more tender and meatier than older beasts (Passmore and Eastwood 1986: 205). Thus not only was the Pharaoh's meat supply plentiful, it was also of good quality. It would be useful to compare meat offerings from the burials of nobles to see whether or not most meat for offerings was butchered young, or whether this was limited to royal burials. Certainly the evidence from Malkata shows that animals were slaughtered at a variety of ages, though the bones of those found in the court's dump were all under the age of four (see below).

Evidence from jars containing meat from both Malkata and, especially, Amarna reveals that some of it came from royal stockyards (Pendlebury 1951: 170, 172; Hayes 1951: 96-7). The evidence from the Amarna jar labels shows that, not only were the royal stockyards the producers of the meat, thereby providing the royal family with fresh meat, but also, the preserved meat was a royal possession regardless of its producers. Many such labels were found in the central city (the Great Temple and Official Palace), while the majority were from the areas of the Maru Aten (Pendlebury 1951: 143, 170) and the magazines between the royal estate and temple (Pendlebury 1951: 211). Thus, its consumers were probably court officials as well as the king. The majority of Malkata labels were from area K, a remnant of a short-lived palace associated with the Birket Habu, and therefore not very revealing in terms of storage areas (Leahy 1978: 2).

Fat jars, containing the by-product of cattle, sheep and geese, in that order, do not seem to come from anywhere but royal holdings (Hayes 1951: 96). This indicates that only in the king's slaughterhouses were a sufficient number of animals killed to make fat that could be used and distributed. It must be remembered, however, that, as labelled jars were a relatively small proportion of all jars (most being unlabeled), they are a biased sample.

Additional information concerning royal eating habits comes from the faunal evidence from Malkata and Amara, as well as from P. Bulaq 18. These will all be discussed below in the larger context of court consumption during the New (and to some degree, Middle) Kingdoms.

Direct evidence of meat consumption by non-royals comes from early periods in Egyptian history in the form of funerary offerings. Again it has proved possible to examine a sample, albeit a small one, of funerary food-offerings from Nag-ed-Deir. Twenty-one graves out of six hundred and thirty-four (i.e. about 4%) graves from the Predynastic cemetery at Nag-ed-Deir contain meat offerings. The results of this examination are summarised in Appendix II and Fig. 45. The presence of meat-offerings is one of the factors indicating status and position[3]. It is interesting to note that the majority of the offerings are from ovicaprids and gazelle;

[3] For a discussion of wealth and status in the cemetery see Podzorski (forthcoming).

there are only five instances where *Bos* is offered. Although this disparity could be blamed on taphonomic processes as well as archaeological recovery, it is more likely that cattle were rare and offered only to very wealthy individuals. Unfortunately the grave goods of the burials containing *Bos* bones provide no real indication and corroboration of wealth due to the fact that many of the graves were disturbed by both humans and animals (Appendix II: ii: **R45-65**). Goats might be predominant at Nag-ed-Deir as *Capra* are easier to take care of than *Bos* as they eat less, require relatively little water, are herbivorous and manage to graze in fairly arid regions, thus being cheaper and easier animals to keep.

Several burials in Gerzeh contained food offerings (**R44-51**). Unlike the Nag-ed-Deir offerings, these were all bovid ribs with the exception of two scapulae and some unidentifiable fragments (Appendix II: **R44-51**; Fig. 62). Whether the bovid is a *Bos* or a species of antelope is unknown, although the bones are more likely to have belonged to the former. A funerary feast in a burial of a high-status woman from an Early Dynastic burial at Sakkara contained two types of poultry (**R22-3**), fish (**R21**), and what have been identified as various cuts of beef (**R25-6**) including the kidneys [**R24**] (Appendix II; Emery 1962: 6). It seems that during this period the preponderance of ovicaprids is limited to Nag-ed-Deir, perhaps reflecting local conditions.

Due to early, more cursory methods of excavation and recording, along with tomb robberies, there is little hard evidence of the types of meat consumed from Old and Middle Kingdom nobles' burials. More information comes from the offering scenes and models in tombs of the wealthier nobles. The tomb of Medunefer is one of the few Old Kingdom example of a tomb with food offerings. It contains both poultry and young beef (Appendix II: **R37-42**). Most Middle Kingdom burials containing food offerings are royal — only one or two (e.g. Senebtisi) were noble (Figs. 60-61). These also contained both poultry and beef, the two items that constituted the mainstay of meat offerings from the Old Kingdom onwards.

The amount of food the nobles consumed depended on their rank within the court, as well as what they inherited (their inherited wealth might have influenced their rank in court). This is elegantly illustrated by a very unusual banqueting relief from the Memphite tomb of Horemheb (Fig. 7) where guests (nobles) of differing wealth and importance are seated before piles of food. The relative importance of these guests is indicated by the elaborateness of their clothing, the amount of jewellery worn, the relative sizes of their paunches of good living, their wigs, and

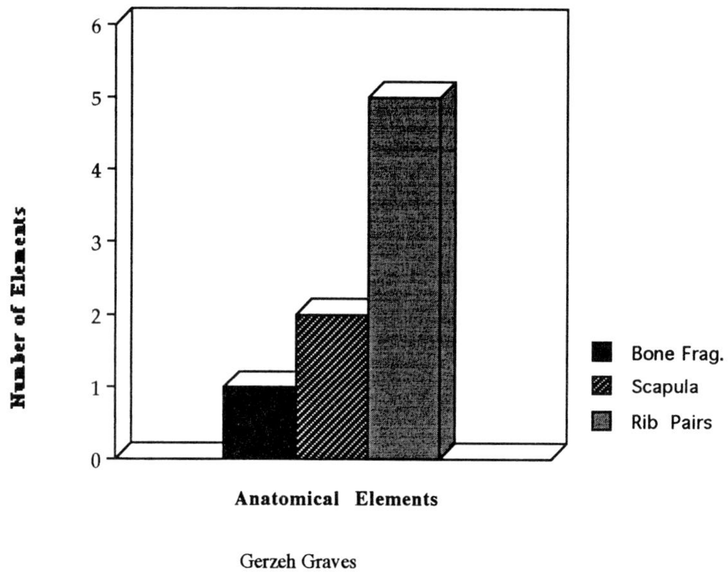

Fig. 62

the presence or absence of incense cones, all of which are marks of wealth and status, and seem to affect the amount and types of food set before them. The more important of these individuals (upper group in Fig. 7) have bigger piles of food before them that include meat and poultry as well as, in one instance, a fish. They also have more bread, beer and wine, as well as vegetables than the nobles of lower status. The guests of lesser importance (lower group in Fig. 7) who are more restrained in their paunches, dress, and adornment have smaller piles of food with fish being the only meat source, although it is possible that one of them does have a joint of meat. Their supplies of vegetables, bread, and drink are also less than those of the wealthier guests. This would indicate that joints of meat and poultry were of higher standing than the fish that was given to the lesser nobles. Of course this is at a banquet — what was consumed by each stratum of nobles at home would be different, primarily depending on their wealth and holdings of poultry and livestock.

Some New Kingdom non-royal tombs contain victual mummies — KV36, KV46, QV46, TT358, MMA1021, and KV60[4] (Fig. 43). Also,

[4] The graves of princes and princesses would be included here as they make up the upper section of this first tier.

the tomb of Kha, a courtier and architect, contains poultry and fish, as well as something that might be meat, but this identification is doubtful. If victual mummies come only from the royal wadis this might be indicative of position and wealth during the Eighteenth and Nineteenth dynasties. Several of these cuts of meat appear to come from the back and rear of the animal, areas of pure meat with no bone. Cuts from the back of the animal are tastier and of higher nutritional value than those of the front. They are very high in protein and B-vitamins and can provide an individual with approximately 83% of his/her protein needs (Paul and Southgate 1978: 94-97).

The inscriptions on jars from Malkata[5] indicate that much of the stored meat came from different officials' stockyards (Hayes 1951: 92), mainly chamberlains (Hayes 1951: 102), and was given to the king for his *heb-sed*. Unless meat was only donated for the *heb-sed* or special festivals (e.g. feasts for the Aten or important days for the king [Pendlebury 1951: 117, 174]) it seems that high-ranking nobles frequently had to provide meat for the king and, by extension, for the whole court. This could have been a form of tax or a route to royal favour much as was done at the court of Louis XIV where 'gifts' were presented to gain favour in the eyes of the king (see above discussion regarding ownership of cattle). It is also possible that the Pharaoh only fed certain nobles, requiring the remainder to supply themselves and their retainers. Certainly in the Tudor court there were 'feed' and 'unfeed' positions, the former benefiting from the generosity of the ruler's table, while the latter had to provide their own nourishment (Starkey 1987: 152-6). This is true to a much lesser extent at Amarna where most of the meat came from the Pharaoh's or the temple's holdings — which in the case of Amarna would definitely be royal property — and more limited amounts from Akhenaten's nobles (Pendlebury 1951: 172). Amarna tombs are rich in depictions of the bringing and slaughtering of cattle for Akhenaten's rituals. There are innumerable scenes showing Akhenaten offering in shrines, with slaughtering scenes on lower registers, as well as scenes showing courtiers bringing oxen to the king. Smith and Redford conclude that "scenes showing officials of various parts of Egypt leading cows probably represent such officials bringing gifts to the king for his *heb-sed*" (1976: 66), which might have been an alternative to providing prepared meat for the king's *heb-sed*, as seems so common in Malkata.

[5] It should be remembered that Malkata is a specialised context largely bound up with the *Sed*-festival.

AREA	E**	J	K	L	P	TOTAL
Bos	7	3	2	1	1	14
Capra	6	1	0	1	0	8
Ovis	3	2	0	0	1	6
Sus	2	4	1	1	1	9
Canis	1	1	0	0	0	2
Equus	1	1	0	0	0	2
Gazella	1	1	0	0	1	3
Trionyx	0	1	0	0	0	1
Crocodilis	0	0	0	1	0	1
Total of Identifiable bone by Species*	371	367	53	53	57	901
TOTAL No. OF BONES	1982	4805	134	271	466	7658

* Includes Ovis/Capra, Birds and Fish.
** One female Homo (sp.) sapiens was found here.

Fig. 63 Malkata: Minimum Number of Individuals (MNI)

AREA	E	J	K	L	P
BUTCHERY MARKS	27	11	6	3	5
TOTAL # OF BONES*	1786	4715	131	254	462

* Postcranial bones only as there are no butchery marks on the cranial bones.

Fig. 64 Butchery Marks on Malkata Bones

	Antelope	Bos	Canis	Capra	Crocodilis	Equus	Gazella	Ovis	Sus	Turtle	Bird	Fish	Tot. Bones/Species*	Total # of Bones
Eº	?	146	6	29	0	2	2	36	26	0	50	16	315	1982
J	?	92	1	6	0	8	3	7	40	1	75	80	313	4805
K	?	38	0	0	0	0	0	0	4	0	11	0	53	134
L	?	27	0	1	1	0	0	0	3	0	11	4	47	271
P	1	31	0	0	0	0	0	1	2	0	14	6	55	466

* Not including Ovis/Capra, or Ovis/Capra/Gazella
º One female Homo sapiens was found here (16 Bones).

Fig. 65 Number of bones for Malkata
(including fragments, and cranial and postcranial bones)

Some written information concerning the distribution of meat in the court of the Thirteenth Dynasty appears in Papyrus Bulaq 18 (Quirke 1990; Scharff 1922, Mariette 1872, Spalinger 1985). This account papyrus deals with income and expenditure at the palace of Thebes and lists provisions and their recipients on different days (Quirke 1990: 17). It points out differences in the ranks of members of the royal family, nobles and other members of the court (Quirke 1990: 37, 42, 60, 108-9; Spalinger 1985: 219, 235). Meat plays only a limited role in the document, and seems to have been given from Royal Largesse in S7 and S75 of Bulaq 18 (Quirke 1990: 105). This leads Quirke to remark that "meat may be regarded as an anomaly in the Egyptian domestic economy, being both a commodity consumed on some scale, like bread, beer and cakes, and a commodity of high value to be spared for special occasions, like wine" [Bulaq 18: S11 and S30] (Quirke 1990: 24). Evidence from this papyrus would suggest that meat was only consumed regularly on some scale by a few people, courtesy of the Pharaoh, and was occasionally distributed by him as a special mark of favour to others. These people would have been free to augment their diet with their own provisions; if they needed more meat they would probably consume their own fish and poultry rather than beef for logistical as well as economic reasons (see below).

Evidence of meat consumption is more difficult to find for artisans and specialists as most tombs tend not to belong to such people, with the exception of those at Deir el-Medina. Here, several tombs contained the meat of quadrupeds cut into joints, though, unfortunately, Bruyère neither specified the type of animal nor the joint (Bruyère 1937: 108, 181-82). It is known through various texts and ostraka that the workmen of Deir el-Medina were given fish regularly (Helck 1961-9: 816, 825) and that they were occasionally provided with beef on feast days (Janssen 1975: 489). In Thebes there were approximately fifty-nine festival days per annum (Nelson 1934: 25; Helck 1964). If meat was distributed to the artisans on every feast day they would have eaten beef as frequently as once a week in terms of the modern calendar. This would have been augmented by other meat that the workmen could provide for themselves. They are known to have owned "small cattle" — i.e. sheep and goats, along with cattle, although it is more likely that the latter were draft animals that were only consumed if and when they outlived their other uses (Janssen 1975: 165-73). The sheep and goat would have provided dairy products as well as meat, although Janssen believes that their chief value lay in their skins being used for leather (Janssen 1975: 401). Although fowl are rarely mentioned in price-ostraka it "is certainly no proof that birds were rare in the village. Their very cheapness was probably the reason for their being so seldom sold" (Janssen 1975: 178); it is more likely they were hunted by the workmen or bought/traded independently from that which the Pharaoh provided through his treasuries and temples in payment. The type and amount of food each worker in the village received depended on his rank. These workers were

Fig. 66

better fed than many other ranks of artisans as they worked for the Pharaoh on one of the most important monuments of his reign — his tomb.

The mummy of Nakht from Deir el-Baheri provides further information about artisans' diets. Nakht was a young man who worked as a weaver in Setnakht's *kny* -temple (Cockburn 1980: 72-4). His body, with the exception of his left knee which had unfused epiphyses and perhaps Harris or growth-arrest lines, showed no signs of malnutrition (Cockburn 1980: 75). Nakht's mummy contained tapeworm ova which suggests that he ate not very well cooked meat — either beef or pork as both can host the tapeworm. Some of the meat he ate definitely was pork as his mummy contained a *Trichinella spiralis* cyst, a parasite that inhabits pork exclusively (Cockburn 1980: 79-80). The pork was also not fully cooked as thorough cooking will rid meat of all parasites. According to Millet (et al) "meat fibers have been found in the intestine of another mummy" (Cockburn 1980: 75). Unfortunately, the type of meat was not identified. Nakht would have obtained more meat than a commercial weaver[6] as he was a temple worker and they, too, were paid in kind (see below).

[6] The debate concerning the probability of private enterprise cannot be entered here; I am working on the assumption that it existed in ancient Egypt.

If the Workmen's Village at Amarna consisted of a similar sort of artisan population as that at Deir el-Medina, then faunal evidence can shed more light on the meat consumption of this class. The range of animals found in Amarna is varied — in addition to the domesticates (with the exception of *Ovis aries*) there is evidence for *Oryx dammah*, *Ammotragus lervia*, and *Hyaena hyaena*, as well as fish and birds (Kemp 1984: 154-5). The pig is the most frequently encountered taxon here, although Hecker cautions that this might have been due to the fact that there are more bones in a pig's skeleton than in that of other animals (Kemp 1984: 156 n.2). However, further work at the village has provided incontrovertible evidence of pig farming at Amarna (Kemp 1984; 1985; 1986; Miller 1990: 125-40). Apart from providing the pigs with water little would have to be done for them as they are omnivorous and would help rid the village of garbage (Miller 1990: 125-6, 137-40). They are economic meat animals, yielding more meat than goats — almost twice as much — as well as more fat. A jar label bearing the legend *mrḥt n š3ỉ*, pig fat, have been found in the village (Kemp 1985: 66)[7]. The majority of the pig remains found here showed that they were slaughtered before they reached two years of age (Kemp 1984: 157).

Other food animals found in the Workmen's Village were fish, birds, some game animals, goats and cattle. There are no jar labels mentioning fish or birds. Presumably the villagers caught them whenever they wanted to augment their diet. This would have also held true for game. Goats were probably raised in and around the village, but to a lesser extent than pigs (Kemp 1984: 159, 161). The *Bos* remains indicate that the meat was being brought into the village from outside. Hecker writes: "the largest number of cleaver-chopped...bones and the highest proportion (12 out of 18) of multi-articulated bones [8] come from cattle, and nine of these were hind-quarters (Kemp 1984: 159). Few meat storage jars were found (Kemp 1985: 66), but they too support the idea that beef was being brought into the village. It is interesting to note that no mutton was brought into the village (Kemp 1984: 162). It is clear that any food provided by Akhenaten was augmented by the independent breeding of livestock as well as hunting, trapping, and fishing. Thus it seems that the artisans at Amarna were relatively well provided with meat, although their meat-staple was pork rather than beef.

[7] This specific jar-label found at Amarna was not a local product, but if the villagers were breeding their own pigs they too could produce pig-fat. In addition to its generally being an uncommon label, perhaps there are so few jars marked pig-fat as it was also being locally produced.

Fig. 67

An Eighteenth Dynasty Faunal Sample: Malkata

Faunal remains from Amenophis III's palace at Malkata also provide us with information concerning the meat consumption of the king, his court, and workers. The faunal evidence from Malkata comes from the 1971-74 excavations of Kemp and O'Connor (Kemp and O'Connor 1974: 101-36) and was analysed by myself in 1991[8]. There is little faunal information recorded from the earlier Metropolitan Museum of Art excavations (Winlock 1912: 184-90). The Kemp and O'Connor excavations were concentrated in the area of the Birket Habu but included area E, the edge of the palace, area J, the South village, area K, remnant of a short-lived palace (?), area L, the mud structures that could possibly have been mangers, and area P[9], the Eighteenth Dynasty town (Kemp and O'Connor 1974: 114-22). Out of these only area J was sieved in lots of five square metres by context, often a room deposit. Only specific features were sieved in other areas. This provides us with a range of areas with

[8] Some work on this assemblage had been done previously by A. J. Legge to whom I am deeply indebted for the use of his work, his help with identifying gazelles, as well as his advice on analytical methods. I am also grateful to Mr. Kemp for permitting me to work on the assemblage.

[9] The deposits at areas P and L were very thin and very close to the surface, thus it is possible that the faunal material was contaminated as well as being badly damaged due to its proximity to the surface.

differing socio-economic groupings. The bones were examined and identified for taxon, anatomical element, part of bone, age, side of body, traces of burning, butchery, taphonomic processes like gnawing, and, whenever possible, gender. They were then entered into a database[10] and run through programmes for calculating the total number and percentage[11] of different anatomical elements, taxa, and ages for the entire site as well as for each area[12], which enables the calculation of minimum numbers of individuals (MNI) for an area along with their ages (Fig. 63). Information concerning butchery-marks (Chapter VI; Figs. 29-30, 64) and consumption is also derived from this study[13].

Area E provides data for what the nobility (and royal family?) consumed (Figs. 44, 63-65). The bones found in this area at the edge of the palace are probably from one of the palace dumps, consequently it includes food for the Pharaoh, his family, and his courtiers. Area E contained the second largest number of bones — the largest number for unsieved areas. The assemblage from this part of the site was the most diverse of all five areas, containing bones of cattle, sheep, goats, pigs, gazelles, birds, dogs, horses, donkeys, and fish. It contained the largest number of *Bos* (123=17.5%) and *Ovis/Capra* (89=12.7%) bones on the site[14] (Fig. 65). The *Bos* MNI was seven individuals, while the *Ovis/Capra* was nine individuals (Fig. 63). These are the highest MNI's from all five areas despite the fact that E, unlike J, was not sieved. Clearly more animals were consumed here than anywhere else in Malkata, thereby indicating that this was a wealthy area, as might be expected. The high MNI of *Bos* corroborates the evidence of victual mummies, insofar as it emphasizes that the Pharaoh and members of his household had more access to meat animals, notably cattle, than any other members of society (perhaps excepting the priesthood). All the animals were fairly young — in the case of *Bos* all the animals were killed before the age of four years — before the epiphyses in the humerus, femur, and tibia were fused[15]. All but one of the *Ovis/Capra* were under three years of age. Fairly young and tender animals were consumed in this part of the city.

[10] I am very grateful to M. Levine for the use of her program and her tutelage.

[11] All percentages in this work are calculated by using all the bones, including fragments, not just completely identifiable remains.

[12] This was accomplished with the help of E. Yannouli to whom I am greatly obliged for explaining the mysteries of SPSS.

[13] All information discussed here is presented in the tables in Appendix IV.

[14] There was a total of 1786 postcranial bones recorded for this area; all percentages depend on postcranial material.

[15] I have used Silver's (1969) ageing method.

Fig. 68

Despite the fact that there are more bones in a pig than in any other domesticate, there were very few *Sus* bones found in this area (7=1%), the MNI being two individuals (Fig. 63). Their age could not be determined although it was clear that they were both older than a year. Several (primarily beef) bones showed evidence of butchery, thereby confirming that these beasts were indeed consumed (Fig. 64).

The faunal remains at Malkata for area J (Figs. 44, 63-65), a workmen's village, do not place the same emphasis on *Sus* remains as do those from Amarna. Out of a total of 4715 bones there were 86 (10.2%) *Bos* bones and only 17 (2%) *Sus* bones. However, there were more *Sus* bones in area J than in all of the other areas combined. The *Bos* MNI is four individuals, while the *Sus* MNI is five individuals (Fig. 63). Thus it seems that both animals were consumed regularly in this village, although there is a suggestion that pork was more available here than in the palace or in any other areas of Malkata. Three of the *Bos* were under four years old, and one was considerably older. The *Sus* in area J were all above one year of age, possibly two years — unfortunately the determining portions of the bones have not survived. Sheep and goats were the second most common species in J; 54 bones (6.3%) were found. Out of these six could clearly be identified as *Capra*, and seven as *Ovis*, with the remainder belonging to either, providing a total MNI of three individuals for ovicaprids. These animals were all older than those found in E, and consequently tougher and less tasty. The variety of taxa found in J was as diverse as that found in E. There were more bird (75=8.9%) and fish (80=10.2%) bones found here. This result is not necessarily because birds and fish were being used to augment the diet (although this is quite possible), but is a result of sieving. The inhabitants of area J had plenty of meat of all types but it was not as young and tender as the meat consumed in the palace. There was definitely more pork consumed here than in any other part of Malkata, although not to the same extent as in other Workmen's villages.

Thus, as can be seen from the Malkata evidence (Figs. 44, 63-65), area E, the area which probably contains debris from the table of the king and his court, had the largest variety of animals, including *Canis* and *Equus*[16] as well as the greatest number of of edible animal bones

[16] According to Legge "the fragmentation of the equid bones at Malkata shows that the species was used as food, though the greater abundance of equid bones in site J, the workmen's village, suggests that it was a food of rather low status. The small size of the bones indicate an ass rather than a true horse (unpublished: 7)." By the time the bones were examined by the author they were too fragmentary to determine whether or not they had been used as food.

with the exception of *Sus*, and perhaps ovicaprids. The bones found here showed the most number of butchery marks (27). Area J contained examples of all types of animals, but only a few *Bos* bones. It was the richest area in *Sus*, birds, and fish. Areas K, L, and P were all very poor in *Bos* and and ovicaprids, but did contain *Sus*, as well as birds and fish. J and L contained the anomalies of turtle carapace (possibly consumed), and L had the additional novelty of a thoracic vertebrae of a *Crocodilus niloticus*. Thus, it would seem that the Palace was where most meat, especially beef and mutton (including goat) was eaten, and the village where beef and pork were mainly consumed, in addition to fish and poultry. Area K was surprisingly poor in most animals, while L and P were only marginally wealthier. Of course, one must remember that the numbers and types of bones found in each area is skewed due not only to the archaeological retrieval methods used, but also to taphonomic processes which varied over the entire site.

It would have been possible for these people to purchase beef. The sale of beef is known from Papyrus Bulaq 11, an account papyrus that records the amount of meat, wine, and bread supplied to various merchants by an unknown supplier, possibly a temple (Peet 1934: 194). It is possible that in this papyrus there is "an example of the manner in which the vast offerings made at the temples were disposed of by the priesthoods in return for gold and silver" (Peet 1934: 194). The joints of meat traded are the *ḏꜣḏꜣ*, the *msdt*, the *swt*, *sms*, *ʿḳꜣ*, *tpt*, the *sbt*, and the *ḳꜣbt*. The most commonly mentioned is the head (*ḏꜣḏꜣ*). If the supplier of the meat was indeed a temple this abundance of heads would be explained as heads were frequently offered (see discussion in Chapter VI), but are infrequently listed as being distributed to people. The reason for this might be that although there is meat on the head, notably in the cheek, it is generally stringy and tough, making it a low prestige joint and one that might not be given to anyone as a reward or even pay. In modern Egypt, as in Britain, the head and even the brain is not regarded as quality meat and is consumed by the poorer classes. As head-meat is nutritionally poor, it is possible that the temple benefited more from selling the heads than from consuming the meat, whereas the other joints would have been deemed more valuable and used as payment, or for consumption as they could have fed more people.

The papyrus shows that different breeds of cattle were traded (*ḳꜣ*, *iwꜣ*, *wnḏw*), and the price of the same item, e.g. the head, varied depending on the type of cattle: "1 head of *iwꜣ* = 1/2 *šʿty*, 5 heads of *ḳꜣ* = 1/2 *šʿty*" (Peet 1934: 194: col.I: 7-8). Apparently *iwꜣ* were the most valuable type of cattle as one head of *iwꜣ* was equivalent to five heads of *ḳꜣ*. The price of *iwꜣ* varied too. In the same papyrus "1 head of *iwꜣ* = Silver, 1/2 a piece, and 2 heads of *iwꜣ* = Gold, 1/2 a piece" (Peet 1934: 194: III: 3, IV: 4). It is possible that one piece of silver was equal to half a piece of gold, and that $_{1/2}$ *šꜥty* = 1/2 of silver, although 5 heads of *ḳꜣ* = 1/2 a piece of Gold, which confuses matters if one tries to correlate this with the price of *iwꜣ* (Peet 1934: 194: V: 4)[17]. However, "an examination of the accounts shows that the joints even of the same animals had no fixed values. No doubt they were valued according to size or weight, and perhaps even quality" (Peet 1934: 195).

The merchants who purchased this meat could have sold it on to households belonging to nobles or artisans who could afford it. Although the nobility possessed large herds of cattle they would not kill an animal every time they needed fresh meat. There are differing estimates as to how many people the meat of one modern animal can feed. If, during the First World War, the Germans could feed between fourteen hundred and two thousand men with the meat of one steer, it would be possible for an ancient Egyptian animal (which would probably be thinner and stringier) to feed one thousand people at least (Posener-Kriéger 1976: 271-2 n. 4). Modern British butchers also estimate that one animal, with the average dressed weight of 300kg, and with the weight of 263kg[18] with the head, etc. removed, would serve up to one thousand or more people, depending on serving size (Green 1992), something that cannot be determined. Ancient and modern cattle are different in their diet, weight, and therefore their nutritional value[19]. The amount of meat on an ox depends on many things: breed[20], diet, exercise, and kill-off age. Some Egyptian oxen were very well fed as they were force-fed on cereal (Davies 1903: pl. XXXI), while other animals belonging to less wealthy people would have been given indifferent feed. The kill-off age and therefore the quality of meat also varies depending on the owner. The victual mummies had all been killed before they had reached the age of

[17] For a discussion of the value of *šꜥty* see Peet 1934 and Janssen 1975.

[18] Out of these 263kg only 71.6% are retailed, i.e., 188.4kg (Green 1992).

[19] The nutritional composition of any foodstuff is influenced by its 'diet' as well as method of preparation (Passmore & Eastwood 1986: 177). It is difficult to calculate how many people an ancient animal could feed as its weight and fat content are unknown. Modern African cattle are much leaner and stringier than their European counterparts. It is possible that many of the ancient Egyptian cattle were the same, however, there are the depictions of fattened cattle whose meat would be much fattier.

[20] For a full discussion on the different Egyptian breeds and rearing of bovids see Ghoneim 1977, Brewer et al. 1995, and Houlihan 1995.

Fig. 69

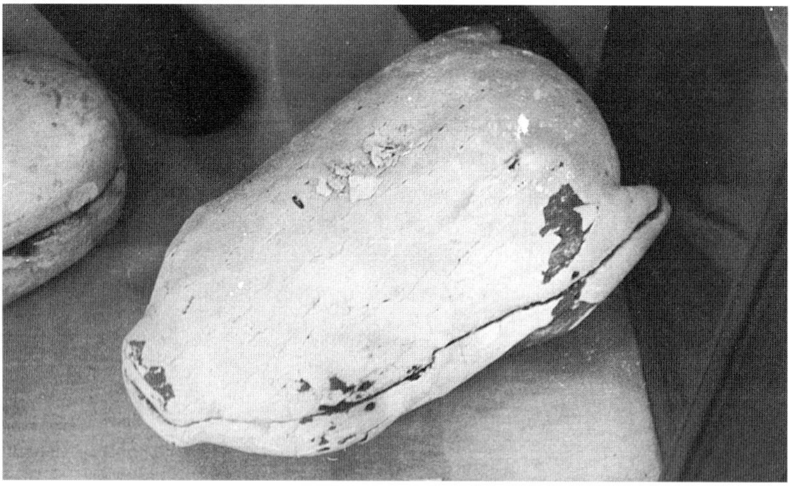

Fig. 70

four years, as had the animals in Malkata's area E, thus providing more and better quality meat per pound. Modern African pastoralists favour the age between four and five years for slaughter of their oxen (Dahl and Hjort 1976: 167), while modern Egyptians tend to favour young *gamoosas* under the age of two. Using the herds of contemporary pastoral societies which do not use chemicals or non-indigenous feeds for their animals one can attempt to deduce the state of ancient herds and methods of husbandry (see Chapter IV.3). "It takes eleven head of cattle to produce one metric ton of beef in the US compared with ten to twenty head for Australia and Argentina. By contrast, it takes fifty-five to two hundred to produce the same amount of beef in Africa" (Dahl and Hjort 1976: 19). Therefore, the estimate that one cow can feed upto one thousand people (Posener-Kriéger 1976: 271-2 n. 4) is based on a modern animal, and does not necessarily hold true for an Egyptian ox. It is difficult to estimate the amount of meat on an ancient ox, but the weight would be much less. Regardless of the weight of an ancient beast, it is doubtful that an ox would be killed each time a family wanted to eat fresh beef. It is far more likely that they would purchase it from an outside source instead of killing an entire animal. If they killed an entire animal they would have to consume the meat by the next day and preserve the rest, or distribute it. This meat might also have been purchased by merchants, bureaucrats, and artisans who were not supported by the Pharaoh but who had enough wealth in goods or services to buy meat rather than slaughter their own animals if indeed they possessed them. It is possible to imagine that donating an ox to a temple could have been a social obligation for the rich, with the meat then distributed to a wider population; we still know too little of the mechanics of piety. Certainly the idea of a wealthy person donating an animal for consumption by poorer individuals is common in modern Egypt. Animals are sacrificed and their meat distributed on the *Eids*, as well as on births, deaths, or when special requests are made to God.

It is quite possible that many people benefited from redistribution of food from temples. Priests would definitely have been recipients as would have the workers at Deir el-Medina (Valbelle 1985: 144-7) and Amarna. The military would probably also have been paid via the temple. It is also likely that any excess of meat that was not allotted to people would have been sold by the temple to merchants or householders (see P. Bulaq 11, above). Redistribution of food and goods would have been performed by the temple in its position as an administrative center (Kemp 1972; Garnsey 1989: 114-27).

Fig. 71

Fig. 72

The first significant source of information for redistribution (Sauneron 1957: 54; Nelson 1934: 48, 53) comes from the archives of the funerary temple of Neferirkare at Abu Sir (Posener-Kriéger 1976). The documents record the daily transportation and distribution of meat and vegetables. Using the list of meat-offerings, Posener-Kriéger calculates that at least one ox was sacrificed every day for the cult of Neferirkare, thirty oxen a month[21], and the meat was divided between the solar and funerary temples of the god, from which it was distributed amongst the priests and other temple employees (Posener-Kriéger 1976: 50 n.3, 508, 519, 611, 625, 634-5). Antelopes were sacrificed, too (Posener-Kriéger 1976: 314). Poultry also made up part of the daily offering, with an estimated consumption of 660 birds a month, equalling 10,000 birds a year (Posener-Kriéger 1976: 634). Generally the poultry consisted of ducks and geese, though there were occasional pigeons and other birds on festival occasions (Posener-Kriéger 1976: 314, 320). The *ḫpš, swt,* and *ỉwʿ* were the most favoured joints and were, after being offered to the god, given to various high ranking priests and palace officials (Posener-Kriéger 1976: 318). Pieces of meat in jars were also distributed to groups (phyles?) of priests: *wʿbs, ḥmw-nṯr*, and *ḫntiw-š*, among others (Posener-Kriéger 1976: 322). The number of persons in each group is unknown, as is the frequency with which the meat was distributed to them. As one ox feeds almost one thousand people, there were, potentially, several thousand people who would benefit from these offerings, unless, of course, everyone received a joint a day. Unfortunately the temple records are not so well preserved that one can precisely determine the number of times certain individuals receive meat, and whether the cut and amount of meat is the same each time. However the records do show that meat was offered to the god and then passed on to temple and palace personnel (perhaps one person fulfilling many duties?). The amount of meat varied with rank within the temple hierarchy. It is not known whether further redistribution occurred with the more important and wealthier priests providing for priests who were junior to them and under their protection. It is also a mystery whether the food they were given was consumed by them during the time of temple service, or whether any of it was passed on to their families. It is possible that the meat was redistributed to other unknown people, or that it was processed and preserved in one (or more) of the ways discussed in Chapter VII.

[21] Sacrificing such a large number of oxen would certainly have a considerable impact on herd size and modes of husbandry as a large number of animals would have to be housed close to the temple for the daily sacrifice.

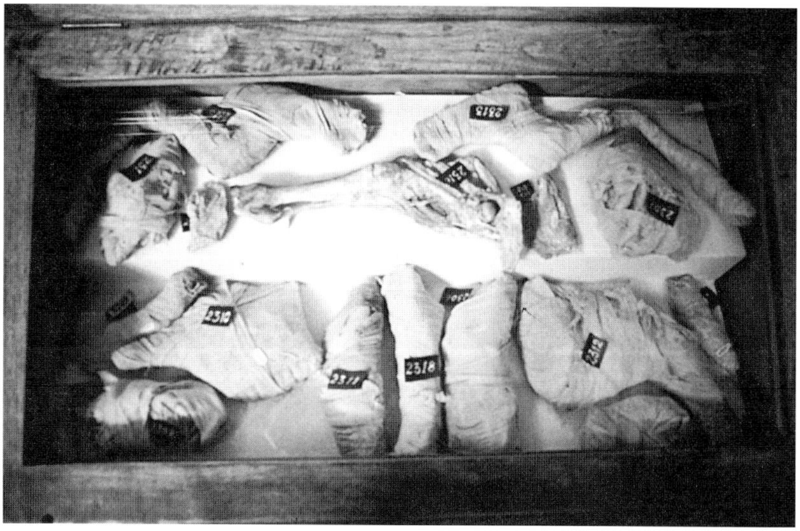

Fig. 73

There seem to be at least four sources of meat that was then distributed: the sun temple, the domain (?), the residence, and the palace (Posener-Kriéger 1976: 611-2). With the channelling of resources from the Pharaoh's domain to the sun-temples, which were, at this time adjuncts of the pyramids and not typical of temples, it is obvious that they were given access to what might be termed secular lands and resources in addition to lands and resources that were granted to them (e.g. the possessions of the sun temple). Perhaps goods from the residence and palace were passed on to the temple only when a successor to the throne demonstrated his piety by ensuring that his father's endowment and cult flourished, and that all its dependents did not starve.

Another example of redistribution comes from the Middle Kingdom tomb of Hepdjefa (1) in Asyut where the revenues given to the deceased's *k3* statue are passed on to the temple and the *k3*-priest (Spalinger 1985: 7-10). What is of note here is that some of the meat-offerings that were passed on to Hepdjefa's statue were those that he was awarded for his position as nomarch when alive. He simply re-directed his revenues from his living position to his mortuary cult, helping to feed his personal priest as well as other temple personnel[22]. The question

[22] For a thorough discussion of this somewhat problematic example of redistribution see Spalinger 1985.

remains as to where the succeeding nomarch obtained his meat. Perhaps the meat was a gift to Hepdjefa specifically, and the new nomarch had other sources of protein (Spalinger 1985: 12). It is also possible that the nomarch was a descendent of Hepdjefa, perhaps his eldest son, who could also play the role of the personal priest, thereby keeping the meat-offerings within the family.

Evidence for the redistribution of meat in the New Kingdom comes from the village of Deir el-Medina where the villagers were paid in kind by the temple of Medinet Habu as well as the other funerary and cult temples that were dotted about Thebes. In Amarna it was not only the workers who benefited from the sacrifices in the temple: an inscription of Pentu states that he consumed various foods "...which [co]me from [the] *Ḥwt-pꜣ-itn* in Akhet[aten]" (Davies 1906: pl. 3). Leprohon writes that "the scene from the tomb of Ahmose [Davies 1905c: pl. 30] where men are seen carrying goods in and out of the doors of the sanctuary reminds us of Panehsy's prayers that he be '...granted a receiving of offerings which came from every festival of the living Aten in the *Ḥwt bnbn*'. Both examples document the reversion of offerings, in which the food, having already served its cultic purpose, was redistributed to the faithful followers of the king for their own offering tables, or simply for their own consumption" (Leprohon 1988: 49). Perhaps with his desire to emphasize the link between temple and state Akhenaten butchered all his cattle as offerings and had the meat distributed from the temple in an obvious manner, rather than having palace officials distribute it[23] as seemed to have been done in P. Bulaq 18. In the case of Amarna it seems that both the royal family and the king's inner circle of Aten supporters benefited from the Pharaoh's offerings[24].

Ethnographic information concerning the current meat consumption of artisans such as carpenters, metal workers, etc. is not necessarily relevant to their ancient counterparts as they no longer work under a Pharaoh's or a noble's patronage. However, in the earlier parts of this century and in the

[23] The source of the meat is unclear in P. Bulaq 18. It is quite possible that it came from temple slaughter houses. There does not seem to be any clear rule as to whether the temple and its dependents or the palace and its dependents (the two not being mutually exclusive) benefited from the slaughter of cattle from its holdings. Indeed it is rarely clear where the slaughtered beasts came from and where exactly they were slaughtered — in the palace slaughter yards or those of the temple? This is occasionally made clear in case of Neferirkare when most of the animals seem to be killed in the sun temple, although the live animals might come from any number of holdings.

[24] Although some of the offerings that pharaoh made might have originated from the nobles as did the *heb-sed* gifts.

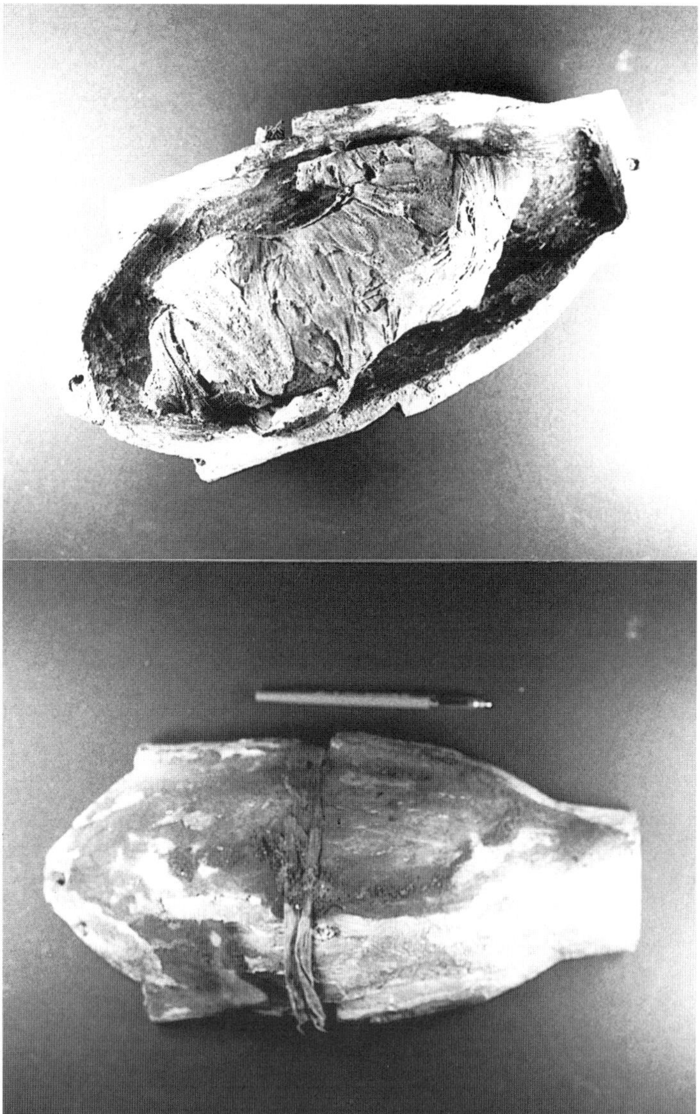

Fig. 74

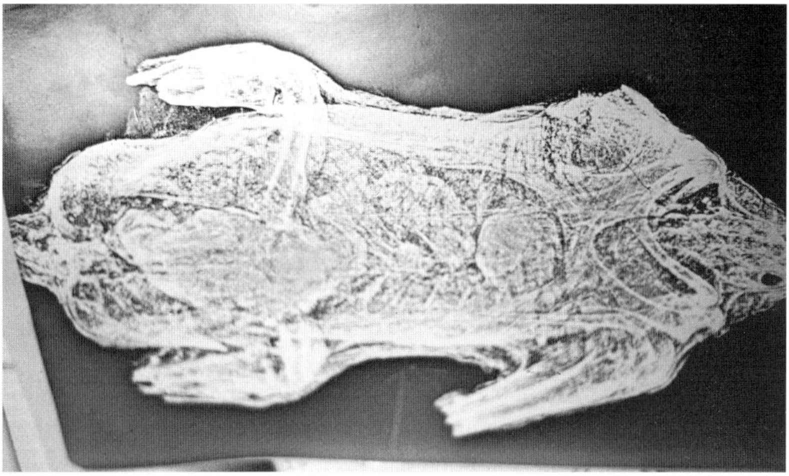

Fig. 75

last century wealthy households frequently employed artisans on a permanent or semi-permanent basis. These people would be better fed than their independent counterparts as their patrons would provide them with meat on a regular basis from their own livestock (Koraiem 1990), supplementing any animals they might have bred or any meat they purchased. Questioning and observing people's eating habits in villages and small cities nowadays (mainly in Upper and Middle Egypt), it transpires that meat is eaten a few times a week and many city-dwellers are better off than the pure agriculturalists. Both urban and rural populations tend to eat beef in the form of *gamoosa*. The *fellahin* eat beef once or twice a week, and poultry twice a week, in addition to the beef, depending on their wealth. Thus meat is consumed three to four times a week by most *fellahin*.

There is limited evidence for meat consumption when it comes to the diet of agriculturalists, herders, fishermen, birdcatchers, quarry workers, potters, and others who made up the lowest strata of society. Obviously the fishermen and birdcatchers would have had access to meat; herders would certainly have access to dairy products, but only to meat if they owned the animals or if the owners distributed meat to them. The only meat available to the agriculturalists and their like would have been what they could have brought down in the hunt or received in distribution[25]. It

[25] Quarry workers and others who might be employed/conscripted by the government would have rations issued to them as would any prisoners of war. Slaves would be fed depending on their masters' and their own rank.

is probable that very few individuals owned animals, and then they would have only been sheep, goats, or pigs. In the case of the former it was more probable that they were used for their dairy products than their meat; the latter could have been consumed, but there is little evidence of this.

In modern Egypt the *fellahin* and other lower-class workers, such as potters, do not consume much mammalian meat. Most of the *fellahin* fish and have their own pigeon houses so they have a stable supply of some meat; most do not own cattle or even sheep or goats; if they do they are used for dairy products. Meat is eaten on a grand scale by virtually the entire population on feast days, such as the *Eids,* when the meat of a slaughtered animal might be distributed through the village by a richer member of the community or when an extended family joins together to purchase an animal for sacrifice. Early in this century, whilst doing some research for diet in prisons, Wilson noted that poorer natives (*fellahin*) probably took meat only once a week on market day, and were mainly, due to lack of choice, vegetarians (Wilson 1917: 5). In some areas of Egypt that have recently become more prosperous market day is twice a week instead of once, and there are more families able to afford meat on both these days — however these are families who have members working in large cities, government jobs, or abroad. Wilson goes on to state that "the town dwellers would purchase about one and a half pounds of meat three times a month for a family of four, the meat being often the waste material (viscera) of the slaughter-house" (Wilson 1917: 23). It is doubtful whether the liver was among the viscera purchased as, though cheaper than cuts from the main body of the animal, liver is the next most expensive portion available. According to Wilson the Bedouin also consumed little meat, especially in summer. They ate meat once in fourteen days (Wilson 1917: 16), surviving on pulses, vegetables, and perhaps occasional pieces of dried and salted meat cooked with their vegetable stews.

In fact, it is possible to survive efficiently without meat as is shown by much of the population of India and the many vegetarians throughout the rest of the world. However, meat was indeed a part of the diet of the Egyptians, along with vegetables, pulses, cereals, dairy products, and fruit. Although meat was but one element of the diet, it was an important one, as it is a major source of protein. Legumes also provide protein, although protein derived from them is harder to absorb. Modern Egyptians believe that legumes are much harder to digest than meat, and given a choice would chose meat over pulses. Given that, on the whole, there were several protein sources available, it is doubtful that diseases of malnutrition were common in Egypt. It is only in times of famine (Vandier 1936; Unas Causeway reliefs) that this might have been a problem. The Harris lines mentioned in connection with Nakht's mummy are indications of food deprivation. However, the class of people who might be expected to suffer from malnutrition would be unable to afford a burial that might be excavated and studied by archaeologists, so it is impossible to make a generalisation for or against the presence of malnutrition in ancient Egypt. It should be noted, however, that all the components of meat (vitamins, minerals, protein) are only relevant to fresh produce. If the meat is preserved in any way then its nutritional value decreases. This is especially the case with fish (Platt 1962: 22; Paul and Southgate 1978: 134-149). Thus, the nutritional value of preserved meat would differ depending on the method of preservation. Any method using fat (e.g. pemmican, *confit*) would be nutritionally much richer than drying or salting. Both these methods, especially the latter, leach meat (be it mammalian, piscian, or avian) of many of its minerals and vitamins.

It is impossible to determine the exact amount of meat consumed and the frequency of its consumption by any ancient Egyptian. As long as other sources of protein were available, meat was not a necessity. From the evidence presented above it would seem that the Pharaoh ate meat of all types regularly, perhaps daily. Pharaohs' mummies show that although they might have had small-pox and other illnesses, lack of protein was not a cause of any disease (Smith and Dawson 1924: 154-62; Cockburn 1980: 29-70). Some priests and nobles would have beef regularly too, perhaps four or more times a week, while others would consume it with lesser frequency. Poorer peasants might only have it on feast occasions, while others would augment festival meats with wild or domesticated poultry or small animals. A further reason for being unable competently to compute meat consumption, even assuming that it was the only source of protein available, is that the basal metabolic rate of any of the population is unknown, and without more knowledge of it, as well as an idea about the physical dimensions of the ancient population, it would be unwise to attempt to hypothesize on the amount of meat protein necessary for an ancient Egyptian.

However, in the early part of this century some work was done on minimum amounts of protein necessary for an Egyptian in 1917 by Wilson. The men in the study were in prison, without labour, so their energy requirements were lower than that of a farmer or other labourer. According to Wilson, only thirty grams of animal protein a day are necessary if not to leach the body's own store (Wilson 1917: 41). This would provide the men with 123cal of protein, working on his rule of 1 g. protein = 4.1 cal. (Wilson 1917: 14). He cautions that "the nearer the intake is to the minimum the longer the organism takes to rebuild any damaged tissues or to make up for a temporary deficiency in the protein supply; a high protein intake appears to increase the defensive powers of the tissue against disease" (Wilson 1917: 41). In summer there is less need for meat as "an excess of protein...increases production of heat...and an increased flow of blood to the skin and an equivalent increase in the discharge of heat...which [must] be dissipated by sweating" (Wilson 1917: 41), thus people tend to eat less lentils and beans (and meat) in the summer. He thought that meat should be given to the prisoners two, or at most, three times a week as when most of them were free they would have consumed it twice a week at most (Wilson 1917: 3). Furthermore, there was no need for more as they were given other forms of protein as well as a balanced diet. Modern day nutritionalists feel that even that small amount of animal protein is not strictly necessary, although it is the most efficient source, providing the people in question were getting other sources of protein and iron (Green 1992; Black 1992).

Currently there are attempts being made to determine of what the ancient diet consisted. A method that has been developed to try to determine this is by isotopic and elemental analysis of human bone. The method consists of analysing samples of human bone for various trace elements (e.g. Sr [Strontium], Mg [Magnesium], and Zn [Zinc]) and for ratios of $^{12}C/^{13}C$ (Carbons) and $^{15}N/^{14}N$ (Nitrogen) of bone collagen. These methods allow for the determining between plants versus animal foods (White and Schwarcz 1989; DeNiro 1987; Byrne and Parris 1987). Such work has been done on the Mayan diet (White and Schwarcz 1989) and Native American diets (Byrne and Parris 1987), and more recently, at Porros, an Iron Age site, in Spain (Subira and Malgosa 1992: 199-204) where the diet of the human population was contrasted with that of the other fauna found at the site. This last study used Zn and Cu (Copper) to test for a carniverous diet, and Ba (Barium), Mg, and Sr to test for a vegetarian diet. Unfortunately no such work has been done on an Egyptian population, although such a study has been performed on a Nubian population by Dr. C. White (Fairgrieve 1992). Although there is always some debate about the validity of the results of such studies, a similar study would be of great value to Egyptologists as, although it might not provide all the answers, it would provide a working hypothesis which can be further explored. It is hoped that such studies will be undertaken in the near future. It would be of special value, for example, if the Nag-ed-Deir collection were so tested since many of the burials contain food offerings of different types and it would be interesting to see the degree of correlation, if any, between the offerings and the actual bones of the deceased. Furthermore, the types of grave goods in addition to the size of the graves determined the wealth of the individual (Podzorski forthcoming), thus providing some information concerning the correlation between wealth, status, and food consumption.

Until such a time as these more quantitative and scientific studies can be carried out, one can only use the evidence available to try to reconstruct what the ancient Egyptians ate. At the moment this evidence points to the existence of a large pool of meat animals in ancient Egypt. These animals would not have been equally accessible to all members of society: wealth would probably have been the determinant in deciding who had the choicest cuts of the most expensive meat — beef. Thus, the Pharaoh was guaranteed all types of meat, from beef to game, including fowl and fish. The latter animals were probably available to his most humble subjects as well. They would only have tasted fresh beef on feast days. It is possible that some form of processed meat, probably salted or dried, was available to them at all times. The rest of society ate what it could afford. Proximity to the Pharaoh probably influenced wealth, so those in his inner circle no doubt benefited from his generosity, and might, like the nobles of Charles I, have been given a "table" (of food) as a mark of favour. Priests would have been better fed than most other sections of the population enjoying a supply of both fresh and preserved meat. If, as Egyptologists have hypothesized, a rota system of priests was in operation, then being well-fed for some parts of the year would have been a privilege accorded to a significant number of the population on a cyclical basis. The temples, through redistribution, would also have provided food for artisans and workers engaged on state business. Thus, it seems that meat of one type or another, preserved or fresh, would have been available universally, albeit to a varying degree. It was by no means the mainstay of the peasants or even the artisans and other professionals, as it probably was for the Pharaoh and his inner circle, but meat was consumed by all, and given its importance, it is no wonder that it continually featured in tomb decoration and funerary texts which always contain a request for "choice cuts".

APPENDIX I

"MEAT BOXES"

Some Old and Middle Kingdom burials contained hollow limestone or calcite boxes carved in the shape of joints of meat or fowl. All of these, unlike the similar wooden boxes dating to the New Kingdom (Appendix II), were found empty, suggesting that they were larger examples of the solid calcite and limestone models (Fig. 66). The Old Kingdom examples come from tombs at Giza and Sakkara, excavated by Reisner, Jéquier, and others. Not all the mastaba numbers are known. The Middle Kingdom examples came from pyramid of Amenemhet III at Hawara, and some from Lisht. The Old Kingdom examples are made of limestone which is sometimes painted. In his excavation records Reisner notes that these cases were "curiously empty" (Reisner 1913), and indeed showed no trace of ever having been used. There are only a few examples of these limestone boxes known to me, and none of those examined shows any indication of ever having contained meat. They are more like three-dimensional offerings, although unlike model offerings that are solid pieces of limestone, wood or cartonnage that are shaped in the form of offerings, these cases can be opened and are potentially able to contain an offering. However no mummified offerings have been found in conjunction with these objects; in fact, not even un-mummified joints have been found close to them. Thus the reason that they are shaped as cases remains mysterious. It is interesting to note that they have only been found in the Memphite area, and, unlike the majority of their Middle Kingdom counterparts (see below) are not limited to Royal burials.

The few Middle Kingdom examples of boxes/sarcophagi for victual offerings are also of stone, although they are only made of alabaster and not of limestone. These only take the shape of duck or geese, and, like the Old Kingdom examples, are empty and appear never to have been used as containers for actual fowl. There are, however, allusions to embalmed funerary meat offerings in the Middle Kingdom: found in the tomb of Princess Nubhoteptikhered (see Appendix II) were "des oiseaux et des pièces de viande de bœuf embaumés" (de Morgan 1895: 108). However he provides no description of these embalmed meats or any mention of cases, so one cannot compare the Middle Kingdom meat mummifying practices, if any, with those of the New Kingdom. Thus, so far as one can determine, the practice of mummifying meat and placing it in the tomb as an offering is chronologically limited to the New Kingdom and geographically limited to the Theban necropolis.

Organization of the Catalogue:

I have assigned an **N** number to each object followed by a museum number (**D** for Dokki Museum of Agriculture, **M** for the Metropolitan Museum of Art, **B** for the Boston Museum of Fine Arts) depending on its location. As CG and JE numbers are sometimes used for objects both in the Cairo Museum and the Dokki Museum of Agriculture, a '**C**' will follow the CG/JE number if the object is in the Cairo Museum, and a display number, where known. Where available the dimensions are given in metric, followed by the object's provenance and a brief description.

N1 **B13.4323a,b** **L**: 43.5 **W**: 23 **H**: 9.5 **Prov**: G2381 Z128
Limestone case for a shoulder [unclear if he meant scapula — possibly scapula and humerus] piece found in the shaft of a mastaba (Reisner 1913), excavation #12-12-128/9. There is a 2 cm high ridge in the centre of the lid. There was a yellow wash inside the case which is not airtight (Fig. 37).

N2 **B13.4310,-12** **L**: 43 **W**: 20 **H**: 9.5 **Prov:** G2385 A
A femur or humerus shaped limestone box found in a mastaba shaft dating from the Sixth Dynasty. It is covered by a yellow wash both inside and out (Fig. 39).

N3 **M37.6.5a,b** **L**:53, 53.5 **Prov:** Giza
A limestone case in the shape of a leg joint of a large animal. It was found empty in the shaft of the mastaba of Yenti.

N4 **CG44622;C** **L**: 40 **Prov**: Giza
A limestone box in the shape of a leg joint. It is covered by a yellow wash and was found empty.

N5 **CG44623-24;C(lid)** **L**: 49,53 **Prov**: Giza
Limestone box in the shape of a joint of meat. It has a faint yellow wash, but was found empty.

N6 **CG44623;C** **L**: 53 **Prov**: Giza
A limestone box in the shape of a leg joint. It is covered by a yellow wash and was found empty.

N7 **D1061-62** **L**: 31.5 **Prov:** Sakkara
A limestone box, painted a reddish brown, in the shape of a joint of meat.

N8 **CG44630-31;C** **L**: 48,45 **W**: 22,21 **Prov**: Giza
A limestone box carved in the shape of ribs with a faint yellow wash like those in the BMFA. It was found in the Giza necropolis by Reisner. This, like all of the other boxes, was found empty with no trace of any objects found within.

N9 **B13.4324,-26** **L**: 55 **W**: 27 **H**: 3 **Prov:** G2385 A
A limestone case in the shape of four ribs. It was found in the shaft of a Sixth Dynasty mastaba (Fig. 67).

N10 **B13.4325** **L**: 57 **W**: 28 **H**: 4 **Prov:** G2385 A
A limestone case carved in the shape of four ribs. It was found in the shaft of a Sixth Dynasty mastaba at Giza.

N11 **B13.4327** **L**: 34 **W**: 27.7 **H**: 4 **Prov:** G2385 A
A limestone case carved in the shape of four ribs. The lid is covered by a yellow wash both inside and out. It was found in the shaft of a Sixth Dynasty mastaba at Giza. It is hollowed, so it would be possible for it to contain something, although the curved shape of real ribs would not fit easily in the box.

N12 **CG49810;C** **L**: 23 **W**: 13 **Prov**: Sakkara
Limestone disk painted reddish brown like a joint of meat.

N13 **CG49809;C** **D**: 20 **Prov**: S. Sakkara
Oval limestone box painted reddish brown like a joint of meat. From the tomb of Ptahwash.

N14 **D1063-64** **L**: 24 **Prov:** Sakkara
Oval limestone box painted a reddish brown in imitation of a joint of meat. Fifth or Sixth Dynasty.

N15 **CG44625** **D**: 27.5 **Prov**: Giza
Shallow limestone oval box, like **N17**, with a yellow wash presumably supposed to represent a (unidentified) joint of meat.

N16 **B13.4320** **L**: 27.7 **W**: 21.7 **H**: 5.5 **Prov:** G2381 Z
The slightly domed lid of an ovoid limestone case, like **N17**, supposedly for a joint of meat. It is covered both inside and out by a yellow wash and dates to the Sixth Dynasty. The box was found in the shaft of a mastaba.

N17 **B13.4375a,b** **L**: 32 **W**: 24 **H**: 11 **Prov:** Giza
This ovoid limestone box found in the shaft of a mastaba is thought to represent an unidentified piece of meat. It resembles some of the ovoid meat determinatives in Gardiner's (1947a,b) *Onomastica*, but other than that have no obvious correlation with an animal. It is possible that it might represent a cake or bread. There is a yellow wash both inside and outside the case.

N18 **M37.6.4a,b** **L**: 26.8,27 **Prov**: Giza
This limestone box is in the shape of a duck with its wingtips and legs removed. The head of the duck is shown in relief on the lid. There are traces of a yellowish wash on the limestone.

N19 **M37.6.3a,b** **L**: 29,29.3 **Prov**: Giza
This Sixth Dynasty limestone box is in the shape of a duck with its wingtips and legs removed. The head of the duck is shown in relief on the lid. There are traces of a yellowish wash on the limestone.

N20 **M37.6.2a,b** **L**: 35.5,36 **Prov**: Giza
This limestone box is in the shape of a duck with its wingtips and legs removed. The head of the duck is shown in relief on the lid. There are traces of a yellowish wash on the limestone, and the bead and eyes show traces of black pigment.

N21 **B13.3484** **L**: 23 **W**: 15 **Prov**: Giza
This limestone case in the shape of a duck with its head carved in relief on the lid dates from the Sixth Dynasty. Previously it has mistakenly been identified as a box for a pigeon. It was found empty.

N22 **B13.3478-9** **L**: 39 **W**: 24.5 **Prov**: G2385 A
A limestone case for a duck or goose with the bird's head carved in relief on the lid (Fig. 68). Both the external and internal surfaces of the case are covered by a yellow wash.

N23 **B13.4316** **L**: 19.9 **W**: 13.6 **H**: 6 **Prov**: G23812
A Sixth Dynasty limestone case in the shape of a bird. The bird has previously been misidentified as a pigeon. The case was found in the shaft of a mastaba by Reisner.

N24 **CG63254a,b;C (lid)** **no dimensions** **Prov**: Giza
Limestone box in the shape of a duck with yellow wash over the whole box. Duck's head, wings, and feet cut off.

N25 **D1084** **L**: 25.5 **Prov**: Sakkara
Limestone box in the shape of a bird with the head in relief on the belly, and the legs and wing tips removed. The beak and eyes are painted black, the rest of the bird covered by a yellow wash. There is an inscription on the box, *ṯrp*, as is found on New Kingdom wooden cases (see Appendix II).

N26 **D1086** **L**: 27.5 **Prov**: Sakkara
Limestone box in the shape of a goose with the head in relief on the belly, and the legs and wing tips removed. The bird covered by a yellow wash. There is an inscription on the box, *r*, as is found on New Kingdom wooden cases (see Appendix II).

N27 **D1088-89** **no dimensions** **Prov**: Sakkara
Limestone box in the shape of a goose with the legs and wing tips removed.

Princess Neferuptah

In his publication of Amenemhet III's pyramid in Hawara, Petrie reports that he found parts of nine or ten "duck bowls, the largest of which was 19X8.5X5cms. and the smallest was 7X3.5X2.5cms. (Petrie 1890: 17). All of these were of alabaster and many bore the name of the king's daughter, Neferuptah. All of these are now in Cairo.

N28 JE36447a; C1028 L: 42
An alabaster duck shaped vessel in two parts, both labelled A. The size is however more indicative of a goose, which it quite possibly is meant to represent as geese were frequently offered. An inscription reading *ȝpd nswt ptah nfrw* is incised on it. The bird's legs and wings are shown in relief. Its head has been cut off, but there is a hole at where the neck would be. The division between the ducks is equal and instead of being a horizontal or lateral cut through the bird, is a vertical cut as if the bird had been divided equally in two by splitting its backbone in half.

N29 JE36447c; C L: 16
Alabaster duck shaped vessel in two parts with an inscription reading *ȝpd nswt ptḥ nfrw* carved on it. It is identical to **N28** save for its size which is more indicative of a small duck. It was broken and repaired in antiquity.

N30 JE36447b; C L: 47
Alabaster goose shaped vessel in two parts, both labelled A. An inscription reading *ȝpd nswt ptaḥ nfrw* is incised on it (Fig. 69). It is identical to its fellows and was broken and repaired in antiquity like **N29** (Petrie 1890: 17).

Food Cases from Lisht (Middle Kingdom)

N31 CG25406; C118 L: 41 **Prov:** Lisht
Large Twelfth Dynasty alabaster vessel in the shape of a goose. The dividing cut between the two halves of the box are vertical, having one wing and one leg on a side, rather than horizontal, the legs and wings on top with the case on the bottom. Perhaps from the pyramid of Sesostris I (Arnold 1988: 70; Dodson Forthcoming: 18).

N32 CG25403; C115 L: 25.2 W: 12.3 **Prov:** Lisht
An alabaster box of the Middle Kingdom in the shape of a duck with a vertical cut forming the division between the two parts of the bird. Perhaps from the tomb of Sesostris I (Arnold 1988: 70; Dodson 1994: 18).

N33 CG25404; C116 L: 32 W: 16.5 **Prov:** Lisht
An alabaster receptacle in the shape of a bird. It too has a vertical cut through the body of the goose like **N31**, and no head. Perhaps from the tomb of Sesostris I (Arnold 1988: 70; Dodson 1994: 18).

N34 No Dimensions **Prov:** Abydos
A. M. Dodson has kindly pointed out that the excavators of the tomb of Sesostris III reported that the canopic chest contained fragments of alabaster dishes "in the trussed duck pattern" (Ayrton et al. 1904: 27). The present location of these fragments is unknown.

New Kingdom

N35 CG24699 **L**: 34cm **W**: 24 cm **Prov:** Biban el-Moluk
Stone box of goose shape made in two parts (Daressy 1902: 177). Neither the actual object nor a photograph was seen by the author, so its resemblance, if any, to the Old Kingdom types is unknown. The only allusion to it is from Daressy and the CG. This also features as **S4** in Appendix II. Stored in in P12 S4 *bas* of the Cairo Museum.

APPENDIX II

PHYSICAL EVIDENCE

The most informative evidence concerning the jointing and preserving of meat comes from actual pieces of meat and bone found in tombs. The study of the different cuts of meat that appear as actual offerings, both mummified and plain, in relation to the cuts of meat featuring on tomb walls and their names (when labelled) is of great help in determining words assigned to specific joints as well as the actual method of jointing favoured by the Egyptians.

From the Predynastic onward tombs have been found to contain desiccated joints of meat or bones that were meant for the nourishment of the deceased's *ka*. These meat offerings usually consist of various cuts of meat, along with other foods such as loaves of bread and fruit, placed at the head of the coffin or in front of the face of the deceased. The remains of meat offerings are of two types: the first are "victual mummies" found in Theban tombs of the New Kingdom, both in and out of coffinets; the second is animal bones (frequently with butchery marks) placed in pottery receptacles or placed on the floor of the tomb. The former consist of joints of meat or of entire birds that are wrapped in bandages. Most of these bandaged meats were placed in individual sycamore-wood 'coffins' shaped to the form and dimensions of the meat, the lids of which were joined to the body by dowels and holes. These boxes were plastered and whitewashed both inside and out. The interior of many of these 'coffins' was coated with resin, which might have been used to seal the boxes and make them resistant to moisture, thus preventing the meat from deteriorating and smelling. It is possible that a roasted appearance (browning) was given to victual mummies by the application of very hot resin on the wrapping which no doubt cooked the exterior surface of the mummy (Lortet and Gaillard 1905: 248). The method of making the 'victual mummies' is unclear, although it has been suggested that the meat was preserved by immersion in resinous natron (Lortet and Gaillard 1905: 307). The author has examined samples from the tomb of Meryetamun (TT358) under a SEM, and found that there is some evidence of crystals that are quite probably salt (Chapter VII). However, one must keep in mind that it is possible for other compounds to have a similar crystalline structure. Other Joints from that tomb as well as MMA1021 were examined under a microscope and showed evidence of salt as well as natron having been used on the meat (see **S374**). Other assemblages (e.g. Imhotep [QV46]) showed evidence of salt to the naked eye. Thus, it would seen that the joints were preserved in either liquid or solid salt (see Chapter VII.3). The cuts of meat can generally be determined from the bones that are included in the joint, or sometimes from the shape of the box. These coffined or wrapped pieces of meat form the first part of the catalogue (Figs. 42-43, 59).

Part Two of the catalogue consists of a selection of the unwrapped meat offerings found in tombs from the Predynastic onward. Unlike Part One, Part Two is not an exhaustive catalogue of all the offerings ever found in tombs throughout Egyptian history. It lists only some of the tombs from each period of Egyptian history that contain funerary offerings. The reasons for this limit are, firstly, that it is meant only to provide a rough idea of the variety of meat offerings that appear throughout Egyptian history; secondly, space constraints, and thirdly, the fact that in the early years of excavation in Egypt (and even until quite recently) faunal material might have been noted but is rarely mentioned and even less frequently adequately discussed in publications.

Organization of the catalogue

Part I:

Each entry of mummified food remains is organized by tomb number and owner. Within the tomb the entries are organized by body part, front to back. The numbering system for each entry is as follows. The **S** number is my number. This may be followed by another number which can be any of the following types: the number assigned to the object in another publication; the excavation number; a museum number. The type will be made clear in the introductory matter for each tomb. This number will be followed by the dimensions (in centimetres or grammes) of the piece, **L**ength, **W**idth, **W**eigh**t**, **H**eight, and **Th**ickness, whenever possible. If the entry comprises the 'coffin' for the victual mummy, its dimensions will be given along with those of the mummy, but with the legend "CS" to separate it from the measurements of the meat. A short description of the object(s) will follow the dimensions. All the remains are bovine remains (*B. taurus*, *B. brachyceros*) unless otherwise stated.

Part II:

Many of these entries have not been examined by the author, and many of the identifications rely on publications as well as her own identifications, based solely on photographs and drawings. **R** numbers are assigned to each entry; sometimes one number is used for an entire tomb, at others, whenever possible, each item in the tomb is given an **R** number. The same abbreviations for dimensions employed in Part I are used therein.

Part One

Amenophis II [KV35] & Tuthmosis III [KV34]

These two tombs have been grouped together as they were found at the same time, excavated by the same person, Victor Loret, and shared a (preliminary and sole) publication (Loret 1898). Previously, the food offerings in both have been studied as a group so that it is impossible to determine which item belonged to which Pharaoh, another reason for grouping them together. There are over 200 meat offerings — beef, calf, goose, duck — that were found in the two tombs. Some were found in the left back annex, room II in Loret's plan, in KV 35 (Loret 1898: 102, pl. 8.), while others were found "dans la seconde pièce à droite…au milieu de la pièce, les ossements d'un boeuf au complet; de quoi monter, en pièce anatomique, le squelette entier d'un taureau datant du 16e siècle avant notre ère" (Loret 1898: 94).

Although all of were studied by Lortet, Gaillard and Daressy, only a few of them (approximately 8%) possess Catelogue Général numbers (CG29565-29577; CG249696-99) (Gaillard and Daressy 1905: 62). There are three main publications that deal with the meat remains from these tombs, Gaillard and Daressy 1905 (referred to in catalogue as **G&D**), Lortet & Gaillard 1905 (referred to in catalogue as **L&G**), and Lortet and Gaillard 1907 (referred to in catalogue as **L&G2**). There is some disagreement between these publications as to the exact number of some kinds of joint found in the tomb. This may be due to the confusion occasioned by the mixing of the finds from the tombs of Amenophis II and Tuthmosis III, or perhaps some items were lost or miscounted at one time or another. Below are listed the types of joints, and then the number of each type, according to the three publications which concentrate on these remains. One should note that Lortet and Gaillard's 1905 publication is the only publication that provides a comprehensive list of the objects found in the tombs. What *is* strange, is that *within* the publication they seem to disagree about the number of each joint found. I provide both lists.

Joint	G&D #p.63-4	L&G2 #p.114
Tibia	21	21
Femur	20	20
Humerus	18	18
Radius & Ulna	18	18
Scapula	14	14
'Legs' without phalanges*	8	no mention
Ribs	5 sets of 4-8 ribs	no mention
Vertebrae	9 portions of 5-8 dorsal; 6 lumbar	1 mentioned
Sternum	11	no mention
Pelvic (?) meat	c. 50	no mention
Liver	no number	no mention
Total	**175**	**92**

Joint	L&G #p.311-14	L&G #p.314
Scapula-alone	12	14
Scapula & Humerus	2	no mention
'Legs' without phalanges	8 [forelegs]	8*
Humerus-alone	17	18
Radius & Ulna	18	18
Vertebrae	10 portions	5 frontal, 8 dorsal
Ribs	7 sets of 4-8 ribs	sets of 4-8
Sternum	11	no mention
Femur	20	20
Tibia	19	24
Pelvic (?) meat	c. 48	50
Liver	5	no mention
Total	**177**	**indeterminate**

* Gaillard and Daressy specify foreleg for "Leg" (Gaillard and Daressy 1905: 63), while Lortet and Gaillard only call it "Leg" (Lortet & Gaillard 1905: 314), and do not define its identifiable components, as do Gaillard and Daressy. Here I am using the Gaillard and Daressy 1905 definition and am interpreting "Leg" as foreleg for all publications. Note that the "Leg" ended at the division between the carpals and metacarpal.

The totals of both sets of authors are not exact except perhaps in the case of Lortet and Gaillard (1905: 311-14), although the duck/goose cases mentioned by Daressy are not all listed. The reason that the joints and the birds do not make a total of 200 is that several of the so-called wrapped offerings were actually bundles of cloth and unidentifiable bones (**L&G** 20, 52-4, 72-4, 110-5, 150-7, 187-8) and in one case a wrapped rat (**L&G** 46). These twenty-three other 'offerings' provide most of those missing from the tombs. The remainder may have been represented by the avian offerings mentioned by Daressy in 1902, but not all included in the other publications. Furthermore, strips of meat might have been destroyed or miscounted, and they might be responsible for the remainder of the offerings.

None of these publications specifically mentions the avian sarcophagi, although Lortet and Gaillard 1905 do mention two *A. albifrons* remains. These avian offerings feature in the volume by Daressy (1902: 177), and are also listed below with their Catalogue Général numbers. These are the only avian offerings from this tomb that can currently be traced.

In their descriptions none of the authors specifies for each item whether or not it was wrapped. From the photographs of some of the items, and from the present appearance of objects in the Agricultural Museum at Dokki, it appears that the majority of the offerings were wrapped in linen bandages, and some had vestiges of resin on them. Generally resin on bandages indicated that they were in a wooden sarcophagus which, after being plastered, frequently had its interior coated with hot resin which would sometimes adhere to the victual mummy. For the most part caseless mummies are not coated with resin, although occasional traces of resin appear on the bandages. From the photographic and physical evidence available one can tentatively conclude that some of the meat offerings for the two pharaohs had cases whose interior was coated with resin, and that virtually all of them (with the notable exception of the strips/pieces of meat) were bandaged in antiquity.

The list below is based on Daressy 1902 and Lortet and Gaillard 1905 and has, whenever possible, additional information from personal observations and research conducted at the Museum of Agriculture, Dokki, and the Egyptian Museum, Cairo. As only objects with the Catalogue Général numbers could be found in the museums, they are the only ones that have been examined by the author. Regrettably, the whereabouts of the other items is unknown as it seems that the museum took a representative sample of each type of joint, but not the whole collection; it is conceivable that they are held in storage in Luxor.

S1 **CG24696**; **JE 32558a,b** **L**:38cm **W**: 26.5cm.
Wooden box in the form of a duck with its wingtips, feet and head cut off. Stuccoed white inside and outside, with the interior covered with resin (Daressy 1902: 177). Stored in the Cairo Museum: P12 S4 Case D *bas* #3055.

S2 **CG24697** **L**: 26cm **W**: 12cm
Wooden box in the shape of a bird similar to the above, but smaller (Daressy 1902: 177).

S3 **CG24698**, **JE32383** **L**: 31cm **W**: 16cm
Like **S2** but the case is more crudely carved (Daressy 1902: 177). Stored in P12 S4 bas of the Cairo Museum.

S4 **CG24699** **L**: 34cm **W**: 24 cm
Stone box of goose shape made in two parts (Daressy 1902: 177). Neither the actual object nor a photograph was seen by the author, so its resemblance, if any, to the Old Kingdom types is unknown. The only allusion to it is from Daressy and the CG. This also features as **N34** in Appendix I. Stored in in P12 S4 bas of the Cairo Museum.

S5 **L&G**177 **no dimensions**
A goose, *Anser albifrons* (?), with the wing tips, feet and head cut off.

S6 **L&G**209 **no dimensions**
Broken *A. albifrons* with bits of humerus, femur, and sternum distinguishable.

S7 **L&G**48 **no dimensions**
Right scapula of a juvenile *B. taurus*.

S8-10 **L&G**92-94 **no dimensions**
Three left scapulae of different sizes (presumably from different animals).

S11-4 **L&G**127-30 **no dimensions**
Four scapulae of *B. taurus* of varying ages (juvenile to adult). Unfortunately the sides cannot be determined.

S15 **L&G**176 **no dimensions**
Left scapula.

S16-8 **L&G**203-05 **no dimensions**
Three scapulae of young/small *B. taurus*. Perhaps a *B. brachyceros*.

S19 **L&G**95 **no dimensions**
Right scapula and humerus of a juvenile *B. taurus*.

S20a **L&G**40 **CG29574**; **D3395** **L**: 50cm **W**: 20cm
This is a very well wrapped piece of meat, the bandages of which show signs of being touched by resin in some places. The card says that it is a radius and humerus of a bovid.

S20b **L&G**40 **CG29574**; **D3396** **L**: 50cm **W**: 16cm
Unwrapped connected right scapula and humerus with a lot of attached meat from a mature bovid. This piece of meat has the same CG number as S20a, but

the object is different. D3396 is the one that appears in L&G, while D3395 does not, although according to the entry in the Dokki museum it has the same CG number and provenance.

S21 L&G43 **no dimensions**
Right scapula, humerus, radius, ulna and carpals of a juvenile animal.

S22 L&G41 **no dimensions**
Right foreleg of a juvenile *B. taurus*, composed of the scapula, humerus, radius, ulna and carpals.

S23 L&G132 **no dimensions**
Right scapula, humerus, radius, ulna and carpals of a B. taurus.

S24 L&G172 **no dimensions**
Right scapula, humerus, radius, ulna and carpals, all in conjunction, of a young *B. taurus*.

S25 L&G116 **no dimensions**
Right scapula, humerus, radius, and ulna of a young animal.

S26 L&G122 **no dimensions**
Right scapula, humerus, radius and ulna.

S27 L&G173 **no dimensions**
Right foreleg of very young *B. taurus*.

S28 L&G33 **CG29567; D4007** L: 27 W: 10
Left humerus covered with meat from a young *B. taurus*.

S29 L&G34 **CG29577; D4006** L: 27 W: 11
Entire left humerus with meat of a young bovid, over three years old. The joint is unwrapped and is blackened in some places, possibly due to a combination of resin and aging, and whitish in others. The white colour seems to come from a fine surface deposit, quite possibly salt (taste) that was used either to cure the meat, or a later deposit from the soil.

S30-31 L&G22-23 **no dimensions**
Two left humeri with meat attached of a small bovid

S32 L&G44 **no dimensions**
Left humerus of a small bovid

S33 L&G77 **no dimensions**
Right humerus of a young bovid.

S34 L&G78 **no dimensions**
Left humerus of a young bovid.

S35 **L&G**82 **no dimensions**
Left humerus.

S36 **L&G**134 **no dimensions**
Left humerus of a small *B. taurus*.

S37-8 **L&G**165-66 **no dimensions**
Two left humeri of a young *B. taurus*.

S39-40 **L&G**125-26 **no dimensions**
Two left humeri.

S41 **L&G**178 **no dimensions**
Left humerus of a small *B. taurus* with meat attached

S42 **L&G**179 **no dimensions**
Right humerus.

S43 **L&G**196-98 **no dimensions**
Humerus of a young *B. taurus*.

S44 **L&G**207 **no dimensions**
Right humerus of a young *B. taurus*.

S45 **L&G**175 **no dimensions**
Left radius and ulna of a juvenile *B. taurus*.

S46-8 **L&G**193-95 **no dimensions**
Three ulnae and radii.

S49-52 **L&G**199-202 **no dimensions**
Four radii of young *B. taurus*.

S53-55 **L&G**162-64 **no dimensions**
Three radii and ulnae from different animals.

S56-7 **L&G**123-24 **no dimensions**
Two radii and ulnae from the left side of a young *B. taurus*.

S58 **L&G**83 **no dimensions**
Radius and ulna with large amounts of meat attached.

S59 **L&G**91 **no dimensions**
Left radius and ulna.

S60 **L&G**24 **no dimensions**
Left ulna and radius with meat attached.

S61 L&G25 **no dimensions**
Left ulna and radius with meat attached.

S62 L&G36 **no dimensions**
Right radius and ulna.

S63 L&G45 **no dimensions**
Six thoracic vertebrae of *B. taurus*.

S64 L&G59 **no dimensions**
Five thoracic vertebrae of a young *B. taurus*.

S65 L&G55 **no dimensions**
Thoracic vertebrae of juvenile *B. taurus* linked together.

S66 L&G96 **no dimensions**
Seven cervical vertebrae of a small *B. taurus*.

S67 L&G97 **no dimensions**
Seven cervical (?) vertebrae of a juvenile *B. taurus*.

S68 L&G137 **no dimensions**
Twelve thoracic vertebrae of *B. taurus*.

S69 L&G208 **no dimensions**
Two lumbar vertebrae and sacrum with meat attached.

S70 L&G138 **no dimensions**
Eight thoracic vertebrae of *B. taurus*.

S71 L&G35 **CG29575; D3399** L: 22 W: 15
Six cervical (?) vertebrae. There is a lot of meat on them, as well as traces of resin.

S72 L&G21 **no dimensions**
Six connected lumbar vertebrae with meat attached, from a young, small bovid.

S73 L&G39 **CG29566; D3398** L: 37 W: 15
The entire sternum of *B. taurus* with cartilage and muscle. On the external face is a deep longitudinal incision to help desiccate the flesh. There are large chunks of resin all over the meat and in the incision. No signs of bandaging, which would indicate that the meat was salted and dried and then covered with resin. Generally if a piece of meat is salted and dried and not bandaged there is no (or very little) evidence of resin.

S74 L&G38 **no dimensions**
Sternum of bovid with left and right costal cartilage. There are longitudinal incisions in the meat, presumably to help it dry.

S75 L&G51 **no dimensions**
Bovid sternum, wrapped.

S76-7 L&G98-99 **no dimensions**
Sternum fragments.

S78 L&G135 **no dimensions**
Sternum and cartilage costal of young B. taurus

S79 L&G136 **no dimensions**
Sternum and costal cartilage of a juvenile B. taurus

S80-2 L&G56-58 **no dimensions**
Three sterna with costal cartilage cut off lengthwise.

S83 L&G169 **no dimensions**
Sternum and costal cartilage of a young B. taurus.

S84 L&G37 **CG29573; D4000** **L**: 36cm **W**: 13cm
Four unwrapped ribs from the front of the right side of a bovid. There is quite a lot of meat on them.

S85 L&G27 **no dimensions**
Part of bovid thorax composed of eight posterior ribs from the right side.

S86-7 L&G65-66 **no dimensions**
Two portions, each composed of six right ribs.

S88 L&G100 **no dimensions**
Six ribs of a small *B. taurus* from the right side.

S89 L&G139 **no dimensions**
Eight ribs from the right side of a young bovid.

S90 L&G180 **no dimensions**
Eight ribs from the right side of a *B. taurus*.

S91 L&G32 **CG29565; D4004** **L**: 22 **W**: 8
Left femur of juvenile bovid with considerable amounts of meat attached. It is unwrapped and there is a whitish deposit, probably salt, on most of the surface.

S92 L&G28 **CG29576; D4005** **L**: 30 **W**: 11
Left femur of a *B. brachyceros* or a juvenile bovid (this cannot be determined as the epiphysis is not visible). The femur is unwrapped and is dark in places and whitish, possibly due to salt, in others.

S93 L&G26 **no dimensions**
Right femur.

S94 **L&G**31 **no dimensions**
Left femur.

S95 **L&G**29 **no dimensions**
Right femur; the meat has lengthwise incisions in line with the bone. Presumably these incisions were helpful for the preservation process of the meat as they allowed for greater absorbency of salt and provided space for air circulation necessary for drying (see discussion in salting).

S96 **L&G**49 **no dimensions**
Left femur, wrapped.

S97 **L&G**42 **no dimensions**
Left femur of a young bovid.

S98 **L&G**75-76 **no dimensions**
Left femur with meat attached.

S99-101 **L&G**117-19 **no dimensions**
Three right femurs with meat attached.

S102-3 **L&G**120-21 **no dimensions**
Two left femurs with meat attached.

S104 **L&G**160 **no dimensions**
Wrapped joint consisting of the right femur of a mature animal.

S105 **L&G**161 **no dimensions**
Left femur with considerable meat attached.

S106 **L&G**206 **no dimensions**
Right femur of a young *B. taurus*.

S107-8 **L&G**79-81 **no dimensions**
Right and left femurs of a juvenile *B. taurus*.

S109-10 **L&G**158-59 **no dimensions**
Two femurs (possibly of different animals) with meat attached.

S111 **L&G**30 **no dimensions**
Left tibia and calcaneum (the latter is an unusual element in these offerings).

S112 **L&G**47 **no dimensions**
Right tibia and tarsals of a juvenile *B. taurus*.

S113 **L&G**131 **no dimensions**
Left tibia and tarsals.

S114 **L&G**133 **no dimensions**
Right tibia and tarsals of a young *B. taurus*.

S115-6 **L&G**167-68 **no dimensions**
Tibia and tarsals (side not mentioned).

S117 **L&G**174 **no dimensions**
Right tibia and tarsals of *B. taurus*.

S118-22 **L&G**189-92 **no dimensions**
Four tibiae.

S123 **L&G**84 **no dimensions**
Right tibia and tarsals of a young *B. taurus*.

S124-5 **L&G**85-86 **no dimensions**
Two left tibiae and tarsals.

S126-7 **L&G**87-88 **no dimensions**
Two tibiae from a juvenile *B. taurus*.

S128-9 **L&G**89-90 **no dimensions**
Two tibiae from a small animal.

S130 **L&G**1-17 **CG29569**; **D4008** **L**: 9 **W**: 5
Two fragment of meat probably removed from the pelvic region. Both have the same dimensions and carry the impression of the string by which they were hung (biltong — see VII.1).

S131 **L&G**1-17 **CG29571**; **D4009** **L**: 9 **W**: 4
Like **S130**, with a trace of the cord that tied the meat for suspension at one end.

S132 **L&G**1-17 **CG29572**; **D4002** **L**: 10 **W**: 5
Meat from the lumbar or pelvic region, with traces of cord marks at one end.

S133-47 **L&G**1-17 **L**: 15-20 **W**: 4-5
Diverse portions of salted and dried meat with traces of the thread that held them while they dried.

S148-52 **L&G**67-71 **no dimensions**
Five fragments of meat.

S153-61 **L&G**101-09 **no dimensions**
Nine fragments of meat.

S162-9 **L&G**142-49 **no dimensions**
Eight meat fragments, probably from the rump.

S170-1 **L&G**170-71 **no dimensions**
Two portions of dried meat, possibly from the pelvic (rump) region of a bovid.

S172-6 **L&G**181-85 **no dimensions**
Five segments of dried meat

S177 **L&G**18 **CG29568**; **D4003** **L**: 9 **W**: 6
Part of a lobe of liver from a bovid.

S178 **L&G**19 **CG29570**; **D4001** **L**: 10 **W**: 9
Portion of a lobe of bovine liver. Large cracks run through it, and a whitish substance, probably salt, covers some areas.

S179 **L&G**50 **no dimensions**
Portion of liver.

S180-1 **L&G**140-41 **no dimensions**
Two portions of liver

S182 **L&G**186 **no dimensions**
Unidentified bones of a juvenile *B. taurus*.

Discussion

Although almost all portions of the animals are represented in the offerings, it is notable that no heads, necks, tails, metatarsals, metacarpals, phalanges, lungs, hearts, or kidneys are offered. Although the pelvis is not offered, it is very probable that the boneless strips or, in some cases, chunks of meat are probably from the pelvic/rump area. Several (c. 30+) of these pieces of meat consist of long thin strips (15-20cm × 4-5cm) with the impression of a string at one end. These are probably biltong and are discussed in Chapter VII.

The incompletely fused epiphyses of the leg bones examined, notably the humerus, femur and tibia, indicate that the animals were relatively young — between two and three and a half years old (Silver 1969 is used for ageing) when they were slaughtered. They were also quite small, probably a small *B. taurus* or *B. brachyceros* (Lortet and Gaillard 1905: 315).

By using the list of joints provided by Lortet and Gaillard 1905 one can estimate the number of animals offered for these two royal burials. According to Lortet and Gaillard, twelve animals were responsible for the sum total of the joints found in these tombs (Lortet & Gaillard 1907: x).

Unfortunately, they do not explain how they arrived at this conclusion. If one were to use the scapulae as the determining bone one would arrive at eleven animals (22 scapulae in toto, divided by two per animal). By the same process the bones from the nether regions indicate (using the femurs) that ten animals were used. If one uses the humerus as the determining bone then one would have at least thirteen animals being offered, if not more (total humeri are 27). Thus, given the evidence available one cannot establish an exact number of animals slaughtered, but one can limit the number between eleven (there are more bones supporting the scapulae-generated number than the femur-generated number) and thirteen. It is quite possible that thirteen animals were sacrificed, but this does not mean that all thirteen were offered in their entirety. There are a number of explanations possible that would explain the fate of the 'missing' offerings. They could have been destroyed in excavation, lost subsequent to excavation, been stolen and devoured before they became official offerings, might have been the (legitimate) payment for the butchers/necropolis workers, or could have provided part of the funeral banquet. Regardless of whether eleven or thirteen animals were sacrificed, it was still a large number (Figs. 42-43, 59). Most of the offerings did indeed go into the tomb covered with meat, with a few portions left unaccounted.

All CG-numbered beef items are in the Agricultural Museum. The avian material is in the Cairo Museum.

Tutankhamun [KV62]

A total of forty-eight boxes of food offerings placed on a reed mat (numbered 62a-vv in Carter's card-catalogue [Carter 1932]) were found in Tutankhamun's tomb under the the Hathor couch [number 73]. The cases containing the meat "are made of wood (probably sycamore); covered with gesso and painted white on the exterior, and the interior varnished with black resin. Their contents [?] ... were much deteriorated and the various joints etc. apparently not preserved before they were wrapped in linen bandages. A number of ideograms [...] written in hieratic upon the exterior (generally underneath) seems to refer to the contents" (Carter 1932). Only twenty-eight boxes were labelled in hieratic (Cerny 1965: 17; Gardiner 1932). The shape of these finely made boxes acts as an additional indicator of the contents. Unfortunately the state of the bandages has not been recorded.

The information found below is from Carter's excavation notes (1932), a letter written to him by Gardiner (1932), who analysed the hieratic inscriptions, as well as from Cerny's remarks on the hieratic (1965: 17-8). Although Carter has identified all the joints, it is uncertain whether or not he unwrapped all of them or if he unwrapped only some of them and extrapolated from there for the rest. Not all his identifications are precise (e.g. shin, etc.) and some of them, like **S216**, identified as tongue, seem unusual as they are the only examples of mummified tongues. One wonders if Carter might have mistaken liver for a tongue, but until the contents of these boxes are re-examined thoroughly one cannot confirm or deny his identifications. Unfortunately, the author has only examined six of these boxes, since the rest are in magazines K and L of the Cairo Museum and were unavailable for study. All the measurements given are for the boxes/cases, not for the actual joints of meat. For illustrations of the inscriptions see Figs. 70-72.

In addition to the mummified meat, "seven ribs, a shoulder blade, two vertebrae, and a leg bone of an ox" (Carter 1932: #179) were found in the tomb. They were, along with a few other objects, "lying on the ground, beside (the) east wall of (the) tomb chamber, near (the) doorway of (the) store-chamber" (Carter 1932: #179). Perhaps these joints were a part of the funerary feast and put into the tomb just before sealing it up? It is possible that this collection of bones was scattered as "some ribs of an ox (were) found near (the) silver trumpet (175) found in (the) south-east corner of (the) tomb-chamber" (Carter 1932: 175b).

S183 **62b** **JE61409** **L**: CS38 **W**: CS20 max
Triangularish box containing a scapula. Inscribed *tp*. (See VI for discussion).

S184 **62d** **JE61407** **L**: CS36 **W**: CS18 max
Triangularish box containing a scapula. Inscribed *tp*. (See VI for discussion).

S185 **62q** **JE61406** **L**: CS37 **W**: CS22 max
Triangularish box containing a scapula. Inscribed *tp*. (See VI for discussion).

S186 **62s** **JE61408** **L**: CS38 **W**: CS22 max
Triangularish box containing a scapula. Inscribed *tp*. (See VI for discussion).

S187 **62a** **JE61402** **L**: CS45 **W**: CS33 **H**: CS22
Five lumbar vertebrae in a somewhat triangular box. Inscribed *iꜣt* (See VI for discussion).

S188 **62dd** **JE61403** **L**: CS44 **W**: CS36 **H**: CS24
Six lumbar vertebrae in a somewhat triangular box. Inscribed *i3t* (See VI for discussion. Fig. 71).

S189 **62c** **JE61430** **L**: CS56 **W**: CS37 **H**: CS29
Oval box containing ribs 1-8.

S190 **62f** **JE61416** **L**: CS37 **W**: CS16 **H**: CS16
An oval box containing ribs 1-3. Inscribed *g*, and discussed in Chapter VI.

S191 **62m** **JE61417** **L**: CS38 **W**: CS15.5 **H**: CS13
An oval box containing ribs 1-3. Inscribed *g*, and discussed in Chapter VI.

S192 **62g** **JE61432** **L**: CS52 **W**: CS37 **H**: CS22
Oval box containing ribs 1-8.

S193 **62y** **JE61433** **L**: CS56 **W**: CS37.5 **H**: CS29
Oval box containing ribs 11-18, side not determined.

S194 **62bb** **JE61431?** **L**: CS56 **W**: CS38 **H**: CS26
An oval box containing ribs 11-18.

S195 **62ff** **JE61419** **L**: CS38 **W**: CS14 **H**: CS12
An oval box containing ribs 1-3. Inscribed, and discussed in Chapter VI.

S196 **62hh** **JE61418** **L**: CS40 **W**: CS15.5 **H**: CS15
An oval box containing ribs 1-3. Inscribed *g*; discussed in Chapter VI (Fig. 72).

S197 **62ee** **JE61421** **L**: CS44 **W**: CS26 **H**: CS19
An oval box containing a sternum. Inscribed *ḳ*, and discussed in Chapter VI.

S198 **62jj** **JE61405** **L**: CS50 **W**: CS40 max
A triangularish box containing a sternum. Inscribed *ḳ*, and discussed in Chapter VI.

S199 **62h** **JE61427** **L**: CS38 **W**: CS15 **H**: CS14
An oval box containing a humerus. Inscribed *ib*, and discussed in Chapter VI.

S200 **62p** **JE61428** **L**: CS36 **W**: CS18 **H**: CS17
An oval box containing a humerus. inscribed *ib*, and discussed in Chapter VI.

S201 **62v** **JE61426** **L**: CS42 **W**: CS32 **H**: CS17
A lozenge shaped box containing a humerus. Inscribed *ib*, and discussed in Chapter VI.

S202 **62qq** **JE61425** **L**: CS45 **W**: CS33.5 **H**: CS20
A humerus in an oddly shaped box. Inscribed *ib*, and discussed in Chapter VI.

S203 **62aa** **JE61422** **L**: CS45 **W**: CS32 **H**: CS20
A radius placed in a box the shape of a squashed triangle. The box is inscribed *s3*, and is fully discussed in Chapter VI.

S204 **62u** **JE61420** **L**: CS46 **W**: CS29 **H**: CS19
An oval box containing a radius. The box is inscribed *s3*, and is fully discussed in Chapter VI.

S205 **62z** **JE61440** **L**: CS46 **W**: CS18 **H**: CS14
An oval box containing "the shin, metacarpal, etc" (Carter 1932). This is an unusual portion of the animal to be included.

S206 **62vv** **JE61435** **L**: CS39 **W**: CS20 **H**:CS14
An oval box containing the patella to the end of the metacarpal. Like **S205**, it is an unusual joint for inclusion.

S207 **62l** **JE61411** **L**: CS45 **W**: CS25 **H**: CS18
A rectangularish shaped box containing a femur. Inscribed *swt*, and discussed in Chapter VI.

S208 **62e** **JE61410** **L**: CS46 **W**: CS23 **H**: CS17
A rectangular box containing the femur and patella. Inscribed *swt* , and discussed in Chapter VI.

S209 **62r** **JE61429** **L**: CS43 **W**: CS33 **H**: CS18
A lozenge shaped box containing the femur and patella.

S210 **62n** **JE61412** **L**: CS50 **W**: CS41 max
Triangular box containing the femur. Inscribed *swt*, and discussed in Chapter VI.

S211 **62kk** **JE61404** **L**: CS52 **W**: CS38 max
A tibia in a triangularish box. Marked *ḫn*, and discussed in Chapter VI.

S212 **62oo** **JE61415** **L**: CS44 **W**: CS27 **H**: CS20
A tibia in a triangularish box. Marked *ḫn*, and discussed in Chapter VI.

S213 **62rr** **JE61414** **L**: CS47 **W**: CS38 max
A tibia in a triangularish box. Marked *ḫn*, and discussed in Chapter VI.

S214 **62t** **JE61413** **L**: CS42 **W**: CS22 **H**: CS17
Rectangular box containing a tibia. Marked *ḫ*, and discussed in Chapter VI.

S215 **62o** **JE61423** **L**: CS35 **W**: CS12 **H**: CS12
An oval box containing a tongue (Carter 1932). The inscription reads *nnšm*, and is discussed in Chapter VI.

S216 **62i** **JE614241** **L**: CS35 **W**: CS13 **H**: CS13
An oval box containing a tongue (Carter 1932). The inscription reads *nnšm*, and is discussed in Chapter VI.

S217 **62tt** **JE61434** **L**: CS38 **W**: 1CS8 **H**: CS12
An oval box containing something that might be flesh.

S218 **62gg** **JE61436** **L**: CS40 **W**: CS18 **H**: CS17.5
Oval box containing three pieces (probably chunks) of flesh.

S219 **62ii** **JE61438** **L**: CS36 **W**: CS18 **H**: CS14
Oval box containing four chunks of flesh.

S220 **62mm JE61443** **L**: CS40 **W**: CS18 **H**: CS17
Oval box containing three pieces (probably chunks) of flesh.

S221 **62cc** **JE61437** **L**: CS40 **W**: CS19 **H**: CS15
A piece of flesh in an oval box.

S222 **62w** **JE61442** **L**: CS40 **W**: CS18 **H**: CS16
Two pieces of flesh in an oval box.

S223 **62ll** **JE61439** **L**: CS40 **W**: CS19 **H**: CS18
Three pigeons in an oval box.

S224 **62nn** **JE 61400** **L**: CS41 **W**: CS24 **H**: CS20
A goose in an oval box with the legs and wings indicated by modelling. The inscription *r*, is correct (Fig. 70).

S225 **62x** **JE61397** **L**: CS40 **W**: CS27 **H**: CS16
A goose in an oval box with some indication of legs and wings.

S226 **62j** **JE61396** **L**: CS40 **W**: CS22 **H**: CS20
A goose in an oval box with some indication of legs and wings.

S227 **62k** **JE61399** **L**: CS40 **W**: CS22 **H**: CS20
A goose in an oval box with some indication of legs and wings.

S228 **62ss** **JE61401** **L**: CS27 **W**: CS16 **H**: CS15
A duck in a box with the legs and wings indicated on the lid.

S229 **62uu** **JE61441** **L**: CS40 **W**: CS18 **H**: CS15
Three pigeons in an oval box.

S230 **62pp** **JE61390** **L**: CS29 **W**: CS18 **H**: CS10
A goose liver(?) in a goose shaped box.

Discussion

It is certain that in addition to eleven birds, at least two entire animals were sacrificed for this burial as is indicated by the number of scapulae, sternums, femurs, humeri and tibiae. The inclusion of metacarpals amongst the offerings is rare as generally, unless the entire leg was offered, the leg elements below the tarsals/carpals were not offered. This collection of food offerings is especially notable for the unusual offering of tongues. However, it is possible that Carter mistook the liver for a tongue. The identity of the tongue/liver is fully discussed in Chapter VII.

The chunks of meat provide an interesting puzzle, Are they identical to the small pieces of air dried meat found in the offerings of Amenophis II, are they large slabs of meat like those found in the tomb of Queen Meryetamun (TT358), or are they larger and longer versions of the air dried chunks of King Amenophis II, and therefore closer in shape and dimension to the modern types of biltong? Unfortunately none of these can be examined and only the dimensions of the boxes are available so the identity of these pieces will have to await further investigation.

The relationship between joints and box shape and label is strange in this assemblage. Objects of the same type appear in boxes of different shapes, and the labels on the boxes frequently bear no resemblance to their contents. It is possible that the boxes were labelled in advance and left to be filled by workers who were unable to read or were lazy or had consumed the required item. Or maybe our reading and interpretation are at fault? The hieratic name-labels on the boxes are of especial interest and are discussed in Chapter VI.

All the Food-Boxes from Tutankhamun's tomb are stored in Cairo Museum magazines K and L, with the exception of those on display.

Tuthmosis IV [KV43]

Food offerings consisting of mummified joints and geese were found in this tomb in the second room on the left, B on Davis' plan (Reeves 1990: 36). Out of these, only the bird mummies have been recorded. All of these are now in the Agricultural Museum in Dokki (**D** number), and have been studied by Gaillard and Daressy (1905), and Lortet and Gaillard (1908), as well as myself. If the allusion to the joints was not a mistake in identification made by the excavator, and there were indeed joints in the burial, they are now lost.

S231 **CG29704** **L**: 30 **W**: 15

A bird, quite well wrapped in linen bandages. The head, neck, tarsals and wingtips are removed as they would be for a roasting bird. The bandages are coated with resin, which might have come from boxes in which the birds were originally stored, or might have been applied to the bandages to better preserve the meat. According to Gaillard and Daressy, the bandages were also coated with an undefined antiseptic (1905: 115). The bird is probably *Anser albifrons*, the white coated goose in Medum frieze that stays in Egypt during the cold season (Gaillard and Daressy 1905: 115).

S232 **CG29705; D2935** **L**: 30 **W**: 13
A. albifrons, well wrapped in "bandages impregnated with some resin and natrons" (Gaillard and Daressy 1905: 115). My examination showed there were only some areas of the bird that were covered with resin, although it is possible that some of the resin became detached from the bird over time. The head, neck, tarsals and wingtips are removed.

S233 **CG29706; D2936** **L**: 19 **W**: 8
A small duck, maybe a teal, with its head, neck, tarsals and wing tips removed. It is bandaged tightly and there are traces of resin on the bandages.

S234 **CG29707; D2937** **L**: 28 **W**: 12
Perhaps *A. albifrons* with its head, neck, tarsals and wing tips removed. It is bandaged, with traces of resin on the bandage in the area of the breast and near the neck.

S235 **CG29708; D2938** **L**: 13 **W**: 6
Probably a small migratory duck with its head, neck, tarsals and wing tips removed. It is bandaged and the bandages are resin coated.

S236 **CG29709; D2939** **L**: 8 **W**: 4
A very small bird that is bandaged after the removal of the extremities and head. Quite probably a quail (Gaillard and Daressy 1905: 116), though it is more probably a pigeon or a dove.

Discussion:

Although this tomb was supposed to contain meat offerings in addition to the birds none of these can be traced. However, this assemblage provides us with a variety of birds and indicates, from the types of birds offered, the time of year the burial might have taken place. The pharaoh was probably buried during the winter as that is when *A. albifrons*, a migratory bird not native to Egypt, is present. However, one might argue that these mummies were not made at the death of the king but that they were a regular part of the funerary equipment and that a convenient stock of mass produced victual mummies was kept in storage and used whenever necessary. Given the limited number of burials containing these food offerings it seems highly unlikely that a stock of these objects was available. It is far more likely that the boxes were prepared in advance at one place and that the food remains were put in at the time of the funeral. One thing is clear, however, and that is that the internal resin coating is generally done just before putting in the mummies as the bandages are frequently stuck to the resin meaning that they were put in when the resin was still warm. If this were done well in advance of the burial there would have been time for the resin to dry before filling it with meat. Also, who would pay for a stock of meat offerings? Meat was dear and not easily available. Perhaps the boxes were made in advance, but the resining and filling would only be done at the time of the burial.

An interesting stylistic detail that helps determine the identity of the wrapped birds comes to light in the study of this collection of food offerings. The legs of the geese are wrapped as separately defined entities, while the legs and wings of ducks, pigeons and other small birds are wrapped as part of the body mass, and are not separately defined.

Tomb of Maherpri [KV36]

This tomb contained eleven boxes of avian and mammalian food offerings in addition to unwrapped offerings of ribs. Some of these offerings were placed at the northern end of the burial chamber, probably near the head of the coffin, while Daressy notes that "dix côtes de veau, enveloppées de bandelettes de toile" (Daressy 1902: 58) were found in the coffin (CG24002) itself. The sycamore wood boxes (Daressy 1902: 22) are finely made with the usual dowel and hole arrangement for firm closure. The boxes are plastered both inside and out. There is no trace of resin on any of the boxes or on any of the bandages. Not all the joints are boxed; those that are unboxed are wrapped in linen and are untouched by resin.

It is interesting to note that this tomb in the Valley of the Kings is a non-royal tomb from the Eighteenth Dynasty (reign of Tuthmosis IV — See Reeves 1990: 146-7 for discussion of dating). These mummies have only been found in royal burials, or in burials of people very closely associated with royalty, like Yuya and Thuyu (see below).

All the offerings found in the tomb are listed below. They are housed in the Cairo Museum (the **C** numbers are institution's exhibition numbers), and have been examined by the author. They were studied by Daressy (1902) and Lortet and Gaillard (1905). The measurements provided are those of the cases.

S237 **CG24047**; **JE33834** **L**: 1.035 **W**: 31-21 **H**: 8-11
This box contains what is possibly the entire wrapped hindleg (femur, tibia,

patella, tarsals and metatarsal and possibly phalanges) of a bovid, well cushioned in its box by linen. The location of this box is unknown, although photographs from Daressy (1902: pl. vi) and the JE have been examined in order to verify the identification.

S238 **CG24048**; **JE33835**; **C4230** **L**: 51 **W**: 23 **H**: 24
An oval wooden box for meat of an undetermined cut, although the shape of the box suggests, given the identified parallels, a leg bone. It is plastered white, both inside and out. There is no evidence for the interior being coated with resin. The bandages are supposedly impregnated with resinous natron (Daressy 1902: 22).

S239 **CG24049**; **JE33836**; **C4231** **L**: 49 **W**: 23 **H**: 14
A high domed oval box containing the left humerus of a calf.

S240 **CG24050**; **JE33837**; **C4236** **L**: 50 **W**: 19 **H**: 14
An elongated wooden box for food containing a "morceau d'une cuisse de veau" (Lortet and Gaillard 1905: 307). It quite possibly contains a leg joint, perhaps the femur.

S241 **CG24051**; **JE33838**; **C4235** **L**: 44 **W**: 22 **H**: 19
A wooden box in the shape of a bird, most probably a goose. In the box the legs and wing tips are carved free of the body of the box. The bird is tentatively identified as a goose due to its size as no other identification is possible given its well wrapped state.

S242 **CG24052**; **JE33839**; **C4232** **L**: 44 **W**: 22 **H**: 19
A box in the shape of a goose, containing *Anser albifrons*. The tarsals and carpals onward are cut off, and the internal organs removed and then wrapped up after replacing the heart, the liver, gizzard, all tied together by string (Lortet and Gaillard 1905: 308).

S243 **L**: 49
Oval case like that of **S242**, although it is bigger than that. No signs of resin.

S244 **CG24053**; **JE33840**; **C4234** **L**: 35 **W**: 17.5 **H**: 16
A wooden box in the shape of a duck or goose. Identified as a duck covered by a resinous and saline substance, whose internal organs were removed, and then the heart and liver replaced into the body cavity. The bone from the tarsals and carpals onward was cut (Lortet and Gaillard 1905: 309).

S245 **CG24054**; **JE33841**; **C4233** **L**: 27 **W**: 13 **H**: 11
A teardrop shaped wooden box for a food offering, possibly a pigeon (Lortet and Gaillard 1905: 309) or a duck.

S246 **CG24055**; **JE33842**; **C4227** **L**: 27 **W**: 14.5 **H**: 11
Wooden case with light white-wash and plaster on exterior. No signs of the internal surface being coated by resin. Identified by Daressy as a pigeon, although given the dimensions of the box this is somewhat unlikely (Daressy 1902: 23).

S247 **CG24056; JE33843; C4229** **L**: 25 **W**: 25 **H**: 15.5
A square box containing meat. The shape of the joint is disguised by the wrapping (Lortet and Gaillard 1905: 310), although it could be the liver of sheep and some pieces of meat with the bandages being soaked in a saline/resin mixture (Daressy 1902: 24).

S248 **CG24056 *bis*, C4228** **no dimensions**
This entry, according to Daressy, should consist of ten sides of veal wrapped in bandages found in coffin 24002 (Daressy 1902: 24); however, on display in the Cairo museum were eight or nine wrapped ribs with no evidence of resin on any part thereof.

Discussion

According to Lortet & Gaillard (1907: x) all the meat was from one animal. This is a feasible conclusion as long as one ignores the ten sides of veal mentioned by Daressy, since his reference makes it unclear whether he is referring to ten ribs or to ten separate packages containing undisclosed numbers of ribs, the latter supposition making the total of the number of animals offered more than one.

It is notable that none of the visible bandages show traces of resin, although some scholars who have directly studied and unwrapped the offerings write that the bandages were soaked in a mixture of resin and salt. Resin soaking would leave the bandages discoloured, so if the bandages were indeed soaked with resin, perhaps it was restricted to only the internal bandages. Resin would not be necessary to preserve the meat as salt would be sufficient preservative on its own. Generally the resin found on the bandages of other offerings is thought to have come from boxes, rather than playing a part in the meat wrapping.

All the boxes are in the Cairo Museum. There are a few objects in the Dokki Agricultural Museum that are incorrectly labelled as coming from this tomb (Darby et al 1977: fig.3.43); the catalogue for the museum assigns them to KV35.

Yuya & Thuyu [KV46]

Supposedly eighteen boxes of provisions were found in the burial chamber of the tomb near the head of the mummies (Reeves 1990: 150, Quibell 1908: 4). However, the Cairo collection contains nineteen boxes allegedly from this burial, all of which are listed below. As yet, this discrepancy between the objects reputed to be found and those located in the museum in inexplicable, although it is possible that a bird belonging to another burial was mistakenly attributed to this burial. The **C** numbers are Cairo Museum display numbers.

All the mummies in this assemblage were boxed. All the cases are of wood and are plastered. However, unlike other such boxes, in some instances the interior and the exterior are covered by resin.

S249 **CG51084**; **JE95334a,b**; **C111** **L**: 72
A "wooden box covered with plaster, then resined inside and out, and containing some animal or joint of meat, dried and wrapped in cloth. Lid and box are generally of about equal size and pegged together with four pegs. The outline of the boxes follows roughly the shape of their contents. It appears that each box and each lid is hewn out of a single piece of wood, but it may be that some joints are hidden by the thick coat of resin. This box appears to contain a leg of veal. Two foot bone are wrapped separately" (Quibell 1908: 46). This looks like a foreleg to me. This case is bituminized, but not at the join.

S250 **CG51085**; **JE95338**; **C115** **L**: 35
An oval box possibly containing the tibia of a calf.

S251 **CG51086**; **JE95341**; **C118** **L**: 21
A box containing a wrapped pigeon or other small bird.

S252 **CG51087**; **JE95339a,b**; **C116** **L**: 40
An elongated triangular box with rounded corners containing a wrapped scapula with a lot of meat attached. The scapula could quite possibly belong to an antelope as Quibell (1908: 47) suggests, though it might belong to a *Bos*. It is impossible to determine its identity for certain as the scapula is well wrapped.

S253 **CG51088**; **JE 95324**; **C101** **L**: 50
A large kidney bean shaped box containing four juvenile bovid ribs.

S254 **CG51089**; **JE95325**; **C102** **L**: 42
An oval box whose contents are not certain. The JE lists the contents as being 'ribs of veal', although the shape of the box suggests a portion of the leg, possibly the tibia. Unfortunately this object was not seen so its identity remains uncertain.

S255 **CG51090**; **JE95326**; **C103** **L**: 38
A wooden box containing food offerings. Unfortunately this case is in storage and was not examined; furthermore there was no information concerning it available in the CG or the JE.

S256 **CG51091**; **JE95327**; **C104** **L**: 43
A wooden box containing food offerings. Unfortunately this case is in storage and was not examined; furthermore there was no information concerning it available in the CG or the JE.

S257 CG51092; JE95336a,b; C113 L: 45
A well wrapped goose in a wooden case with the lid carrying carved details of the legs and wings (with the feet and wing tips cut off). The bottom part of the box consists of an oval.

S258 CG51093; JE95333; C110 L: 30
A wooden case for a duck. The legs and wings are carved on the bos.

S259 CG51094; JE95332; C109 L: 25
A box containing a small duck, or possibly a pigeon or dove.

S260 CG51095; JE95335a-c; C112 L: 46
A wooden box for a goose. The carving on the lid of the box has legs and wings indicated. The bird's wing tips and feet have been cut off, and the legs have been wrapped separately away from the rest of the carcass. The bandages are heavily coated with resin.

S261 CG51096; JE95328; C105 L: 44
A large wrapped goose with the carpals and tarsals removed.

S262 CG51097; JE95329; C106 L: 35
A wooden box containing food offerings, the contents of which are unknown. It could be a leg bone.

S263 CG51098; C108 L: 37
Box containing ribs of bovid.

S264 CG51099; JE95330; C107 L:42
Foot of an ox in a box. This was wrapped and consequently not examined closely. It is highly unlikely that this box contains the foot of an ox as feet (metacarpals, metatarsals, and phalanges) were almost never offered, and if they were offered it was only as part of the entire leg and never independent of it.

S265 CG51100; JE95340a,b; C117 L: 45
This box contains a mature *Bos* humerus, probably from the right side of the body. The JE entry states that this number should be assigned to a poultry food offering.

S266 CG51101; JE95337a,b; C114 L: 65
Shoulder of veal (?) I believe it is the scapula and humerus of a young bovid. The resin on the inside of the box has gotten on the bandages quite badly. The joint is very well wrapped and thickly coated with resin so the identification is not secure.

S267 JE95333a,b L: 30
A case containing a duck? Probably from another burial.

Discussion

Unfortunately there are several identifications in this assemblage that are not quite secure making it difficult to determine exactly how many animals were offered. However the evidence indicates that a calf and portions of a mature *B. taurus* were probably offered. The assemblage is unusual in the possible inclusion of an antelope scapula. The scapula in question (**S252**) does look like that of an antelope, and as antelopes of various sorts were consumed, the appearance of a part of one as a funerary offering is not impossible, although rare. The workmanship of these boxes was poorer than that of any other collection examined. These boxes were especially unusual in the generous amount of resin utilized on the interior and exterior.

Most of these victual mummies are on display in the Cairo Museum, while a few are in storage in the cases.

Isitemkheb D

The burial of Isitemkheb D, the wife of Pinudjem II, was one of the five primary Twenty-first Dynasty burials in the Deir el-Bahari cache excavated in 1881. Part of her funerary equipment consisted of reed baskets containing wrapped meats and birds as funerary offerings. These victual mummies are like those of the Eighteenth Dynasty in that they consist of joints of meat or entire birds that are wrapped in linen bandages and given as offerings to the deceased, although these Twenty-first Dynasty examples were not placed in individual wooden coffinets as were so many of their earlier counterparts. Furthermore they were not coated with resin, a feature common to many of the earlier examples of victual mummies with the notable exception of Maherpri. Numbers C2311-20 (JE 46899-908) were all found in the basket JE46880 as part of the Deir el-Baheri cache of 1881. C2305-10 (JE46893-8) were found in the basket JE46880, also in the Deir el-Baheri cache. All the victual mummies from her burial are in the Cairo museum (Fig. 73).

S268 **JE46868; C2298** L: 29
Wrapped head of an animal, probably a calf — a unique example.

S269 **JE46879; C2291** L: 25
Wrapped duck, feet and wings cut off. There is some problem with the JE entry as next to this number it says, "tête de veau dans une toile (ou cuissot)", while the photograph illustrating it clearly shows the object in the vitrine, a wrapped duck.

S270 **JE46875; C2292** **L**: 30
A wrapped food offering. Again a problem with the JE entry which reads: "Cuissot? enveloppe dans une toile." The object does not look like a humerus or femur as the shape is wrong; it more closely resembles a wrapped fowl, or even a well wrapped lobe of calf's liver.

S271 **JE46885; C2293** **L**:36
A wrapped duck, feet and wing tips cut off.

S272 **JE26263; C2299** **L**: 22 **W**: 30 **H**: 21
Box made of reeds and papyrus said to contain a duck or goose. The bird is unwrapped and appears quite dry. It could have been roasted and then put into the box, or it might have been cleaned, salted, and then air dried before being placed in the box.

S273 **C2305** **L**: 15 **W**: 7
A tightly wrapped piece of meat that is quite small and rather lumpy, although the lumps might be due to the bandages.

S274 **JE46894; C2306** **L**: 30 **W**: 10
Very well wrapped victual mummy, possibly a tibia or humerus (more likely) of a sheep or goat.

S275 **JE46895; C2307** **L**: 20 **W**: 14
This victual mummy is very securely bandaged so that the contents are invisible. However, the shape of the package suggests that it contains the mummified liver of a bovid, as it's shape is similar to that of other packages containing beef liver.

S276 **C2308** **L**: 20 **W**: 8
A small mummified bird, possibly a duck.

S277 **JE46897; C2309** **L**: 12 **W**: 5.5
Remains of some sort of meat.

S278 **JE46905; C2310** **L**: 17 **W**: 20
A small well wrapped bundle. The shape intimates that it is a pelvis of a calf.

S279 **JE46899; C2311** **L**: 28 **W**: 8.5
Wrapped food offering.

S280 **C2312** **L**: 41 **W**: 20
Wrapped pelvis and femur (?) of young bovid. Meat attached.

S281 **JE46901; C2313** **L**: 51
A well wrapped right foreleg (humerus, radius, ulna) of a mature sheep or goat.

S282 **JE46902; C2314** **L**: 39 **W**: 8
Wrapped leg, probably hindleg (tibia and fibula are clear), but the bundle is too well wrapped to be absolutely certain, of a sheep or goat.

S283 **JE46903; C2315** **L**: 49
The entire left foreleg (scapula, humerus, radius, ulna, carpals, metacarpals and phalanges) of a small framed animal, either a goat or a gazelle.

S284 **C2316** **L**: 42 **W**: 13
Half of the pelvis and the femur and tibia (fibula?) of a sheep or goat. The animal could conceivably be a gazelle, but one would need a closer examination before this can be determined.

S285 **JE46898; C2317** **L**: 15 **W**: 7
An unidentifiable piece of flesh of no particular shape or cut.

S286 **C2318** **L**:31.5 **W**: 7.5
Wrapped hindleg joint of a bovid. It includes the calcaneum, tarsals, metatarsal, and maybe the phalanges.

S287 **JE46907; C2319** **L**: 22 **W**: 10
Wrapped piece of meat, no resin in evidence on the bandages. It is either a scapula or pelvis of a small framed animal, ie: sheep, goat, or gazelle.

S288 **JE46908; C2320** **L**: 16 **W**: 4
A well wrapped chunk of meat.

Discussion

These are the only known royal food offerings from the Twenty-first Dynasty. They are markedly different from both royal and non-royal food offerings from previous dynasties in that there are no wooden coffinets for any of the offerings and that resin was not used on any of them. It is also notable that no mature ox was offered — only a calf was offered. The inclusion of small mammals as offerings: sheep, goats, gazelles, is unique. Obviously food offerings were not on as grand a scale in the Twenty-first Dynasty as they were in the Eighteenth, although one cannot make any sweeping generalizations, given the limited amount of information available. The fact that this collection of offerings is from a queen's burial rather than a kings might also influence the information.

All offerings are held in the Cairo Museum.

Imhotep [QV 46]

Another non-royal burial containing victual mummies is that of Imhotep, an early Eighteenth Dynasty vizier buried in the Valley of the Queens. Most of his beef and bird offerings are in wooden coffinets, some showing traces of plaster on the exterior, while the interior are resin coated, although a few are simply wrapped in bandages, some showing traces of resin on their exterior, perhaps from missing coffinets? (Donadoni Roveri 1988: 47; Schiaparelli 1924: 25, figs. 24-26). Unlike Kha's burial (see below), no salted and unwrapped birds or fish were found in this burial.

The numbers following the **S** number are excavation numbers; **T** signifies Turin; the letters CS before a measurement signifies the measurements of a case; the number following it is the measurement for the piece of meat in the case.

S289 **T5084** L: CS27 W: CS16.5
Wooden box in the shape of a duck or similar sized bird with its head cut off. The case was covered, both inside and out with plaster and paint, and then the interior was painted with resin. This case was firmly closed, so one can only assume that it contains some kind of fowl. There is no detailing on the case to suggest wings or feet.

S290 **T5085** L: CS21 W: CS14.3
Wooden box in the shape of a duck or similar sized bird with its head cut off. The case was covered, both inside and out with plaster and paint, and then the interior was painted with resin. This case was firmly closed, so one can only assume that it contains some kind of fowl. There is no detailing on the case to suggest wings or feet.

S291 **T5083** L: CS25 W: CS14.3
Wooden box in the shape of a duck or similar sized bird with its head cut off. The case was covered, both inside and out with plaster and paint, and then the interior was painted with resin or resin or resin. This case was firmly closed, so one can only assume that it contains some kind of fowl. There is no detailing on the case to suggest wings or feet.

S292 **T?** L: 22.5 W: 10
An indeterminate chunk of meat, most probably bovine. No signs of bandaging, though the other joints in this assemblage were bandaged. It is impossible to determine from which part of the animal they came. Large salt crystals are visible on much of the surface of this boned chunk of meat.

S293 **T5104** L: 28 W: 10.5
Six vertebrae (cervical or the beginning of thoracic), with bits of their transverse processes broken off. The vertebrae come from a bovid which is between two and three years old. The age is determined by the fusion of the epiphysis (Silver 1969). However one cannot be secure as only one of the ephiphysis is visible as the other five are covered by meat. The bones are not too big, so that would also suggest a younger though mature animal. The meat is cut off both sides of the spines of the vertebrae, much as it was in **S404**. The meat was removed in antiquity, and it is quite possible that it was consumed then. There are no bandages visible on the vertebrae, although the majority of the other members of this group of funeral offerings were wrapped.

S294 **T5105** **L**: 33.6 **W**: 15.2
Four half-ribs (costa to halfway down the shaft) from the right side of a young (about two years as the epiphyses are not completely fused) *Bos taurus*. These ribs are well wrapped in a coarse cloth which has been painted over by a resin-like substance. The bandages do not seem to be salted, while it is more than probable that the meat was.

S295 **T?** **L**: 40 **W**: 14
Three ribs from the left thoracic cavity of a young bovid. Some of the meat is gone from the bone, but it was all wrapped in cloth. Salt is visible on the bones. The entire ribs were wrapped and offered.

S296 **T?** **L**: 35 **W**: 24
Six ribs from the left side of the thoracic cavity of a small bovid. The ribs are wrapped and then the bandages are covered with resin. The meat shows salt crystals on its unwrapped portions.

S297 **T?** **L**: 38 **W**: 33
Seven or eight ribs (the number of ribs is difficult to determine as they are well wrapped and covered by meat) from the right side of the thoracic cavity of a young bovid. They are wrapped in bandages (some portions of which have come off) which were covered by resin. There are small areas (3-5 cms.) of the bandages which are covered by thicker smears of resin. Some of the exposed areas of the ribs show traces of resin on the bone and the meat.

S298 **T5106(?)** **L**: 37 **W**: 17
Right femur of a mature (but quite young) bovid. There is a lump of meat attached to it which might or might not actually belong to it. It is wrapped on, but it does not quite look as if it goes with the bone. The whole was wrapped in linen which was coated with resin.

S299 **T5106(?)** **L**: 26 **W**: 16.7
Right humerus of a bovid. It has several areas where salt is visible; the remainder is wrapped in bandages that were painted over with resin. There is quite a lot of meat on this joint, especially on the distal end.

S300 **T5106(?)** **L**: 37 **W**: 12
Left tibia of a mature bovid. There is meat only on the proximal end of the bone around and below the epiphysis. Then the meat was removed somewhat crudely. After that the entire joint (meat on the proximal end, with the remainder being bone) was wrapped in a bandage and painted over with resin.

S301 **T5106(?)** **L**: 37 **W**: 12
Left femur of a mature bovid. There was once some meat on the proximal end of the bone, but it has fallen off since excavation. The meat only covered half of the bone, or so itr seems, as the lower half of the bone is tightly wrapped and meat-less.

S302 **T5086** L: CS48.5 W: CS21
An empty wooden case, plastered and painted on the exterior and bituminized on the interior. The shape, an unequal oval, suggests that it might once have contained a humerus or other leg bone of a bovid.

S303 **T5087** L: CS53 W: CS18.5
A thick crescent shaped wooden box, plastered on the exterior and bituminized on the interior. The dowels and dowel-holes that kept the lid in place are well made. Small crumbled pieces of meat (?) lie at the bottom and sides of the case. Clearly the case once held a much larger joint, possibly a leg joint (femur or tibia).

S304 **T5078** L: CS54.4 W: CS26.8
A wooden case that is wider at one end than the other. The interior has been painted with resin and the exterior has been left plastered and painted.

S305 **T5079** L: CS45 W: CS15.7
An oval wooden case with the exterior plastered and painted, and its interior covered by resin. Through a small hole in the side of the case one can see that it contains a wrapped object, presumably meat, but the unfortunately the object is not identifiable.

S306 **T5077** L: CS93 W: CS34
A plastered wooden case in the shape of a foreleg. The dark dribbles along the opening of the case indicate that the interior was covered by resin. Unfortunately the case is shut so one cannot determine whether a whole leg was wrapped up and placed therein, or whether it was divided into pieces, each piece wrapped, and then all of them placed in the case. It is also possible that something entirely other was put in a case of this shape, although the museum record reports that the contents are a leg of beef.

S307 **T5081** L: CS31; 21 W: CS18; 8
This kidney-bean shaped wooden case is plastered and painted on the outside, with the inside being covered with plaster and coated with resin. It contains a very tightly wrapped piece of meat (like **S378**). It is clear that the resin was still warm when the victual mummy was put inside the case as parts of the mummy's bandages have adhered to it and a fingerprint is visible on it. The resin has seeped onto the mummy. It is probable that the out of case joints of meat once had cases and that they themselves were not painted with resin, but that the resin attached itself to the mummies from the case.

S308 **T5088** L: CS21; 14 W: CS14; 7.5
A wooden case in the shape of a bird without a head. The exterior is plastered and painted, and the interior was bituminized over the plaster. No details such as head, wings or feet are depicted on the box. The mummified bird in the case is quite small — a little too small for a duck. It has been beheaded, but one cannot tell whether or not its feet have been removed as it is tightly wrapped.

S309 T5099 No dimensions
Preserved duck. Not examined as lost.

S310 T5100 No dimensions
Preserved duck. Not examined as lost.

S311 T5101 No dimensions
Bovid ribs, now lost.

S312 T5102 No dimensions
Bovid ribs, now lost.

S313 T5103 No dimensions
Part of bovid vertebral column, now lost.

Discussion

Unfortunately one cannot give the exact museum number for each piece. Although a detailed study of all the available pieces was impossible due to technical difficulties with the museum cases, only the pieces of meat on display and not in their coffinets could be identified. At least one animal was sacrificed for burial. Based on the fact that all the limb bones were fused with their epiphysis, one can say that all the joints were from a three to four year old ox and seemed to have been preserved in the same way as were similar offerings found in other Eighteenth Dynasty tombs — they were preserved with salt as can be attested by salt crystals found on them and then wrapped in bandages some of which were also coated by salt. Then most of the offerings were placed in individual wooden coffinets whose interiors were coated with resin. In a few cases the pieces of meat without coffinets had traces of resin on their bandages, but a very insignificant amount.

Amenemhat Q [MMA1021: Tomb Field number 7A 1248]

This Eighteenth Dynasty prince's tomb was excavated in 1918-19 by the Metropolitan Museum of Art, New York. All eighty-one of the meat mummies found in the main chamber of this tomb were taken to the Metropolitan Museum where they remained until the 1950s when many of them were deaccessioned and scattered throughout North American collections. I have a complete list of the meat remains from this tomb and I have studied the victual mummies in virtually all the collections that purchased these items from the museum. However, sometimes I find myself unable to match the victual remains in the Boston Museum of

Fine Arts (**B** number), the Oriental Institute, Chicago (**O** number), the American Museum of Natural History, New York (**A** number), and the California Academy of Sciences (**CA** number), with the relevant Metropolitan Museum (**M** number) number; in such cases, I have listed the holdings from the other museums at the end with possible Metropolitan numbers in italics. In some entries I know the Metropolitan Museum number and to which museum the object had gone, but I do not have the relevant museum number. Where this situation exists, I have put the museum initial after the **M** number. If the object remains in the Metropolitan Museum then only the **M** number is supplied. Case measurements are preceded by CS.

Most of the avian mummies from this tomb held in the American Museum of Natural History were unwrapped and studied by Dr. S. M. Goodman (1987: 67-77) whose results are incorporated into the catalogue below, indicated by an * following the **A**. Unless otherwise stated the birds were butchered and prepared for mummification in the following way, hereinafter referred to as the 'standard manner': "feathers removed from the skin; the head, portion of neck, leg at the knee (femur/tibiotarsus joint) severed; and all major muscle masses, pygostyle, and oil glands left in place" (Goodman 1987: 68).

S314 **M19.3.212; A*** L: 34.5
Mummified goose, *Anser anser* (greylag goose) or *A. fabalis* (bean goose), wrapped, with one leg missing. The bird was prepared in the standard way. The heart had been left in the bird, although, when it was found the heart had come out of the body from a hole in the back of the bird. The gizzards with attached intestines and a ball of cloth were also present in the mummy.

S315 **M19.3.213** L: 35
A well wrapped mummified goose, probably *A. anser*. Probably like all the other birds in this collection of offerings it was butchered, plucked, and had its head, portion of neck, leg at the tibiotarsus joint severed. The presence or absence of the viscera is unknown.

S316 **M19.3.214/.215; A*** L: 31.5
Mummified *A. anser* or *A. fabalis* prepared in the standard manner. Only a small portion of the viscera was removed; the liver, gizzard, intestines, heart, lungs and kidneys all remained. The interior also contained a large ball of linen. A large clot of blood was also evident in the abdominal cavity.

S317 **M19.3.216z; A*** L: 27.5
Fragmentary wrapped remains of *A. anser* or *A. fabalis*.

S318 **M19.3.217** L: 30
Mummified *A. anser* or *A. fabalis* prepared in the standard manner with both legs missing.

S319 **M19.3.218**; **A*** L: 28.5
A. albifrons (white-fronted goose) or *A. erythropus* (less white-fronted goose) prepared in the standard manner. The viscera was removed from the body and then replaced with the addition of a ball of linen.

S320 **M19.3.219**; **A*** L: 39
A wrapped *A. albifrons* prepared in the standard manner with all the viscera removed and then replaced with the intestines, gizzard, liver, lungs and heart, along with a wad of linen.

S321 **M19.3.220**; **A*** L: 28.5
A. albifrons or *A. erythropus* prepared in the standard manner with the viscera partially removed and replaced. The heart and lungs were in situ, and and five attached cervicals and a large wad of cloth remain in the cavity. In this goose an incision (30 × 20mm) in the skin of the ventral surface for the removal of viscera was visible.

S322 **M19.3.221**; **B?** L: 28
Probably a wrapped *A. albifrons* prepared in the standard manner.

S323 **M19.3.222**; **O18275** L: 33 **W**:15.8
A mummified goose, *A. albifrons*; the neck is unusually long and the tibiae of both legs are missing. Prepared in the standard manner. It is probable that the box, M19.3.285, contained this goose.

S324 **M19.3.223**; **B?** L: 25
A. albifrons wrapped and prepared in the usual way. Both legs and one wing have fallen off.

S325 **M19.3.224**; **A*** L: 32
A goose, either *A. anser*, *A. fabalis*, or *A. albifrons*, wrapped and prepared in the standard manner with the viscera partially removed and replaced. The abdominal contents are the intestines, both gizzards, liver, heart and a potion of the aorta, and a large ball of linen cloth.

S326 **M19.3.225**; **A*** L: 29.5
Mummified *A. albifrons* prepared in the standard manner with the heart, lung, intestines and section of cervicals replaced in the abdominal cavity after removal.

S327 **M19.3.226**; **A*** L: 27.5
Wrapped *A. anser*, *A. fabalis*, or *A. albifrons* prepared in the standard manner with the viscera removed and partially replaced. The present viscera include the intestine, spleen, gizzards, portion of liver, heart and attached lungs, section of the cervicals and the usual ball of linen. A blood clot was found in the remaining portion of the oesophagus, some of the blood having dripped into the abdominal cavity. The outer bandages surrounding these areas were soaked through by blood.

S328 **M19.3.227**; **A*** **L**: 29.5
A. albifrons, wrapped and prepared in the standard manner with the viscera removed and partially replaced. The abdominal cavity contains the heart, lungs, section of trachea and a large wad of linen.

S329 **M19.3.228**; **A*** **L**: 28.2
An *A. albifrons* prepared in the standard manner with the viscera removed leaving only a section of cervicals and a large ball of cloth in the abdominal cavity.

S330 **M19.3.229**; **C389-2672** **L**: 18.0 **W**: 9.9 **Wt**: 257g
A duck, probably *A. acuta*.

S331 **M19.3.230**; **A*** **L**: 19
A pintail duck, *A. acuta*, prepared in the standard manner. The viscera was removed and four cervicals left attached to the trunk. The abdominal cavity contains the upper cervicals, several large blood clots, and a wad of linen. Some of the wrappings were bloody as blood was found at the distal portion of the neck and in the abdominal cavity.

S332 **M19.3.231**; **A*** **L**: 21
A. acuta prepared in the standard manner and completely eviscerated, although there are two blood soaked balls of cloth in the abdominal cavity.

S333 **M19.3.232**; **CA389-2673** **L**:16.9 **W**:10.5 **Wt**:148.7g
A small well wrapped water fowl, probably a pin-tailed duck. In some places the bandages are dark because of resin. The breast bone is clearly defined.

S334 **M19.3.233**; **A*** **L**: 33
A. albifrons, wrapped and prepared in the standard manner with the viscera partially removed and replaced. The abdominal cavity contains the intestines, lungs, kidney, gizzards, portion of the cervicals and three balls of linen.

S335 **M19.3.234**; **A** **L**: 14
A well wrapped small bird, perhaps a garganey, *Anas querquedula*, prepared in the usual manner.

S336 **M19.3.235**; **A*** **L**: 14
A garganey, *A. querquedula*, prepared in the standard manner with the viscera removed and partially replaced. The abdominal cavity contains the heart, portion of liver, gizzard and two balls of cloth.

S337 **M19.3.236**; **CA389-2675** **L**: 17.0 **W**:7.7 **Wt**: 160g
A very small and narrow bird, possibly a pigeon. It definitely had a case at one point as there is a thick coagulation of resin on the dorsal side.

S338 **M19.3.237**; **A*** **L**: 15
Wrapped and standardly prepared *A. querquedula* with the viscera removed and partially replaced with six cervicals left attached to trunk. The abdominal cavity

contains the heart, portion of the trachea, and a small ball of linen. A large clot of blood which had permeated the wrappings was found in the neck and throat area.

S339 **M19.3.238; A** **L**: 13.8
A small bird, perhaps a garganey, wrapped and prepared in the usual manner.

S340 **M19.3.239; B** **L**: 10
A wrapped and eviscerated pigeon or dove.

S341 **M19.3.240; B** **L**: 10
A wrapped and eviscerated pigeon or dove.

S342 **M19.3.241; A*** **L**: 10
A turtle-dove, *Streptopelia turtur*, wrapped and prepared in the standard manner with the visceral partially removed; the heart and lungs are *in situ* and a small ball of cloth is added.

S343 **M19.3.242; A*** **L**: 10
Mummified *S. turtur* prepared in the standard manner with the viscera partially removed and four cervicals left attached to the trunk. The abdominal cavity contains a ball of cloth and the lungs and kidneys *in situ*.

S344 **M19.3.243; A*** **L**: 9
A *S. turtur* that is well wrapped and prepared in the standard manner with the viscera partially removed and replaced. The surface bandages were heavily coated and discoloured with resin. The abdominal cavity contains the gizzard, heart, lungs and kidney *in situ*, and a ball of linen.

S345 **M19.3.273; CA389-2688** **L**: 9 **W**: 4.6 **Wt:** 51.4
Completely wrapped in fine linen, possibly a mummified bird.

S346 **M19.3.275** **L**: 33
A wooden case in the shape of a headless duck that is stuccoed white. The two halves are pegged together and the resin coating the interior of the case has sealed it. The two halves are bound together with diagonal crossed bandages. The container probably contains a duck.

S347 **M19.3.276** **L**: 15
A small stuccoed wooden case representing a headless bird, probably a garganey. The interior coating of resin has spilled out between the two halves of the coffinet that are pegged together. It is further secured by having bandages wrapped around the middle of the case.

S348 **M19.3.280** **L**: CS51 **W**: CS21.3
Half a plastered bird shaped wooden case with the inside resined. It contains a well wrapped goose, possibly *A. anser*, whose feet and wing-tips were removed.

S349 **M19.3.281** L: CS41.5 W: CS20.5
A wooden box in the shape of a headless bird with the legs and wings carved out in both the lid and the body. The box is stuccoed both inside and out, with the interior surface covered by a layer of resin. It probably contains a goose.

S350 **M19.3.282** L: CS40 W: CS19
Half of a wooden case in the shape of a bird; it is plastered and the internal surface is painted with resin. Probably for a goose, *A. anser*.

S351 **M19.3.283a,b; B37.553** L: CS36 W: CS17
A wooden box in the shape of a headless bird with the legs carved out in both the lid and the body. The box is stuccoed both inside and out, with the interior surface covered by a layer of resin. The case is closed and the top and bottom halves are transversely bound together by a thin strip of linen bandage. Radiographs show that it contains a duck which was prepared in the standard manner (Figs. 74-75).

S352 **M19.3.284a** L: CS32.7 W: CS15.5
The lid of a wooden box shaped like a headless bird containing a mummified duck.

S353 **M19.3.285; O18275** L: CS33.5 W: CS16
Part of one half of a stuccoed wooden case in the shape of a bird with the interior painted with resin. The legs are broken off. It is quite probable that the goose, **S323** was stored in this case.

S354 **M19.3.286a,b** L: CS32 W: CS16.5
A wooden box in the shape of a headless bird with the legs carved out in both the lid and the body. The box is stuccoed both inside and out, with the interior surface covered by a layer of resin. The case is closed and the top and bottom halves are transversely bound together by a thin strip of linen bandage. It probably contains a duck.

S355 **M19.3.287a,b; CA389-2674** L: CS30.5 W: CS15
The upper half of a duck shaped wooden case. This lid was plastered both inside and out, and then the inside was covered with resin.

S356 **M19.3.288** L: CS32.5 W: CS15
Half of a duck shaped stuccoed wooden case, resin-covered on the inside.

S357 **M19.3.289a,b** L: CS21 W: CS11
Wooden box containing the remains of a quail, pigeon or dove. The box is plastered both inside and out, with the internal surface being painted with resin. Both the lid and the body of the case is carved in the shape of a headless bird with the legs sticking out at one end.

S358 **M19.3.290a,b; B37.551** L: CS18.5 W: CS10.5
Wooden box in the shape of a small bird, plastered both inside and out, with the internal surface being painted with resin. The box contains a wrapped quail,

pigeon or dove on whose neck and wing are smears of resin from the box. As is usual with these boxes the lid is kept on by pegs fitting into holes.

S359 **M19.3.291a,b; B37.552** **L**: CS18.8 **W**: CS9.5
A headless bird shaped wooden box containing the tightly wrapped remains of a quail, pigeon or dove. The box is plastered both inside and out, with the internal surface being painted with resin.

S360 **M19.3.292; A95-3163a-c**
The right foreleg, including: scapula, humerus, with the radius, ulna, carpals, metacarpals, and first and second phalanges all joined by cartilage, with a third phalange separate. The scapula has bits of naturally dried meat and some hide on it. The broad end of the scapula has bits of salt clinging to it. No knife cuts are visible on it. However there are some cut marks on the frontal surface of the radius indictating that the meat had been removed. None of the other bones has butchery marks on it, yet almost all of the meat has been removed. What is especially curious is that the bones are still articulated as they are held together by cartilage and that the metacarpals and phalanges are included in the offering. They are never seen independently in offerings of this sort.

S361 **M19.3.247** **L**: 35
This wrapped package was x-rayed and contains the right scapula of a mature bovid with meat still on the bone (Figs. 34-36). The entire package is in a wooden case with the interior surface covered by resin, **S362**.

S362 **M19.3.278** **L**: 46.5 **W**: 30
A large wooden case stuccoed white and bituminized inside containing a well wrapped package of meat (**S361**).

S363 **M19.3.246; A95-3160A** **L**: 32.1 **W**:12.4 **Th**: 6.3 **Wt**: 2kg+
Humerus with meat attached to it. The animal is fairly young as its proximal epiphysis is unfused, thereby making it under three and a half years old. The meat from this joint is very black — more so than that of any previously studied joints. It is heavily salted and the salt has even gotten into the bone where it is unfused. There are no obvious marks of jointing on the exposed proximal end of the bone. Parts of it are very heavily encrusted with resin, and are responsible for its dark colour. One cannot be sure if all the meat from the humerus is still attached to the bone. Pieces of what seem to be hardened clay (maybe mixed with resin in some places) are attached to some portions of the bandages.

S364 **M19.3.244; A?** **L**: 91
A wrapped hind leg of a bovid with the hoof cut off. Unfortunately the location of this piece is uncertain.

S365 **M19.3.245; CA389-2676** **L**: 44 **W**: 15
The x-rays show that this cut consists of the tibia, astragalus, tarsals, and metatarsals. The meat is salty, though the bandages are not. This is similar to **S360**. Dermestid beetles have grazed on the cartilage and parts of the meat.

S366 **M19.3.248**; **A?** **L**: 45
Wrapped ribs from a bovid. Not examined.

S367 **M19.3.249**; **A?** **L**: 36
Supposedly a cut of four ribs whose bandages are dilapidated. Not examined.

S368 **M19.3.274**; **A95-3159** **L**: 23.5 **W**: 15 **Th** 4.4 **Wt**: c.2700
Mummified liver of a bovid with much resin on top of the bandages, colouring them black in places. Microscopic examination of the liver shows that it was cut out of the body, immersed in salt, and then wrapped.

S369 **M19.3.250**; **A95-3162a** **L**:16.2 **W**:7.5 **Th**: 1.8 **Wt** 176.5g
A slab of meat, possibly steak from the rump of the cow. Some chunks of the meat are missing. Along one side of the meat runs a 1.2mm band of fat. Salt crystals and powder along with bits of resin are visible on the meat.

S370 **M19.3.267**; **A95-3162b** **L**: 14.4 **W**: 7.5 **Th**: 1.6 **Wt**: 184.5
Perhaps a piece of steak from the rump or the flank of a cow. Dark resin shows under the microscope. A piece is broken off on top.

S371 **M19.3.251**; **A95-3162c** **L**: 13.5 **W**: 8.4 **Th**: 1.7 **Wt**:173.5g
Perhaps a piece of steak from the rump or the flank of a cow. Top and bottom chunks of the piece are missing. The meat is covered by salt and resinous bandages.

S372 **M19.3.254**; **A95-3162d** **L**:15.5 **W**: 7 **Th:** 1.9 **Wt**: 146.5g
Perhaps a piece of steak from the rump or the flank of a cow; most of the blackened wrapping and one tip of the meat have vanished.

S373 **M19.3.257**; **A95-3162e,f**
Fragments that do not fit any of the other members of this group. They are both wrapped and salty. One weighs 48.3g and is half a steak piece.

S374 **M19.3.260**; **A95-3161a** **L**: 18.4 **W**:11.3 **Wt**: 259.0g
A large piece of boneless meat the whole of which was once bandaged; now only portions of the bandage remain. Part of the meat has broken off and in the break the meat has a woody texture and looks as if it might have been eaten by dermestid beetles. There are definite resinous deposits on the meat. The bandages are very dark from the resin. It is possible that the resin comes from boxes, although there is no evidence for boxes for meat in this tomb. Furthermore, the resin-coating of the bandages is much more even than seen on other offerings, thereby suggesting that they were actually painted over with resin as a sealant for preservation. Under the microscope the areas not covered with resin show a white powder tasting and looking like natron (a bag of natron from Lisht [AMNH95-2809] in the collection was studied under the microscope and tasted for comparison) or fine salt.

S375 **M19.3.263**; **A95-3161b** **L**: 15.7 **W**:7.2 **Wt**: 200.0g
A lumpy irregular piece of meat, possibly brisket or rump, although one cannot be sure. Most of the wrappings have fallen away. Large salt crystals are obvious due to their size; smaller powdery grains are visible under the microscope.

S376 **M19.3.262**; **A95-3161c** **L**: 15.4 **W**: 4.2 **Th**: 4.3 **Wt**: 146g
A tightly wrapped almost crescent shaped piece of meat. No bones are visible and the wrappings are quite dark possibly due to resin coating, although no resin was apparent under the microscope.

S377 **M19.3.252**; **A95-3161d** **L**: 15.5 **W**: 9 **Th**: 2.6 **Wt**:132.5g
No bone is visible anywhere in this victual mummy. This piece might be from the shoulder, but this is impossible to determine. A corner of it has been cleanly cut with a chopper. Most of the slab of meat is covered by dark bandages and salt crystals.

S378 **M19.3.265**; **A95-3161e** **L**: 21 **W**: 8.5 **Th**: 5.4 **Wt**: 318.5g
X-rays and CAT-scans reveal this to be a long boneless piece of meat that is coiled around itself before being well wrapped and covered by resin. The length of the piece of meat suggests that it is from the flank.

S379 **M19.3.266,A95-3161f** **L**: 18.9 **W**: 7.8 **Th**:5.6 **Wt**: 271.5g
A boneless chunk of meat that is well wrapped and covered by resin. There are holes in the meat bored by dermestid (?) beetles. Driblets of resin have made a few very dense, dark splotches on the wrappings. Of course, they could be unrelated to the processing of these meats and have fallen on them by accident, although given the presence of resin on other offerings from this assemblage that is fairly unlikely.

S380 **M19.3.264**; **CA389-2682** **L**: 16.5 **W**: 10 **Wt**: 194.4g
A lumpy piece of meat that is quite well wrapped and therefore cannot be identified.

S381 **M19.3.256**; **A?** **L**: 8 **W**: 8
A small wrapped piece of steak. Not seen.

S382 **M?**; **A95-2455a** **L**: 12.7 **W**: 11 **Th**: 2.5 **Wt**: 154.0g
A wrapped boneless victual offering purchased 1958, provenance unknown, although there is a reference to a MMA collection which leads one to think that it might be from this tomb. The bandages are coloured blackish grey.

S383 **M19.3.253**; **CA389-2677** **L**: 17 **W**: 8 **Wt**: 237.6g
A steak-like piece of meat that is unwrapped and covered with salt.

S384 **M19.3.272**; **CA389-2687** **L**: 14.4 **W**: 5.7 **Wt**: 208g.
Very well wrapped up piece of meat. It is long and thin, and x-rays show it to very similar to **S378**, a piece of flank steak that was coiled around itself.

S385 **M19.3.270**; **CA389-2685** **L**: 19 **W**: 7.9 **Wt**: 208.2g
An extremely tightly wrapped up piece of meat. It is long and thin and could possibly be, like the similar piece x-rayed in New York, **S378**.

S386 **M19.3.258**; **CA389-2679** **L**: 19.8 **W**: 14.6 **Wt**: 465g
Bits of resin on the top and side of this steak. It is probably from the box which is now lost.

S387 **M19.3.259**; **CA389-2680** **L**: 21.8 **W**: 12 **Wt**:377.2g
Dermestid beetle activity has made some holes in this steaky piece. Small chunks of resin cling to the bandages and in places have even touched the meat.

S388 **M19.3.268**; **CA389-2683** **L**: 26 **W**: 11 **Wt**: 703.4g
This piece is unusually heavy for its dimensions. It is well wrapped in coarse linen with the bandages missing in places. It was salty, the bandages were not.

S389 **M19.3.271**; **CA389-2686** **L**: 17 **W**: 6 **Wt**: 119.4g
This piece is quite small, long and knobbly, and is wrapped in extremely fine linen.

S390 **M19.3.261**; **CA389-2681** **L**: 12.2 **W**:9.7 **Wt**: 150.6g
This piece is broken off from a larger piece, the rest of which is not in this collection. The bandages are dark and deeply impregnated by resin.

S391 **M19.3.269**; **CA389-2684** **L**: 23 **W**: 8.4 **Wt**: 403.2g
An unidentified piece of bovine flesh well wrapped in fine linen almost of the same quality as that of **S389**.

S392 **M19.3.255**; **CA389-2678** **L**: 11.7 **W**: 5 **Wt**: 53.5
This relatively small piece of meat is wrapped in very dark bandages of coarse linen. The bandages are not salty.

S393 **M19.3.277**; **CA389-2689** **L**: 20.8,CS28.8 **W**: 7.8,CS16.1
A large and very desiccated piece of meat stuck into half of the bottom portion of a wooden case. The bandages are orangey and coarse. The meat is marginally salty — far less so than other joints I have tasted. Some portions of the bandages are salty too. The meat is extremely dry and has shrivelled away from the bandages. The interior of the case is coated with black stuff which microscopic study shows to be resin. In the bottom of the box, some of the resin has permeated through the bandages causing the meat piece to stick to the case. The exterior of the case shows evidence of plaster and maybe white paint. The dowels and dowel holes are visible.

S394 **M19.3.279** **L**: CS39 **W**: CS23
Half of a rectangular wooden case with rounded corners. It is plastered and the internal surface is painted with resin. This case was probably used to hold a beef offering.

Discussion:

About twenty geese, thirteen ducks, nine pigeons/doves, twenty-six pieces of meat, two sets of ribs, one liver, one humerus, two scapulae, one foreleg and two hindlegs were offered. Meat offerings come from one cow, while the not inconsiderable remainder of the offerings are avian. There is considerable variety in the quality of linen used — perhaps this is due to the fact that all sorts of different discarded linen was used for mummification.

Queen Meryetamun [TT358]

This tomb in Deir el-Baheri was excavated by H. E. Winlock for the Metropolitan Museum of Art (MMA) in 1928-29. The tomb belongs to an Eighteenth Dynasty queen, probably the wife of Amenhotep I, although there has been considerable debate as to her dating; for architectural evidence of her early date, see Wysocki 1984. It is a rich source of victual mummies containing over twenty-five victual mummies. The tomb was disturbed and restored in antiquity so that when Winlock found it "the embalmed meat was simply thrown out in the rubbish in the corridor or swept up in a basket in which it scarcely could have belonged" (Winlock 1932: 28). The meat was simply wrapped up in bandages and placed in the tomb — no joint shaped wooden coffinets were found save one duck-shaped box. The bandages of the victual mummies are all stained a dirty gray or black by the preserving materials used on them, while three of them, supposedly hearts, are wrapped in cloths stained a bright orange color. Winlock writes that "it seems likely that most, if not all, of the meat was put into rectangular boxes, of which the tomb contained at least two and perhaps more as we found one lidless box, the lid of an exactly similar box, and a seal knob perhaps from still a third. Apparently three boxes of the size of the one found (L: 44, W: 25.5, Ht: 22, Th: 1.5; JE55152-3 [Winlock 1932: 73]) would have held all the meat discovered in the tomb and would have gone excellently with the duck-shaped case. Like the latter they were made of sycamore wood, were coated with a thick white stucco inside and out, and thickly smeared with black resin inside. While this resin was still soft, something wrapped in linen, like the embalmed meat, had been put in them. They had then been closed, with the result that the linen had stuck to the bottoms, and the lids had stuck to the boxes" (Winlock 1932: 28-9) much as it had in joint shaped cases from other burials. When the tomb was 'restored' in antiquity many of the mummies were placed in a clothes basket (MMA30.3.17), while others were replaced in the one intact wooden box that was found. One or two remained scattered through the entrance corridor of the tomb.

Most of these mummies are now in the American Museum of Natural History (AMNH), their accession numbers being here prefixed with an **A**. Unfortunately there are no specific **M** numbers to attach to the **A** numbers as the MMA did not number the individual pieces of meat, but treated them as a group assigning them the number 30.3.22. In the **A** number sequence I have given each mummy a letter when they are just given a general number as none of the pieces are individually labelled.

S395 **A95-2900a** **L**: 12.3 **W**: 10 **Wt**: 169.5g
A very tightly wrapped, quite dense, very smelly, piece of meat. The bandages are dark brown and yellowy.

S396 **A95-2900b** **L**: 11.7 **W**:10.3 **Wt**: 171.2g
A rectangular piece of meat wrapped in yellowy and black bandages. X-rays showed it to be boneless.

S397 **A95-2900c** **L**: 13.2 . **W**: 6.1 **Wt**: 97.5g
A steaky chunk of meat whose wrapping looks as if it has been covered with a dark resinous material. The meat was cut with a fairly blunt knife as the cut surface, part of which is visible, shows signs of jagged cutting.

S398 **A95-2900d** **L**: 12.7 **W**: 6.4 **Wt**: 119.1g
A chunk of meat covered by dark bandages. It has whitish spots that, under the microscope, look like salt.

S399 **A95-2900e** **L**: 12.5 **W**: 8.2 . **Wt**: 118.6g
A very salty piece of meat that seems quite hard all the way through. The meat itself is blackish and yellowish, much like the bandages. This could be the result of time, resin, or/and salt.

S400 **A95-2900f** **L**: 11.5 **W**: 7.4 **Wt**: 89.8g
A chunk of meat that is wrapped in bandages. The meat is less salty. It is lighter than other similar sized pieces of meat, perhaps due to less salt?

S401 **A95-2900g** **L**: 12.3 **W**: 8.1 **Th:** 2.5 **Wt**: 244.0g
Exposed sections of this wrapped piece of meat look as if they are covered by salt crystals. The meat does indeed taste salty. As with the other similar pieces of meat the cut is impossible to determine, although it probably comes from the shoulder or the rump. X-rays and CAT-scans show that it is boneless. Its heaviness is extraordinary and inexplicable as it is boneless. Perhaps the weight comes from resin?

S402 **A95-2900h** **L**: 12.6 **W**: 8.9 **Wt**: 100.4g
This piece of meat is partially unwrapped. At one end a clean cut all the way through the piece of meat made with a sharp knife is visible. This cross section shows that the outer edges of the meat are darker than the centre. This could be due to cooking or a level of impregnation of liquid. Under the microscope it definitely seems covered by sprinklings of a white powder which is quite possibly salt/natron crystals. Tastes salty.

S403 **A95-2900i** **D**: 13.5 **Wt**: 226.5g
This almost circular piece of meat could be from anywhere at all. It is small, but reminiscent of onomastica meats ending in a circle, or the Giza and Sakkara

round limestone cases which might have been used for meat interment. It is well wrapped, and x-rays showed it to be boneless (Figs. 31-33).

S404 **A95-2900j** **L**: 31.8 **W**: 26.3 **Th**: 9.8 **Wt**: Over 2kg.
Winlock called this the shoulder of beef. X-rays reveal that this package actually contains six thoracic vertebrae of a full-grown bovid (Figs. 31-33).

S405 **A95-2900k** **L**: 14.5 **W**: 8.6 **Th**: 3.0 **Wt**: 181.99g
This piece is completely unwrapped and appears to be boneless. It is covered by a thin encrustation of dark resin and salt. A smooth cut shows the area where this piece was cut from a larger piece, possibly the rump. Under the microscope one could see that the entire piece was covered by white dust and white crystals as well as some amber coloured crystals which doubtless are resin. It is unusually heavy for its size.

S406 **A95-2900l** **L**: 11.1 **W**: 6.5 **Th**: 2.7 **Wt**: 99.1g
This piece apparently contains no bone and is completely and tightly wrapped.

S407 **A95-2900m** **L**: 11 **W**: 5.2 **Th**: 2.4 **Wt**: 93.4g
This piece is like **S406**, but partially unwrapped. In a cross section the 'outer layer' of meat is darker than the inner. Salt powder and crystals abound, as does resin.

S408 **A95-2900n** **L**: 11.3 **W**: 8.8 **Th**: 1.9 **Wt**: 123.8g
Partially unwrapped boneless piece of meat. Could be from the rump or shoulder Both salt and resin present.

S409 **A95-2900o** **D**: 11.8
Like **S403**. It is extremely tightly wrapped in its dark yellowish-brown bandages.

S410 **A95-2900p** **L**: 11.2 **W**: 8.17 **Th**: 2.6 **Wt**: 113.5g
Tightly wrapped in bandages that are yellow and black and heavily impregnated by resin. The bandage does not taste at all salty, so it is probable that the meat was coated with salt rather than the bandages. X-rays show the meat to be boneless so it is impossible to determine from which part of the animal it originated.

S411 **A95-2900q** **L**: 13.4 **W**: 9.3 **Th**: 2.2 **Wt**: 117.0g
Like **S410**, tightly wrapped and very flat. Unsalty bandages.

S412 **A95-2900r** **L**: 10.8 **W**: 9.5 **Th**: 2.2 **Wt**: 85.9g
Avery badly preserved piece of meat which is extremely brittle. Salt and resin is present on the meat, with a preponderance of resin on the the bandages. The meat was cut with a sharp knife.

S413 **A95-2900s** **L**: 11.2 **W**:8.1 **Th**: 2.3 **Wt**: 151.1g
This piece is completely and tightly wrapped in yellowy orange and black bandages which are not salty; the boneless meat pokes out in one place and is salty. This is one of the so-called beef hearts of Winlock.

The AMNH only definitely records having nineteen pieces of meat from this tomb. However, there are a few other pieces of wrapped meat in the collection that are unprovenanced and purchased from the MMA at the same time as the Meryetamun joints; it is quite possible that they are the missing pieces that somehow got separated from the collection at the time of transfer between the two museums. The other numbers that I believe belong to this collection are:

S414 **A95-2455b** **L**: 15 **W**: 8.2 **Th**: 2.6 **Wt**: 120.0g
A chunk of meat wrapped in black and orange bandages, possibly one of Winlock's hearts (Fig. 33)?

S415 **A95-3165a-d**
A registrar entry listing wrapped food offerings with no information as to their provenance except the fact that they were g Gift from the MMA in 1965. This could belong to either Meryetamun or Prince Amenemhet. Only 'a' was examined as the remainder of the pieces cannot be found.

S416 **A95-3165a** **L**: 17 **W**: 5.6 **Th**: 2.6 **Wt**: 180.0g
A very well wrapped piece of meat. This is not in neat bandages, but in large swatches of cloth deeply impregnated by resin and salt. The cloth sports a bright orange hue on most of its surface — perhaps this is the third heart that Winlock mentions.

S417 **M30.3.21a,b** **L**: 23.5 **W**: 14.5
A sycamore box in the shape of duck covered with white stucco outside and in, and covered with resin on the inside. A leg of a duck rather than an entire duck is associated with this coffinet.

Discussion

This tomb provides a variant on the joint shaped coffinets found in other tombs and suggests that some of the other boxless meat offerings really were not boxless, rather that they were stored together in a square or rectangular box rather than one mimicking the shape of the joint buried therein. All the meat offered came from one cow. There would not have been too much left-over meat as much of the cow's choicest and most meaty pieces were interred.

Eighteenth Dynasty Female Burial [KV60]

A private tomb first cleared by Carter and reported to contain some mummified geese (Carter 1903: 176). It was subsequently cleared by Dr. D. Ryan who found, throughout the tomb, and in a side chamber (B-A), several joints of mummified meat in addition to the mummified geese (Ryan 1989; 1990). The information below is based upon photographs and notes kindly provided by Dr. Ryan. The number and letter sequence following the **S** number is the excavation number.

S418 **60-89-C21-B60** **L**: 32 **W**: 13
Mummified meat fragment.

S419 **60-89-B-O** **no dimensions**
Avian neck vertebrae.

S420 **60-89-B-2** **no dimensions**
Two mummified avian tibiae and femurs.

S420 **60-89-B/BA** **no dimensions**
Mummified avian vertebrae found within **S426**.

S421 **60-89-C4-B14** **L**: 40 **W**: 22
Well wrapped joint that is probably between four and six of the thoracic vertebrae of an ox as it is very similar in shape to **S404**.

S422 **60-89-B-A-117** **L**: 92
Mummified right foreleg of an ox under three and a half years old. The limb consists of the scapula, humerus, radius, and ulna.

S423 **60-89-C4-B15** **L**: 40 **W**: 14
Mummified joint, possibly the humerus.

S424 **60-89-C4-B16** **L**: 32 **W**: 17
A well wrapped joint, possibly from the leg. The way it is wrapped is reminiscent of the shapes of the limestone boxes of meat, notably **N2**.

S425 **60-89-C4-B17** **L**: 60 **W**: 22
A very well wrapped victual mummy. The size and shape suggest that it is part of a leg.

S426 **60-89-B/BA-1** **L**: 58 **W**: 42
Roundish joint consisting of at least six ribs.

S427 **60-89-C3-B12** **L**: 34 **W**: 13
Very well wrapped piece of meat. Might contain a bone.

S428 **60-89-C21-B53** **L**: 10 **W**: 5
Very well wrapped piece of meat.

S429 **60-89-B3-F1-1** **L**: 16 **W**: 8
Very well wrapped piece of meat.

S430 **60-89-B3-F1-2** **L**: 28 **W**: 6
Very well wrapped piece of meat similar to **S378**.

S431 **60-89-C9-B-31** **L:** 23 **W:** 8
Very well wrapped piece of meat similar to **S378**.

S432 **60-89-B/BA-2** **L:** 26 **W:** 10
Very well wrapped piece of meat.

Discussion

At least three birds and one steer were offered. It is possible that some of the mummified joints have been lost, or that only half a carcass was given as an offering. Were the remainder eaten by the mourners, or was only half an animal purchased? Alas, one can only speculate.

Queen Isit [QV51]

In 1905, Schiaparelli excavated the tomb of the mother of Ramesses VI, and found that it contained mummified birds as food offerings (Lortet & Gaillard 1908: 145). They all had their heads, necks, tarsals and carpals removed. The abdominal cavity contained the liver, gizzard(s), and heart. The birds were wrapped in linen bandages and then placed in tear shaped sycamore boxes, almost identical to those of Maherpri.

S433 **no dimensions**
Goose or duck, prepared in the manner described above.

S434 **no dimensions**
Identical to **S433**.

S435 **no dimensions**
Identical to **S433**.

S436 **no dimensions**
Identical to **S433**.

Part II: Unwrapped Funerary Offerings

R1 **Seti I [KV 17]**
According to Belzoni, the discoverer of this tomb, the left side room off the burial chamber contained a bull carcass "embalmed with asphaltum" (Belzoni 1821: 235). It is unknown if this was a later burial or not. There is very little information on it apart from Belzoni's report; unfortunately he did not pay too much attention to such ephemeral and base items as long dead meat products, so with the given evidence it is difficult to determine if this bull was an offering of funerary meat or whether this was the burial of a sacred and revered animal.

R2 Amenophis III [WV22]

"Boxes containing 209 well preserved alimentary offerings, of which about 200 from young cattle or calves, were found in the tombs of Tuthmosis III and Amenhotep III" (Ruffer 1919: 11). However, there is no evidence regarding this statement in the publications dealing with the tombs and Ruffer gives no footnote for this information. Subsequent literature concerning the burial, including Reeves (1990: 323-5), who lists everything that Carter found, does not allude to this find, so it seems very likely that Ruffer made a mistake, or that this was a misprint for Amenhotep II. That said, new excavations in the tomb of Amenophis III by Jiro Kondo of Waseda University, Tokyo, have revealed a joint of meat in the foundation deposit of the tomb (personal communication 1991). However, this joint is yet to be examined and identified. Perhaps further investigation of the tomb will reveal some mummified joints of meat belonging to Amenophis III.

R3 KV51

An Eighteenth dynasty tomb was excavated that contained three mummified ducks (Davis 1908: 18). There were other mummified animals therein, including an ibis and a baboon, so it is unclear whether or not the ducks were victual mummies or not, although I know of no sacred mummified ducks. Unfortunately there are no photographs of these mummies in the publication and their current location is unknown, so it is difficult to tell whether they were food offerings or not.

Mentuhotep II

Food offerings were found by Carter inside the door of the king's cenotaph at Deir el-Bahari, the Bab el-Hosan (Carter 1901: 201-2). The present location of these objects is unknown, and consequently they could not be examined.

R4 no dimensions

On the floor of the passage was the head of a calf.

R5 no dimensions

Unidentified portions of the leg, also found in the passage.

R6 no dimensions

A large chamber containing "two forelegs of a calf which still had on them the dried up flesh, were lying on the floor" (Carter 1901: 201-2).

R7 no dimensions

Skeletons of two ducks, also found in the large chamber.

Discussion:

It is notable that the head of an animal was offered. The only other example of an head being offered is in the Lady Isitemkheb D's burial.

R8 Princess Nubhoteptikhered [Dahshur L.LVIII Tomb 2]
This mid-Thirteenth Dynasty burial in a reused shaft in the enclosure of Amenemhet III is one of the few Middle Kingdom burials where there is mention of mummified meat offerings. The tomb lies next to that of King Hor (de Morgan 1895: 107). The tomb was untouched, and found next to the eastern wall of the burial chamber, just north of the coffin, were offerings of mummified poultry and joints of beef (de Morgan 1895: 108). Alas, as the whereabouts of these offerings is unknown and there is no photograph of them, comparisons between the Middle and New Kingdom practices of presenting food offerings cannot be made.

Princess Ita [Dahshur, enclosure of L.LI]

This Middle Kingdom burial of Amenemhet II's daughter (see Dodson 1988: 130-1 for dating) included an offering chamber adjoining the burial chamber. The offering chamber contained a shallow bowl holding the bones of several water fowl of all sizes. Scattered about the floor in the doorway between the burial and offering chambers were bones of a bovid (de Morgan 1903: 48). Unfortunately the excavators did not identify these offerings, and the analyses are based on the photographs and drawings of the tomb.

R9 no dimensions
Two humeri of a bovid.

R10 no dimensions
A few ribs.

Princess Khnumet [Dahshur, enclosure of L.LI]

This burial is identical to that of Princess Ita. However in this burial none of the bird and bovid bones are placed in bowls, instead they are scattered on the ground between the doorways of the two chambers (de Morgan 1903: 50).

R11 no dimensions
From the drawing it seems that a humerus was one of the bones offered.

Senebtisi [Lisht, enclosure of L.LX]

This Twelfth Dynasty (see Dodson 1988: 130-1 for references to dating) burial at Lisht also includes food offerings. In the south corner of the tomb were two shallow bowls containing food offerings. Identifications are based on the photographs in the publication (Mace and Winlock 1916).

R12 no dimensions
Bowl containing two whole waterfowl as well as some fragmentary bones from ducks as well as doves or pigeons.

R13 no dimensions
Bowl containing one scapula and two humeri from adult bovids.

R14 Ashayet [DBXI.17]
This Middle Kingdom burial of one of Mentuhotep II's ladies contained bones from an immature bovid, some of which were in pots, others of which were scattered on the floor of the tomb (Winlock 1942: 39,73).

Tomb of Wah [MMA 1102]

Wah, a retainer of Meketre's, was buried in the courtyard of the tomb of his master (TT280) at Thebes. One meat offering was found in his tomb. This is a rare example of a foreleg, which was, ironically, in representations, the traditional offering joint.

R15 M20.3.258 a-d no dimensions
Right foreleg of cow, from scapula to phalanges, including hoof. Never preserved as the presence of hoof and skin attest. The dry desert climate has caused the meat to dry out on the bone.

Kha [TT8]

This Eighteenth Dynasty tomb in Deir el-Medina, datable to the reign of Amenophis III on the basis of its equipment, was excavated by Schiaparelli. Although no mummified food remains were found several non-mummified bovine, piscian and avian offerings were found in the tomb. These are listed below and were examined by the author in Turin.

R16 T? no dimensions
A bowl labelled "pieces of meat". On closer inspection the bowl's contents turned out to be a mixture of edible objects, including pieces of salted (?) birds (the bones provided the necessary identification). The birds were most probably doves or pigeons. The bits were disarticulated joints as well as portions of the breasts and legs and wings of the birds. Some of the objects in the tray might be pieces of meat of an uneven thickness and are quite black or dark brown and are heavily encrusted with dust, some plant fibers, small round ball-like accretions and in many places, large salt (?) crystals of a rhomboidal shape. These cannot be identified securely as meat unless chemically tested. The only firm identification of the objects on the tray is that of the birds. Schiaparelli in his publication does securely identify pieces of meat mixed with and almost immersed in a large quantity of solid salt as the contents of a large amphora (Schiaparelli 1927: 159). However the amphora containing the meat was not studied in the

Turin museum; these fragments of meat on a tray seem to be the only meat pieces presumably from it that are readily available.

R17 **T8354** **L**: 9-13
A bowl containing pieces of dried fish that were originally found scattered amongst the bread on the tables found in the tomb. There are no bones visible, and most pieces have had the scales removed. The fish was definitely sun-dried. Without chemical analysis one cannot determine whether or not it was salted prior to drying.

R18 **T8251** **L**: 9-12 **W**: 9-11
Birds of different types were placed in this bowl. All are quite small, and some are fragmentary. Two feet are present. The majority of the birds (those that have some bones and are therefore identifiable) seem to be some type of dove or pigeon. One of them is almost complete — it has been prepared for preservation by a ventral slit, the removal of any internal organs, and has been opened up. This is not the usual method by which the Egyptians spatch-cocked birds. All the other examples show the birds to be spatch-cocked where their back bones have been cut, rather than vice versa. This bird has its vertebrae in place. A large salt crystal is visible on this bird, while there are other smaller crystals visible on many of the other birds, thereby indicating that salt was used for their preservation.

R19 **T8526** **H:** 58 **D**: 11.8 (mouth); 30 (body)
A large amphora with a lid made of a plate held down by a circular piece of clay decorated with an eight petalled flower. It contains desiccated and salted birds (Fig. 47). The birds still have their feet (in all the visible instances). They seem to be opened and emptied of any internal organs. Unfortunately it was impossible to closely examine the amphora and its contents. However, one could determine that the birds had been beheaded, eviscerated and salted. It is probable that the birds had been placed in the vessel after being dried and salted. It is doubtful that they were put in brine of any kind. Again it is difficult to prove this without testing the material of the vessel. It is interesting to note that the other identically shaped amphorae found in the tomb of Kha were all labelled as containing different items. Only three (including this one containing the birds) have had their contents examined. The other two amphorae were labelled as wine jars. One of these actually contains the lees of a red wine while the other contains a dense yellowy vegetal material which has yet to be identified (Valtz 1991).

R20 **Deir el-Medina Funerary Offerings**
Bruyère reports that many tombs contained food offerings. Unfortunately he was not overly scrupulous in recording them. He mentions pieces of meat as well as unidentified joints. Unfortunately very few tombs have specific mention of meat offerings. One sepulchre which he specifically mentions is tomb 1318, which contained three pieces of meat (Bruyere 1937: 181-82).

Sakkara, tomb 3477

This Archaic tomb contained the remains of an entire funerary meal (Emery 1962: 6). Currently the whereabouts of the meal are unknown, although they might be stored in a magazine at Sakkara. The **E** numbers following the **R** numbers are from Emery's publication (1962).

R21 **E20**
A cooked fish (cleaned and dressed with the head removed) on a pottery dish.

R22 **E22**
Pigeon stew on a pottery dish. It is unclear how Emery identified this as stew rather than roasted or otherwise prepared pigeon.

R23 **E18**
A cooked quail on a pottery dish. It is cleaned and dressed with the head tucked under one wing.

R24 **E15**
Two cooked kidneys on a pottery dish

R25 **E19**
Ribs and legs of beef on a pottery dish. Unfortunately Emery did not elaborate on how many ribs, and which parts of the legs.

R26 **E35**
Identity uncertain, but containing cut ribs of beef.

Discussion

This assemblage contains all three types of meat available to the Egyptians: mammalian, avian, and piscian. It is interesting to note the appearance of a cooked fish as well as the quail and kidneys. According to Herodotus, quail could be consumed uncooked (see Chapter VIII).

Dahshur L.LVIII [Amenemhet III], tomb 7

A Twelfth Dynasty burial of a young man, possibly royal, containing six joints of meat from bovids in the burial chamber. These were identified by Boessneck and von den Driesch (1988: 120-1), and the **BD** numbers following the **R** number are theirs.

R27 **BD1** **no dimensions**
The right scapula and humerus from a large *B. taurus*.

R28 **BD2** **no dimensions**
The right scapula and humerus from a small *B. taurus*, perhaps a cow.

R29 **BD3** **no dimensions**
The left femur and patella of a large *B. taurus*, perhaps a steer or an ox.

R30 BD4 no dimensions
The right femur and patella of a small *B. taurus*, perhaps a cow.

R31 BD5&6 no dimensions
Eight ribs from the same side divided into two.

Discussion:

All animals were between two and four years old. The offered joints and the different sizes of the bones indicate that between two and four animals were sacrificed (Boessneck & A. von den Driesch 1988: 121).

Dahshur L.LVIII, tomb 10

This Twelfth Dynasty royal burial contains a meat offering that consists of various joints of meat from oxen and other large mammals, perhaps antelope (?) (Boessneck, J. and A. von den Driesch 1988: 124). The following information is based on Boessneck and von den Driesch (1988), and the abbreviations are as in tomb 7.

R32 BD1 no dimensions
A right scapula and humerus from a middle sized *B. taurus*.

R33 BD2 no dimensions
The left scapula and humerus from a large mammal.

R34 BD3 no dimensions
The right femur and patella from a middle sized *B. taurus*.

R35 BD4 no dimensions
The left femur and patella from a middle sized *B. taurus*.

R36 BD5&6 no dimensions
Eight ribs from one side.

Discussion

The joints offered in this tomb originate from two animals, one bovid and one other unidentified animal.

Balat, Tomb of Medunefer[5]

This mastaba dating from the Sixth Dynasty contained preserved and processed avian and bovine remains. It contained a few jars of preserved waterfowl — the exact number of jars is unknown as they were badly damaged, although the excavator believes that there were definitely three, if not more, jars for poultry. The birds represented in this assemblage are: *Anas acuta*, *Cygnus*, *Aythya fuligula*, *A. querquedula*, and *A. crecca*. (Valloggia 1986: 208). The tomb also contained remains of *B. taurus*, some in jars, and some just lying on the ground.

R37 no dimensions
Head of an adult animal.

R38 no dimensions
Two skulls of calfs who were between the ages of two and three months.

R39 no dimensions
One tibia from an ox who was less than three years old.

R40 no dimensions
Two sets of metacarpals in conjunction with the first phalange.

R41 no dimensions
Two metacarpals in conjunction with their respective phalanges and sesamoids, from an adult animal.

R42 no dimensions
A metatarsal and the three phalanges, with the sesamoids, all connected, of a mature animal.

R43 no dimensions
A metatarsal and first phalange in a jar [#9] (Vallogia 1986: 206).

Discussion

Most of these joints are of very little nutritional value and have miniscule amounts of meat on them. As has been pointed out elsewhere, it is rare to find metapodials and heads as offerings. Louis Chaix suggests that these portions were put into the tomb as a symbol of the entire animal(s) slaughtered for the event, while the rest of the animal(s) were consumed by the living (Vallogia 1986: 206). Most other tombs from the Predynastic, Middle and New Kingdoms infrequently put these joints among the offerings for the deceased. This might be a practice peculiar to the oasis where meat might have been scarcer than in the valley. However, there is little evidence from the Old Kingdom, both in the Nile Valley and the Oases to come to a definite conclusion.

Gerzeh

In eight graves, dating from the Predynastic/Early Dynastic period, animal bones were found. Petrie states that they were of ox or deer. It is highly unlikely that they were of deer; unfortunately these bones are not available for study, so one cannot identify the animal. The **R** number is followed by the grave number.

R44 Grave 209 no dimensions
One scapula.

R45 Grave 16 no dimensions
One scapula.

R46 Grave 10 no dimensions
A pair of ribs.

R47 Grave 20 no dimensions
A pair of ribs.

R48 Grave 33 no dimensions
A pair of ribs.

R49 Grave 109 no dimensions
A pair of ribs.

R50 Grave 110a no dimensions
A pair of ribs.

R51 Grave 116 no dimensions
Unidentified fragments of bone

Discussion

Only one element was placed in each tombs, though ribs always came as a pair (Fig. 62). Ribs were most favoured — perhaps because there are so many. They are not very fleshy, thus it seems that this is again a case of less valuable meat being offered, while the more nutritious joints were consumed by the living (Petrie et al 1912: 7). This selection of joints is very different from that of the Predynastic cemetery at Nag-ed-Deir (see below). Unfortunately it is not clear what type of animal these bones were from.

R44 Sakkara, pyramid of Horus Sekhemkhet
Skulls and bone fragments of oxen, rams, goats, gazelles, pigs and other animals were found in the (portcullis?) shaft of the unfinished Third Dynasty pyramid of

Horus Sekhemkhet. Some of the more complete joints were wrapped in linen (victual mummies?); these animal remains are thought to have been a Twenty-sixth Dynasty reuse of the tomb. Some skulls were split along the median basi-cranial axis to permit access to the brain (Moustafa 1964: 256), a feature that is common in ox crania in Kush. This is the only known occurrence of victual mummies being found outside Thebes.

Nag-ed-Deir

The Predynastic cemetery N7000 of Nag ed-Deir was excavated by Albert Lythgoe under the auspices of Phoebe Hearst and the University of California between 1902 and 1904. The site is located "along the east side of the 'South Wady'... on a nearly level ground terrace which runs back into the limestone plateau to the north and northwest, directly behind the village of Naga-ed-Der" (Lythgoe and Dunham 1965: xiii). The area on which the cemetery is located is a gravel terrace, about fifteen metres above the wady floor and approximately sixty-five metres long and thirty-five metres wide (Lythgoe and Dunham 1965: xiii). The surface of the cemetery had been "untouched in modern times" (Lythgoe and Dunham 1965: xiii) indicating that the plundering of certain graves dated far back in antiquity.

The majority of the objects, and in this case, the faunal material from the site is housed in the Phoebe Hearst Museum of Archaeology and Anthropology (Formerly the Lowie Museum of Anthropology — LMA) in Berkeley, California, and whose curators have generously permitted the use of these objects in this work. Below is a list of the graves from cemetery N7000 that contained food offerings, the type of meat offered, and the grave goods found therein (Fig. 45). The chronology and some of the information concerning gender derives from the work of Patricia Podzorski (1990; forthcoming).

LMA preceding the number signifies the Lowie Museum of Anthropology (now the Phoebe Hearst Museum) number assigned to the objects in question. **N** numbers are grave number, followed by date, gender and the presence of gravegoods.

R45 N7172 Naqada I-IIa female grave goods present
A short cow's horn was found between the woman's hands and her skull (Lythgoe and Dunham 1965: 100; field notes). Not in the Phoebe Hearst collection.

R46 N 7235c Naqada IIc female plundered
Left scapula of *Ovis aries* under 6-8 months of age LMA 6-12032 (supposed to be two, according to the notes).

R47 N7266 Naqada IIc male grave goods present
"Under pot 5 against the wall of the grave was a small leg bone of some animal (offering)" (field notes). A fragment of a right humerus of an immature mammal. LMA 6-12036.

R48 N7290 Naqada IIc unknown adult grave goods present
A goat (*Capra hircus*) horn was found in the burial (Lythgoe and Dunham 1965:166). Not in the Phoebe Hearst collection.

R49 N7296A Naqada IIb male (c.18 years); grave goods
Bones of two animals seem to be in evidence. An almost complete skeleton of what might be an immature gazelle (*Gazella dorcas*/*G. gazella*) or possibly an immature sheep/goat, was in the tomb, as well as fragments of a dog. The gazelle bones: 1st phalange, 2nd phalange, 3rd phalange and thirty-five vertebrae, and one humerus of a young dog (n.b. there is also another dog [*Canis familiaris*] burial outside the wooden lined grave of N7418A). LMA 6-12040.

R50 N7304 Naqada IId unknown grave goods present
Portion of right scapula of a sheep under the age of eight months (*Ovis aries*). LMA 6-12042.

R51 N7421 Naqada IId male, about 50 grave goods present
"Two small pieces of horn like that of a steer or goat, so badly eaten by worms that their shape could not be determined" (Lythgoe and Dunham 1965: 256) were found in front of the stomach area (field notes). Remains are not in the Phoebe Hearst Museum.

R52 N7454A Naqada IIc male grave goods present
An immature gazelle right femur, tibia, 2 (one might be a carpal) tarsals, the metatarsal, calcaneum, and astragalus. There are also 4 vertebrae which might belong to the gazelle, as well as a metacarpal. LMA 6-12070. Also associated with this burial is a right scapular of an immature animal which could belong to either a sheep/goat or a gazelle. It was found inside the pot (#3). LMA 6-12067.

R53 N7461 Naqada IIc adult & child (6-8) grave goods present
"Single leg bone of a young animal, perhaps a sheep or goat, as has been found in other graves in this cemetery" (Lythgoe and Dunham 1965: 288). Unfortunately this bone is not in the Phoebe Hearst collection.

R54 N7481 Naqada IIb-c female plundered
"Mixed up with plundered bones behind the head were the shoulder blade and leg bone of a small animal (Either a gazelle or a goat. E.S)" (Lythgoe and Dunham 1965: 301). Not in the Phoebe Hearst Museum.

R55 N7484 Naqada IIb female grave goods present
"Against [the] grave wall were the ribs and leg bones of a small lamb or goat" (Lythgoe and Dunham 1965: 302). Not in the Phoebe Hearst Museum.

R56 N7497 Naqada IIb female? plundered
Right humerus of an immature goat. LMA 6-12093.

R57 N7519 Naqada IId adult multiple grave goods present
The grave contained "three ribs of a large animal like an ox, and a shoulder blade of a smaller animal was found in the filling" (Lythgoe and Dunham 1965: 330). Not in the Phoebe Hearst Museum.

R58 N7521A Naqada IId unknown grave goods present
Found in the grave was the "lower jaw and a few other bones of a small animal (goat?)" (Lythgoe and Dunham 1965: 333). Though these bones are not in the Phoebe Hearst Museum, they are visible in the photograph of the tomb. They do look like goat bones.

R59 N7525A Naqada IId female & 5 subadults grave goods present
"Part of a cow's horn" (Lythgoe and Dunham 1965: 339). Not in the Phoebe Hearst collection.

R60 N7539 Naqada III subadult grave goods present but tomb plundered in antiquity
The northeastern corner "shows a small group of animal bones lying under pots 38-41" (Lythgoe and Dunham 1965: 359). These are not in the Phoebe Hearst, but the photograph seems to show the ribs, six or seven vertebrae, a tibia, femur, and humerus of what could quite possibly be a gazelle (Lythgoe and Dunham 1965: fig. 159d).

R61 N7583 Naqada IIa-b male grave goods; tomb plundered in antiquity
In front of the face lay the "skull of a goat with short, slightly twisted horns" (Lythgoe and Dunham 1965: 381). Not in the Phoebe Hearst Museum, but the goat skull is visible in fig. 171b (Lythgoe and Dunham 1965) and shows a mature animal.

R62 N7595B Naqada IIa unknown grave goods present
No mention in notes or the book, but there are remains listed from this grave in the museum. From a young goat (*C. hircus*) come: the right radius (with butchery marks on the proximal end), the right ulna, the metacarpal, the first phalange, and the left horn-core. (LMA 6-12117). Also from this grave comes the right horn core and the patella of a young goat. (LMS 6-12116). If these do not come from this tomb they might come from: N 7481 or 7484.

R63 N7597 Naqada I-IIb male matting in grave
From an immature goat there were found: 4 ribs, (one with a butchery mark at the costal end), a left scapula, the right radius and ulna, the right metacarpal, a first phalange (with some fur attached) and some skull fragments, including

deciduous teeth (D1,2, and 3). All are probably from the same animal. LMA 6-12124.

R64 N7621 Naqada I-IIa 6-8 year old child plundered
"A twisted horn of a mountain goat (?), the companion to which was in the north end of N7624, about 4 metres north of this" (Lythgoe and Dunham 1965: 405) was found. This object, not in the Phoebe Hearst Museum, visible in fig. 183d, seems to be the horn of a screw horned goat, *Capra hircus girgantania.*

R65 N7624A Naqada I-IIa female grave goods present
"A twisted horn of a mountain goat (?0, the companion to which was in north end of N7621" (Lythgoe and Dunham 1965: 408). If this is indeed the twin to the horn found in N7621, as fig. 184 a-d would indicate, it too would be the horn of a screw horned goat, *C. hircus girgantania.* It too is not in the Phoebe Hearst Museum.

Discussion:

The total number of graves in the cemetery is 634, out of which 21 graves dating mainly from the Naqada II period contain food offerings. This is only about 4% of the graves. The reasons for this could be: 1) only very few wealthy/important people had food offerings of meat (there are many more examples of grain and bread); 2) the bones were lost through time and taphonomic processes; 3) the bones were lost due to the recovery method of the archaeologists; 4) they were simply not recorded or recognised by the excavators. Reasons 2, 3, and 4 are all quite possible, but if so, nothing can be done about it. However, if one were to study the existing faunal assemblage from this site with relation to grave wealth as defined by Podzorski (1993), one might find correlations between the wealth of the individual and the types and amounts of the food-offerings found in the tombs. Certainly, virtually all the people possessing food offerings had significant amount of grave goods. It is interesting to note that food offerings occur in as many male burials as in female (six of each), as well as in two child burials. The appearance of the bones in the child burial suggests that either the children were of important families, or that status and offerings are not necessarily linked.

A cursory study was conducted on the contemporary Ballas cemetery by the author. It showed that 3.4% of its occupants had food offerings, a number quite close to that of Nag-ed-Deir, so it seems more likely that these percentages show the number of people with food burials, rather than an archaeological bias, though this does not rule out taphonomic processes.

It is interesting to note that the main source of food was goat — further evidence for the prevalence of goats in this area is the number of goat skins that were used for clothes as well as "shrouds". Bovid horns were only found in very limited instances (three) in both male and female graves. There are only two instances of offerings of oxen, thereby indicating that they were more highly prized, rarer, and a more significant status food. Gazelles were also eaten as well as being used for their hide. There is a mention of sheep skin being used for bags and it could have been used for food, but due to the immaturity of all of the bones found (butchery marks on some of the young animals means they were sacrificed, while some of the marks suggest that the meat was removed) it is more difficult than usual to determine the differences between sheep and goat.

TABLES

Although references are given to PM, the scene may not be listed in PM, although the individual tomb publications will include it.

Table I: Butchery Scenes

Reference	Tomb	Tombowner	Dynasty	Location	Place
PMIII:447	S3509	Hetepka	5/6	Several	Sakkara
PMIII: 454	20/D41	Sekheminkh-ptah	5	offering room	Sakkara
PMIII: 455	21/D40	Ankhmare	5	offering room	Sakkara
PMIII:455*	22/D39	Kapure	5	Several	Sakkara
PMIII:460	37-8 E1,2&H3	Sabu and Ptahshepses	5	Several	Sakkara
PMIII:468	57/D23	Kaemnefert	5	Several	Sakkara
PMIII:468	58/E6	SnefrunuferI	5/6	Chapel	Sakkara
PMIII:470	60/D22	Ty	5	Several	Sakkara
PMIII:481	67/D16	Ankhmaka	5	Entrance	Sakkara
PMIII:481	69/D15	Sopduhotep	5	Chapel	Sakkara
PMIII:483	75/D10	Tepenankh I	5	Offering Room	Sakkara
PMIII:483	76/D11	Tepenankh II	5	Offering Room	Sakkara
PMIII:485	78/D1	Neteruser	5	Room 4	Sakkara
PMIII:486	79/D2	Kaemnehu	5	Chapel	Sakkara
PMIII:487*	80/D3	Raemka	5	Several	Sakkara
PMIII:492	D72/ LS15	Pehnuika	5	Room 2	Sakkara
PMIII:493	LS 6	Methen	5	Chapel	Sakkara
PMIII:495	LS 16	Rashepses	5	Pillared Hall	Sakkara
PMIII:497	S 913	Perneb	5	Offering Room	Sakkara
PMIII:498	S915	Nikauhor	5	Offering Room	Sakkara
PMIII:499	-	Ptahshepses I	6	Chapel	Sakkara
PMIII:506*	-	-	5/6	block	Sakkara
PMIII:510	-	Khentika I	6	Several	Sakkara
PMIII:513	-	Ankhmahor	6	Room 4	Sakkara
PMIII:516	-	Usahateti	6	Room 3	Sakkara
PMIII:518	-	Ptahshepses	6	Offering Room	Sakkara
PMIII:518	-	Mereri	6	Inner Hall	Sakkara
PMIII:519	-	Wernu	6+	Hall	Sakkara
PMIII:520	-	Meru Terisonb	6	Inner Hall	Sakkara
PMIII:524	LS10	Kagemni	6	Room 7	Sakkara
PMIII:529	-	Mereruka et al	6	Several	Sakkara

PMIII:539	-	Sitintetei	6/FIP	Niche	Sakkara
PMIII:541	-	Kaemsenu	6	Chapel	Sakkara
PMIII:558	-	Shema	6/FIP	False Door	Sakkara
PMIII:569	-	Wazet-ipi	6?	Fragment	Sakkara
PMIII:575	LS17/ H2	Manufer	5/6	Several	Sakkara
PMIII:577	D45	Person	5	Offering Room	Sakkara
PMIII:579	LS22/ C11	Iyzefa	5	North Room	Sakkara
PMIII:581	D47	Nenkheftka	5	Offering Room	Sakkara
PMIII:581	D51	Ptahhotep	5	Passage	Sakkara
PMIII:582*	D52	Senenuankh	5/6	Several	Sakkara
PMIII:583	D55	Neferirtnef	5	Chapel	Sakkara
PMIII:594	D 60	Hetepherakhti	5	Several	Sakkara
PMIII:595	D59	Duahap	6	Offering Room	Sakkara
PMIII:596	-	Sekhemka	5	Offering Room	Sakkara
PMIII:597	D62	Ptahhotep I	5	Offering Room	Sakkara
PMIII:599	D64	Akhethotep & Ptahhotep	5/6	Several	Sakkara
PMIII:607	-	Ptahhotep	5/6	Offering Room	Sakkara
PMIII:609	-	Neferkhuptah	6	Offering Room	Sakkara
PMIII:614	-	Seshemnefer	6	Offering Room	Sakkara
PMIII:615	-	Unis	MK	Room 1	Sakkara
PMIII:616	-	Iynefert	6	Room 1	Sakkara
PMIII:618	-	Idut	6	Several	Sakkara
PMIII:621	-	Mehu	6+	Several	Sakkara
PMIII:625	-	Nebt	6	Room 5 & 6	Sakkara
PMIII:627	-	Niankhtap	6	-	Sakkara
PMIII:628	-	Nebhauhor	6	Several	Sakkara
PMIII:705*	-	Merymery	18	-	Sakkara
PMIII:631	-	Kairer	6	Room 4	Sakkara
PMIII:634	E 17	Akhtihotp	6	Several	Sakkara
PMIII:639	-	Irukaptah	5/6	Offering Room	Sakkara
PMIII:640	-	Nufer	5	Offering Room	Sakkara
PMIII:642	-	Niankhnum & Khnumhotep	5	Several	Sakkara
PMIII:646*	-	Methethi	5	block	Sakkara
PMIII:653	LS31	Ptahhotep	5	Room 2	Sakkara
PMIII:671*	-	Ipi	6	block	Sakkara
PMIII:696*	-	Niankhnesut	6	block	Sakkara
PMIII:699	-	Werirniptah	5	Chapel	Sakkara
PMIII:762*	-	No name	5	block	Sakkara
PMIII:827*	-	No name	6	Fragment	Sakkara
PMIII:891*	2	Insnefru-Ishtef	5/6	block	Dashur
PMIII:4	F 7	Kaseuza	5	-	Abu Rawash
PMIII:5	F 21	Hekaf	5/6	Interior Chapel	Abu Rawash
PMIII:49	LG20	Person	5	Chapel	Giza
PMIII:51	-	Abdu	6	Offering Room	Giza

PMIII:53	G1029	Sekhemka	5/6	Chapel	Giza
PMIII:71	G2100-I annex	Merib	4/5	Chapel	Giza
PMIII:73	G2101	Nensdjrka	5	Chapel	Giza
PMIII:76	G2136	Kahif	6	Chapel	Giza
PMIII:78	G2155	Kanensut I	5	Offering Room	Giza
PMIII:81	G2184	Akhmerutnesut	5/6	Offering room	Giza[
PMIII:82	G2196	Iasen	5/6	Chapel	Giza
PMIII:85-7*	G2370	Senedjemib	5	Rm. 2; chapel	Giza
PMIII:88	G2378	Senedjemib	5/6	Chapel	Giza
PMIII:95	G2430	Hetepniptah	6	Chapel	Giza
PMIII:96	G3008	Snefruhotp	6	Several	Giza
PMIII:105	-	Khnemhotep II	6	Chapel	Giza
PMIII:115	D117	Wehemka	5	Chapel	Giza
PMIII:119	-	Mereruka	6	Offering Room	Giza
PMIII:124	G4121	Ankhemre	5/6	Chapel	Giza
PMIII:132	G4561	Kaemankh	6	Several	Giza
PMIII:133	G4611	Niuty	5/6	Chapel	Giza
PMIII:135	G4651	Kapunesut	5	Chapel	Giza
PMIII:142	G4920	Thenti	5	Chapel	Giza
PMIII:148	G5110	Duaenre	I5	Chapel	Giza
PMIII:149	-	Khentkaus	6	Room 3	Giza
PMIII:149	G5150	Seshethotep	5	Several	Giza
PMIII:154	G5170	Seshemnefer III	5	Offering Rm.	Giza
PMIII:161	-	Setka	6	Offering Room	Giza
PMIII:161	G5370	Djaty	5+	Chapel	Giza
PMIII:162	G5470	Rawer II	5	Several	Giza
PMIII:165	G5550	Nufer	6	Chapel	Giza
PMIII:174	G6020	Iymery	5	Room 3	Giza
PMIII:177*	-	No name	4	blocks	Giza
PMIII:201	-	Kaimsekhem	4	Chapel	Giza
PMIII:185	G7101	Meryrenefer	6	Offering Room	Giza
PMIII:190	G7150	-	5/6	Exterior Room	Giza
PMIII:193	G7391	-	5	Chapel	Giza
PMIII:197	G7530	-	4	Chapel	Giza
PMIII:200	G7550	Dwanhor	4	Chapel	Giza
PMIII:201	G7650	Akhtihotp	4	Offering Room	Giza
PMIII:202	G7660	Kaemsekhem	4	Offering Room	Giza
PMIII:204	G7760	Minjedef	4	Chapel	Giza
PMIII:205	G7810	Djaty	4/5	Chapel	Giza
PMIII:208	G7948	Pakhaefankh	5	Chapel	Giza
PMIII:209	LG63	Kaemneferet	5/6	Chapel	Giza
PMIII:209	LG64	Nesemnaw	5/6	Chapel	Giza
PMIII:217	-	Nunetjer	6	Chapel	Giza
PMIII:221	G5III	Sekhemka	5/6	Offering Room	Giza
PMIII:226	LG53	Seshemnefer I5	5/6	Offering Room	Giza
PMIII:231	LG86	Nebemakhet	4/5	Room 2	Giza
PMIII:236	LG90	Debhen	4/5	Room 2	Giza

PMIII:240	-	Nekhetka	5+	Offering Room	Giza
PMIII:243	-	Kaseuza	5+	Chapel	Giza
PMIII:248	-	Niankhkhnum	6	False doors	Giza
PMIII:248	-	Kakaiankh	5	Offering Room	Giza
PMIII:249	-	Seshemnefer	5+	Offering Room	Giza
PMIII:251	-	Smaankh	6	Chapel	Giza
PMIII: 257	Tomb H	Female	5+	Entrance	Giza
PMIII:252	-	Itisen	5+	Offering Room	Giza
PMIII:255	LG95	Khuwiwer	5	Offering Room	Giza
PMIII:263	-	Kaemnefert et al.	5+	Several	Giza
PMIII: 609		Ishethi	6/FIP	Niche	Sakkara
PMIII:266	-	Rawer	5	E. Passage	Giza
PMIII:272	-	Sekheman-khptah	6	Chapel	Giza
PMIII:272	-	Shepseska-fankh	5/6	Chapel	Giza
PMIII:275	-	Ankhtef	6	Offering Room	Giza
PMIII:278	-	Nesutpunuter	5	Pillared hall	Giza
PMIII:280	-	Ireru	5/6	W. Chapel	Giza
PMIII:289	LG81	-	26	Main Room	Giza
PMIII:299	-	Iynefert	5/6	Chapel	Giza
PMIII:303*	-	-	5/6	Several	Giza
PMIII:309*	-	-	5	Block	Giza
PMIV:52	-	Khesuwer	12	Hall	Kom el-Hisn
PMIV:93*	16B	Itet & Nefermaat	4	Facade	Medum
PMIV: 90*	6	Rahotep	3/4	Doorjamb	Medum
PMIV:125	-	Pepyankh	6	Hall	Kom el-Ahmar
PMIV:138	14	Niankhpepy	6+	Side Room	Zawyet el-Amwat
PMIV:143	2	Amenemhet	12	Several	Beni Hasan
PMIV:147	3	Khnemhotep III	MK	Hall	Beni Hasan
PMIV:157	17	Khety	MK	Several	Beni Hasan
PMIV:160	29	Bakt I	MK	Hall	Beni Hasan
PMIV:160	33	Bakt II	MK	Hall	Beni Hasan
PMIV:189	2	Werirni	5/6	Outer Hall	Sheikh Said
PMIV:190	4	Wiu	12	hall	Sheikh Said
PMIV:244	8	Ibi	6	Hall	Deir el-Gebrawi
PMIV:245	12	Djaw	6	Hall	Deir el-Gebrawi
PMIV:250	B1	Senbi	12	Hall	Meir
PMIV:251	B2	Ukhhotep	12	Niche	Meir
PMIV:253	B4	Ukhhotep	12	Several	Meir
PMIII: 539		Sitin	6/FIP	Niche	Sakkara

PMIV:253	B4	Ukhhotep	12	Hall	Meir
PMIV: 79	-	Antefoker	-	S. Hall	Lisht
PMIV:255	D2	Pepyankhhirib	OK/FIP	outer hall	Meir
PMIV:262	I	Hepdjefa	MK	Shrine	Asyut
PMIV: 215	4	Meryre I	18	Pillared Hall	el-Amarna
PMIV: 212	1	Huya	18	Shrine	el-Amarna
PMIV: 213	2	Meryre II	18	Hall	el-Amarna
PMIV: 218	6	Panehsy	18	Several	el-Amarna
PMIV: 229	25	Ay	18	Hall	el-Amarna
PMIV: 244	8	Ibi	6	Hall	Deir el-Gebrawi
PMIV: 245	12	Djau	6	Hall	Deir el-Gebrawi
PMII:388	C	Saded	MK	E.Wall	Deir el-Baheri
PMI:16	7	Ramosi	19	Chapel	Thebes
PMI:27	15	Tetiky	18	Chapel	Thebes
PMI:33	19	Amenmosi	19	Hall	Thebes
PMI:35	20	Mentuhirkhop-shef	18	Passage	Thebes
PMI:46	29	Amenemopet	18	Passage	Thebes
PMI:50	33	Pedamenopet	Saite	Portal	Thebes
PMI:57	34	Mentuemhet	TIP	Chapel C	Thebes
PMI:69	38	Djserkaresoneb	18	Hall	Thebes
PMI:74	39	Puimre	18	Central Chapel	Thebes
PMI:83	42	Amenmosi	18	Passage	Thebes
PMI:85	45	Djhout	19	Hall	Thebes
PMI:88	48	Amenemhet	18	Portico	Thebes
PMI:94	49	Neferhotep	18	Pillars	Thebes
PMI:99	52	Nakht	18	Hall	Thebes
PMI:104	53	Amenemhet	18	Inner Rom	Thebes
PMI:107	55	Ramosi	18	Several	Thebes
PMI:114	57	Khaemhet	18	Several	Thebes
PMI:121	60	Antefoker	12	Several	Thebes
PMI:128	64	Hekerneheh	18	Hall	Thebes
PMI:137	69	Menna	18	Several	Thebes
PMI:142	72	Re	18	Several	Thebes
PMI:144	74	Thanuny	18	Several	Thebes
PMI:151	77	Ptahemhet	18	Inner Hall	Thebes
PMI:152	78	Haremheb	18	Several	Thebes
PMI:156	79	Menkheper	18	Several	Thebes
PMI:158	80	Djhutnefer	18	Several	Thebes
PMI:178	87	Minnakht	18	Several	Thebes
PMI:180	88	Pehsukher	18	Hall	Thebes
PMI:181	89	Amenmosi	18	Hall	Thebes
PMI:184	90	Nebamun	18	Hall	Thebes
PMI:188	92	Suemnut	18	Several	Thebes
PMI:190	93	Kenamun	18	Several	Thebes

PMI:198	96	Sennufer	18	Hall	Thebes
PMI:212	100	Rekhmire	18	Several	Thebes
PMI:217*	103	Dagi	11	fragment	Thebes
PMI:218	104	Djutnefer	18	Inner room	Thebes
PMI:226	109	Min	18	Hall	Thebes
PMI:230	112	Menkheperre-soneb	18	Hall	Thebes
PMI:236	122	Amenemhat	18	Chapel B	Thebes
PMI:261	151	Hati	18	Hall	Thebes
PMI:276	162	Kenamun	18	Hall	Thebes
PMI:277	165	Nehemawy	18	Hall	Thebes
PMI:279	172	Mentiywy	18	Several	Thebes
PMI:281	175	-	18	Hall	Thebes
PMI:336	251	Amenmosi	18	Hall	Thebes
PMI:338	253	Khnemmosi	18	Shrine	Thebes
PMI:341	256	Nebenkemet	18	Hall	Thebes
PMI:344	260	User	18	Hall	Thebes
PMI:408	341	Nektamun	19	Hall	Thebes
PMI:415	347	Hori	19-21	Hall	Thebes
PMI:429	366	Djar	11	Hall	Thebes
PMI:430	367	Paser	18	Several	Thebes
PMI:439	388	-	Saite	Hall	Thebes
PMI:447	A2	Tomb of Dancers	17	Hall	Thebes
PMI:448	A4	Siuser	18	Several	Thebes
PMI:519	11	Ramesses III	19	Room C	Thebes
PMI:540	17	Seti I	19	Several	Thebes
PMI:385	308	Kemsit	11	Burial chamber	Thebes
PMII: 388	C	Sadeh	11	Shrine	Thebes
PMI: 819*	-	-	18	Fragment	Thebes
PMI: 530	14	Tauesert & Sethnakhte	19	Corridor H	Thebes
PMI:225	108	Nebseny	18	Hall	Thebes
PMV:8	-	Kakhent & Ifi	4+	Hall	el-Hammamieh
PMV:27	-	Desher	OK	Rear wall	Nag-ed-Deir
PMV:121	-	Thauti	6	Hall	el Qasr
PMV:162*	-	Iti	6	-	Gebelein
PMV:180	3	Paheri	18	Hall	el-Kab
PMV:185	10	Sebknakht	12	Outer hall	el-Kab
PMV:197	-	Haremkhauef	MK	Hall	Hierakonpolis
PMV:231	25	Mekhu	6+	Hall	Elephantine
PMV:231	26	Sabni	6+	Hall	Elephantine
PMV:235	-	Khunes	6	Hall	Aswan
PMV:138	14	Niankhpepi	6	Side room	Zawyet el-Amwat
PMV:170	-	Ankhtify	FIP	Hall	Moalla

PMIV:181	2	Djhutihotep II	12	Shrine	el-Bersha
PMI: 263	155	Antef	18	Pillar	Thebes
PMI: 391	318	Amenmosi	18	Hall	Thebes
PMI: 338	254	Mosi	18	Chapel	Thebes
PMIV: 122	-	Iteti	5/6	Hall	Deshasheh
PMIII: 710*	-	Patenmheb	18	Chapel	Sakkara
PMIII: 142	G4940	Seshemnefer I	5	Chapel	Giza
PMIV: 236	-	Royal Tomb	18	Room Alpha	el-Amarna
PMIII: 655	-	Horemheb	18	2nd Courtyard	Sakkara
Martin 1987: #79	-	Khaemwast	19	-	Sakkara

* In a museum collection.

Table II: Poultry Processing Scenes

Reference	Tomb	Tombowner	Dynasty	Location	Place
PMI: 32	18	Baki	18	Hall	Thebes
PMI: 38	22	Wah	18	Hall	Thebes
PMI: 101	52	Nakht	18	Hall	Thebes
PMI: 156	79	Menkhepra-soneb	18	Hall	Thebes
PMI:180	88	Pehsukher	18	Hall	Thebes
PMI: 210	100	Rekhmire	18	Hall	Thebes
PMI: 242	127	Senemioh	18	Hall	Thebes
PMI:303	200	Dedi	18	Hall	Thebes
PMI: 316	217	Ipuy	19	Hall	Deir el-Medina
PMI: 459	C.8	Nakht	19	Hall	Thebes
PMV: 179	3	Paheri	18	Hall	el-Kab
PMIII: 516	-	Uzahateti	6	Room 2	Sakkara
PMIV: 254	D2	Pepyankh-hirib	FIP/MK	Outer Hall	Meir
PMIII:69	G2091	Kapi	5/6	Passage	Giza
PMIII:172	G6020	Iymery	5	Room 2	Giza
PMIII:198	G7530+750	Meresankh III	4	Room 1	Giza
PMIII:243	-	Kaseuza	5	Chapel	Giza
PMIII:132	G4561	Kaemankh	6	Bur. Cham.	Giza
PMIII: 448	53509	Hetepka	5/6	Offering Room	Sakkara
PMIV:180	2	Djhutihotep	MK	Hall	El Bershah
PMIII: 593*	D60	Hetepher-akhti	5+	Offering Room	Sakkara/ Leiden
PMIII: 468	60/D22	Ti	5	Offering Room	Sakkara
PMIII: 755*	-	-	18	block	Sakkara

* In a museum collection.

Table III: Fish Processing Scenes

Reference	Tomb	Tombowner	Dynasty	Location	Place
PMI:60*	34	Montuemhat	26	block	Thebes
PMI:143	73	Amenhotep	18	Hall	Thebes
PMI:174	85	Amenemhab	18	Inner Room	Thebes
PMI:164	82	Amenemhet	18	Hall	Thebes
PMI:193	93	Kenamun	18	Passage	Thebes
PMI:210	100	Rekhmire	18	Hall	Thebes
PMI:236	123	Amenemhet	18	Hall	Thebes
PMI: 277	165	Nehemaway	18	Hall	Thebes
PMI: 303	200	Dedi	18	Hall	Thebes
PMI: 316	217	Ipuy	19	Hall	Deir el-Medina
PMI:353	276	Amenemopet	18	Hall	Qurnet Murai
PMI:358	279	Pabasa	26	Pillar C	Thebes
PMIII: 475	59/E7	Snefrunefer	5	Inner Hall	Sakkara
PMIII:482	69/D15	Sopduhotep	5	Chapel	Sakkara
PMIII: 488*	80	Raemka	5	Chapel	Sakkara
PMIII: 492	D70	Pehnuika	5	Room I	Sakkara
PMIII: 498	S915	Nikauhor	5	Offering Room	Sakkara
PMIII:518	-	Mereri	6	Outer Hall	Sakkara
PMIII:523	LS10	Kagemni	5	Room 4	Sakkara
PMIII:583	-	Neferitnef	5	Chapel	Sakkara
PMIII:594	D60	Hetepherakhti	5	Offering Room	Sakkara
PMIII:601	D64	Ptahhotep	6	Offering Room	Sakkara
PMIII:607	-	Ptahhotep Iyniankh	5/6	Offering Room	Sakkara
PMIII: 608	-	Mereri	6	Offering Room	Sakkara
PMIII: 628	-	Nebkauhor	6	Vestibule	Sakkara
PMIII :631	-	Kairer	6	Room 1	Sakkara
PMIII :635	-	Akhtihotep	5/6	Chapel	Sakkara
PMIII :642	-	Niankhnum	5	Vestibule	Sakkara
PMIII: 644	-	Irenkaptah	5	Offering Room	Sakkara
PMIII:695*	-	Niankhnesut	6	blocks	Sakkara
PMIII:753*	-	-	5	block	Sakkara
PMIII:754*	-	-	5	block	Sakkara
PMIII:173	G6020	Iymery	5	Room 2	Giza
PMIII:230	LG86	Nebemakhet	6	Room 1	Giza
PMIII:309*	-	-		Block	Giza
PMIV:91	6	Rahotep	4	Hall	Maidum
PMIV:143	2	Amenemhet	12	Hall	Beni Hasan
PMIV:188	2	Werirni	5/6	Outer Hall	Sheikh Said
PMIV:244*	8	Ibi	6	Entrance	Deir el-Gebrawi
PMIII:419	-	Unas	6	Causeway	Sakkara
PMI: 72	39	Puemra	18	Hall	Thebes

PMIII:755*	-	-	18	block	Sakkara
PMI:155*	78	Horemheb	18	block & hall	Thebes
PMIII: 518	-	Mereri	6	Outer Hall	Sakkara
PMIV: 160	29	Bakt I	11/12	Hall	Beni Hasan
PMIV:180	2	Djhoutihotep	12	Hall	el-Bersha
PMIV: 122	-	Inti	6	Hall	Deshasheh
PMIII: 230	LG86	Nebemakhet	4	Room 1	Giza
PMI: 69	38	Djeserkere-soneb	18	Hall	Thebes
PMIII: 608*	-	Mereri	6	block	Sakkara
PMIII: 475	60/ D22	Ti	5	Inner Hall	Sakkara
PMI: 445	405	Khenty	6	Wall F	Thebes
PMIII:700[8]	-	Werininptah	5	-	Sakkara
PMIII:528[26]	-	Mereruka	6	Room 4	Sakkara
PMIII:207	G7948	Rakhaankh	5+	Chapel	Giza
PMIII:309*	-	-	5/6	block	Giza
PMIV:134	-	Khons	5	-	Zawyet el-Amwat
PMIII: 492	D70/ LS15	Pehnuika	5+	Room 1	Sakkara

* In a museum collection.

Table IV: Pigs

Reference	Tomb	Tombowner	Dynasty	Location	Place
PMIII: 522	LS10	Kagemni	6	Pillared Hall	Sakkara
PMIV: 155	17	Khety	12	Hall	Beni Hasan
PMV:183	7	Renni	18	Hall	el-Kab
PMI:160	81	Ineni	18	Portico	Thebes
PMI:237	123	Amenemhat	18	Hall	Thebes
PMI:257	145	Nebamun	18	Hall	Thebes
PMI:41	124	Nebamun	18	Hall	Thebes
PMI:258	146	Nebamun	18	Hall	Thebes
PMV: 180	3	Paheri	18	Hall	el-Kab

* In a museum collection.

BIBLIOGRAPHY

Allam, S. (1985) *Everyday Life in Ancient Egypt*. Cairo.

Allen, T. G. (ed.) (1960) *The Egyptian Book of the Dead: Documents in the Oriental Institute Museum at the University of Chicago*. Chicago.

—, (trans.) (1974) *The Book of the Dead or Going Forth by Day*. Chicago.

Ammar, H. (1954) *Growing Up in an Egyptian Village*. London.

Anderson, J. (1902) *The Zoology of Egypt*, 2. London.

Anderson, J. R. (ed.) (1980) *Muir's Textbook of Pathology*. London.

Andersson, E. (1912) 'La Denomination Égyptiènne des Bœufs sans Cornes', *Sphinx* 16: 145-64.

—, (1913) 'Notes sur la Brochure de M. Gaillard: Les Tatonnements des Egyptiens de l'Ancien Empire a la Recherche des Animaux à Domestiquer', *Sphinx* 17: 77-82.

Arnold, D. and J. Settgast (1966) 'Zweiter Vorbericht uber die vom Deutschen Archaologischen Institut Kairo im Asasif unternommenen Arbeiten', *MDAIK* 21: 47-61.

Arnold, D. (1962) *Wandrelief und Raumfunktion in Agyptischen Tempeln des Neuen Reiches*. Berlin.

—, (1979) *The Temple of Mentuhotep at Deir el-Baheri*. New York.

—, (1988) *The Pyramid of Senwosret I*. New York.

Arnold, D. (1982) 'Keramikbearbeitung in Dahschur', *MDAIK* 38: 25-66.

Ashtor, E. (1970) 'The Diet of Salaried Classes in the Medieval Near East', *Journal of Asian History* 4.1: 1-24. Wiesbaden.

Ayrout, H. H. (1963) *The Egyptian Peasant*. Boston.

Ayrton, E. R. et al. (1904) *Abydos*, III. London.

Badawy, A. (1958) 'Architectural Provision Against Heat in the Orient', *JNES* 17: 122-8.

—, (1954) *A History of Egyptian Architecture*, 1. Cairo.

Baedeker, K. (1902) *Egypt* . Leipzig.

Baer, K. (1962) 'The Low Cost of Land in Ancient Egypt', *JARCE* 1: 25-46.

Baker, H. S. (1966) *Furniture in the Ancient World*. London.

Balcz, H. (1933) 'Die Gefassdarstellungen des Alten Reiches', *MDAIK* 4: 18-36.

Banks, A. and T. P. Hilditch (1933) 'A Note on the Composition of Some Fatty Materials Found in Ancient Egyptian Tombs', *The Analyst* 58: 265-9. London.

Barguet, P. (1962) *Le Temple d'Amon-Re à Karnak*. Cairo.

Barta, W. (1963) *Die Altagyptische Opferliste*. Berlin.

Bates, O. (1910) *Ancient Egyptian Fishing*. Cambridge, Mass.

Beattie, W. A. (1980) *Beef Cattle Breeding and Management*. Sydney.

Behrensmeyer, K. and A. Hill. (1980) *Fossils in the Making*. Chicago.

Behrensmeyer, K. (1990) *ICAZ Experimental Taphonomy Workshop,* (Smithsonian Institution). Washington, D.C.

Bellion, M. (1987) *Egypte Ancienne: Catalogue des Manuscrits Hieroglyphiques et Hieratique et des Dessins, sur Papyrus, Cuir ou Tissu, Publiés ou Signales.* Paris.

Belzoni, G. B. (1821) *Narrative of the Operations and Recent Discoveries within the Pyramids, Temples, Tombs and Excavations in Egypt and Nubia.* London.

Bietak, M. (unpublished) ISIS Fellowship Lecture, 27 April 1991. London.

—, (1979) *Avaris and Piramesse.* Oxford.

Binford, L. (1978) *Nunamiut Ethnoarchaeology.* London.

—, (1981) *Bones: Ancient Men and Modern Myths.* London.

Birch, S. (1876) *Papyrus Harris,* 1-2. London.

Black, A. (1992) Personal Communication.

Blackman, A. M. (1915) *The Rock Tombs of Meir,* 2. London.

—, (1922) 'A Study of the Liturgy of the Aton', *Recueil d'Études Égyptologiques Dedié à la Mémoire de Jean-Francois Champollion*: 505-27. Paris.

—, (1924a) *The Rock Tombs of Meir*, 4. London.

—, (1924b) 'The Rite of Opening the Mouth in Ancient Egypt and Babylonia', *JEA* 10: 47-59.

—, (1945) 'The King of Egypt's Grace Before Meat', *JEA* 31: 57-73.

—, (1953a) *The Rock Tombs of Meir,* 5. London.

—, (1953b) *The Rock Tombs of Meir*, 6. London.

Blackman, W. S. (1927) *The Fellahin of Upper Egypt.* London.

Bleiberg, E. L. (1981) 'Commodity Exchange in the Annals of Thutmose III', *JSSEA* 11: 107-10.

Boessneck, J. and A. von den Driesch. (1987) *Tuna el-Gebel I: Die Tiergalerien.* Hildesheim.

—, (1988) 'Tierknochenfunde vom Tell Ibrahim Awad Im Ostlichen Nildelta', *The Archaeology of the Nile Delta*, 117-22 (ed. E. C. M. van den Brink). Amsterdam.

—, (1982) *Studien an Subfossilen Tierknochen aus Agypten.* Munich.

Boessneck, J. J. (1953) *Die Haustiere in Altagypten.* Munich.

—, (1976) *Tell el-Dab'a III: Die Tierknochenfunde.* Vienna.

Bonnet, C. (1986) *Kerma.* Cairo.

Borchardt, L. (1907) *Das Grabdenkmal des Königs Ne-user-Re.* Leipzig.

—, (1909) *Das Grabdenkmal des Königs Nefer-ir-ke-Re.* Leipzig.

—, (1910) *Das Grabdenkmal des Königs Saꜣhu-Re*, 1. Leipzig.

—, (1913) *Das Grabdenkmal des Königs Saꜣhu-Re*, 2. Leipzig.

Boreux, C. (1925) *Études de Nautique Égyptiènne.* Cairo

Boulenger, G. A. (1907) *Zoology of Egypt: The Fishes of the Nile.* London.

Boulus, E. A. (1991) Personal Communication.

Bourriau, J. (1981) *Umm el Gaab.* Cambridge.

—, (1991) Personal Communication.

Brain, C. K. (1981) *The Hunters or the Hunted?* Chicago.

Breasted, J. H. (1906) *Ancient Records of Egypt*, 1-4. Chicago.

Breasted Jr., J. H. (1948) *Egyptian Servant Statues.* Washington, DC.

Brewer, D. J. and R. F. Friedman. (1989) *Fish and Fishing in Ancient Egypt.* Warminster.

Brewer, D.J. et al. (1995) *Domestication of Plants and Animals in Ancient Egypt.* Warminster.

Brewer, D. J. (1986) Personal Communication.
Brillat-Savarin, J. A. (1982) *Physiologie du Gout.* Paris.
Briskey, E. J. (1974) 'Meat and Meat Packing', *Encyclopaedia Britannica,* 11: 745-53. Chicago.
Brodrick, A. (ed.) (1972) *Animals in Archaeology.* London.
Brothwell, D. and A. T. Sandison (1967) *Diseases in Antiquity.* London.
Brothwell, D. and P. (1969) *Food in Antiquity.* London.
Brunton, G. (1928, 1930) *Qau and Badari,* 2-3. London.
Brunton, G. and G. Caton-Thompson (1928) *The Badarian Civilisation and Predynastic Remains Near Badari.* London.
Bruyère, B. (1929) *Rapport Sur les Fouilles de Deir el Medineh, 1928.* Cairo.
—, (1933) *Rapport Sur les Fouilles de Deir el Medineh, 1930.* Cairo.
—, (1934) *Rapport Sur les Fouilles de Deir el Medineh, 1931-32.* Cairo.
—, (1937) *Rapport Sur les Fouilles de Deir el Medineh, 1934-35.* Cairo.
—, (1939) *Rapport sur les Fouilles de Deir el Medineh (1934-35).* Cairo.
—, (1952) *Rapport Sur les Fouilles de Deir el Medineh, 1945-47.* Cairo.
—, (1953) *Rapport Sur les Fouilles de Deir el Medineh, 1948-51.* Cairo.
Budge, E. A. W. (1922) *Hieroglyphic Texts from Egyptian Stelae in the British Museum,* 6. London.
Burckhard, J. L. (1819) *Travels in Nubia.* London.
Butzer, K. W. (1976) *Early Hydraulic Civilization In Egypt.* Chicago.
Byrne, K. and D. C. Parris (1987) 'Reconstruction of the Diet of the Middle Woodland Amerindian Poppulatiojn at Abbott Farm by Trace Element Analysis', *American Journal of Physical Anthropology* 74: 373-84. Washington, DC.
Cailliaud, F. (1826) *Voyage a Meroe au Fleuve Blanc,* 1. Paris.
Calverly, A. M. et al. (1933) *The Temple of King Sethos I at Abydos,* 1. London and Chicago.
Caminos, R. (1954) *Late Egyptian Miscellanies.* Oxford.
—, (1964) 'The Nitocris Adoption Stela', *JEA* 50: 71-101.
Carter, H. (1901) 'Report on the Tomb of Mentuhotep I', *ASAE* 2: 201-05.
—, (1903) 'Excavations at Biban El-Moluk', *ASAE* 4: 176-77.
—, (1923) 'An Ostracon Depicting a Red Junglefowl', *JEA* 9: 1-4.
—, (1927) *The Tomb of Tutankhamen,* 2. London.
—, (1932) Notes on meats in the tomb of Tutankhamun. Griffith Institute, Oxford.
—, (1933) *The Tomb of Tutankhamen,* 3. London.
Casson, L. (1986) *Ships and Seamanship in the Ancient World.* Princeton.
Castle, E. W. (1992) 'Shipping and Trade in Ramesside Egypt', *Journal of Social and Economic History of the Orient* 35.3: 239-77.
Caton-Thompson, G. and E. W. Gardner (1934) *The Desert Fayum.* London.
Černy, J. (1935) *Catalogue des Ostraca Hieratiques Non Litteraires de Deir el Medineh, No. 1-113.* Cairo.
—, (1937-38) 'Deux Noms de Poisson du Nouvel Empire', *BIFAO* 37: 35-40.
—, (1954) 'Prices and Wages in Egypt in the Ramesside Period', *Cahiers d'Histoire Mondiale* 1.4: 903-21. Paris.
—, (1965) *Hieratic Inscriptions from the Tomb of Tutankhamun.* Oxford.
Chabas, F. (n.d.) *Le Calendrier des Jours Fastes et Nefastes de l'Année Egyptienne.* Paris.
Chaix, L. (1990) Personal Communication.

Charles, R. (1991) Personal Communication.
Chassinat, E. (1905) 'Note Sur le Titre', *BIFAO* 4: 223-8.
Chevrier, H. (1951) 'Rapport Sur les Travaux de Karnak', *ASAE* 51: 549-64.
Christensen, M. and P. Walter (1991) 'Physico-Chimie en Traceologie: Le Cas des Couteaux Égyptiens', *La Pierre Préhistorique*: 149-71 (M. Menu and P. Walter, ed). Paris.
Christensen, M., P. Walter and M. Menu (1992) 'Usewear Characterisation of Prehistoric Flints with IBA', *Nuclear Instruments and Methods in Physics Research* B64: 488-93. Holland.
Christensen, M. (1992) Personal Communication.
Christophe, L-A. (1967) 'Le Ravitaillement en Poissons des Artisans de la Necropole Thebaine a la Fin du Regne de Ramses III', *BIFAO* 65: 177-99.
Churcher, C. S. (1980) 'Dakhleh Oasis Project — preliminary Observations on the Geology and Vertebrate Palaeontology of Northwestern Dakhleh Oasis', *JSSEA* 10.4: 379-95.
Clark, J. D. (1971) 'A Re-Examination of the Evidence for Agricultural Origins in the Nile Valley', *Proceedings of the Prehistoric Society* 37: 34-79. Cambridge.
Clutton-Brock, J. (1981) *Domesticated Animals from Early Times*. London.
—, (1987) *A Natural History of Domesticated Mammals*. London.
—, (1989) *The Walking Larder*. London.
Cockburn, A. (1973) 'Death and Disease in Ancient Egypt', *Science* 181.4098: 470-81. Washington, DC.
Cockburn, A. and E. (ed.) (1980) *Mummies, Disease, and Ancient Cultures*. Cambridge.
Cottevieille-Giraudet, R. (1931) *Les Reliefs d'Amenophis IV*. Cairo.
Cranstone, B. (1969) 'Animal Husbandry: The Evidence from Ethnography', *The Domestication of Plants and Animals* (P. J. Ucko and G. W. Dimbleby, eds.). London.
Crawford, D. J. (1971) *Kerkeosiris: An Egyptian Village in the Ptolemaic Period*. Cambridge.
Crawford, R. J. M. (1962) 'Cheese and Cheese Starters', *Recent Advances in Food Science* (J. Hawthorn and J. Leitch, eds.). London.
Crum, W. E. (1939) *Varia Coptica*. Aberdeen.
Dahl, G. and A. Hjort (1976) *Having Herds*. Stockholm.
Darby, W. J. et al. (1977) *Food: The Gift of Osiris*, 1-2. San Francisco.
Daressy, G. (1902) *Fouilles de la Vallée des Rois*. Cairo.
—, (1933) 'Un Casse-tête Préhistorique en Bois de Gébelein', *ASAE* 22: 17-32.
Darwin, C. (1839) *Voyages of Adventure*. London.
Daumas, F. (1964) 'Quelques Rémarques sur les Representations de Pèche à la Ligne sous l'Ancien Empire', *BIFAO* 62: 67-86.
D'Auria, S. et al. (1988) *Mummies and Magic*. Boston.
David, A. R. (1973) *Religious Ritual at Abydos*. Warminster.
Davide, D. (1972) 'Survey of the Skeletal and Mummy Remains of Ancient Egyptians Available in Research Collections', *JHE* 1.2: 155-60.
Davies, N. de G. (1900) *The Mastaba of Ptahhotep and Akhethetep*. London.
—, (1901) *The Rock Tombs of Sheikh Said*. London.
—, (1902) *The Rock Tombs of Deir el Gebrawi*, 1. London.

—, (1905a) *The Rock Tombs of El Amarna,* 1. London.
—, (1905b) *The Rock Tombs of El Amarna,* 2. London.
—, (1905c) *The Rock Tombs of El Amarna,* 3. London.
—, (1906) *The Rock Tombs of El Amarna,* 4. London.
—, (1908a) *The Rock Tombs of El Amarna,* 5. London.
—, (1908b) *The Rock Tombs of El Amarna,* 6. London.
—, (1913) *Five Theban Tombs.* London.
—, (1917) *The Tomb of Nakht at Thebes* . New York.
—, (1920) *The Tomb of Antefoker and his Wife Senet.* London.
—, (1922) *The Tomb of Puyemre,* 1. New York.
—, (1923a) *The Tomb of Puyemre,* 2. New York.
—, (1923b) *The Tombs of Two Officials of Tuthmosis the Fourth.* London.
—, (1924) 'A Peculiar Form of New Kingdom Lamp', *JEA* 10: 9-14.
—, (1927) *Two Ramesside Tombs at Thebes.* New York.
—, (1929) *The Townhouse in Ancient Egypt.* New York.
—, (1930) *The Tomb of Ken amun at Thebes.* New York.
—, (1933) *The Tombs of Menkheperrasonb, Amenmose, and Another.* London.
—, (1935) *Paintings from the Tomb of Rekh-Mi-Re.* New York.
—, (1941) *Tomb of the Vizier Ramose.* New York.
—, (1943) *The Tomb of Rekh-Mi-Re,* 1-2. New York.
Davies, N. de. G. and A. H. Gardiner (1926) *The Tomb of Huy.* London.
Davies, N. M. and N. de G. (1940) 'The Tomb of Amenmose (No. 89) at Thebes', *JEA* 26: 131-6.
Davis, T. M. (1908) *The Tomb of Siptah.* London.
—, (1907) *The Tomb of Iouiya and Touiyou.* London.
Davisson, W. I. and J. E. Harper (1972) *European Economic History,* 1. New York.
Dawson, W. R. (1924) 'The Mouse in Egyptian and Later Medicine', *JEA* 10: 83-6.
—, (1925) 'Studies in Ancient Materia Medica', *American Drugist* December: 22-4. New York.
—, (1929) *Magician and Leech.* London.
—, (1953) 'Egypt's Place in Medical History', *Science, Medicine and History* (E. A. Underwood, ed.). London.
de la Roque, M. F. B. and J. J. Clere (1928) *Rapport sur les Fouilles de Medamoud-1927.* Cairo.
—, (1929) *Rapport sur les Fouilles de Medamoud-1928,* Cairo.
—, (1930) *Rapport sur les Fouilles de Medamoud-1929.* Cairo.
de la Roque, M. F. B. (1926) *Rapport sur les Fouilles de Medamoud-1925.* Cairo.
—, (1927) *Rapport sur les Fouilles de Medamoud-1926.* Cairo.
de Morant, H. (1973) 'L'Alimentation Chez les Egyptiens' *Archeologia* 61: 64-71. Paris.
de Morgan, J. (1897) *Ethnographie Prehistorique.* Paris.
—, (1895) *Fouilles à Dahchour en Mars-Juin 1894.* Vienna.
—, (1903) *Fouilles à Dahchour: 1894-95.* Vienna.
Debono, F. and B. Mortensen. (1988) *The Predynastic Cemetery at Heliopolis.* Mainz.

Debono, F. (1948) 'El Omari, Exposé sur les Campagnes des Fouilles 1943-1944 et 1948', *ASAE* 48: 561-9.

—, (1954) 'La Necropole Predynastique d'Heliopolis', *ASAE* 52: 625-52.

DeNiro, M. J. (1987) 'Stable Isotopy and Archaeology', *American Scientist* 75: 182-91.

Derchain, P. (1962) *Rites Egyptiens*, 1. Brusells.

Desroches-Noblecourt, C. (1954) 'Poissons, Tabous et Transformations du Mort', *Kemi* 13: 33-42.

Desrosier, N. W. and J. N. (1982) *The Technology of Food Preservation*. Conneticut.

Drummond, J. C. and A. Wilbraham (1958) *The Englishman's Food*. London.

Dixon, D. M. (1972) 'Masticatories in Ancient Egypt', *JHE* 1: 433-49.

Dodson, A. M. (1988) 'The Tombs of the Queens of the Middle Kingdom', *ZÄS* 115: 123-36.

—, (1994) *The Canopic Equipment of the Kings of Egypt*. London.

Donadoni Roveri, A. M. (ed.) (1988) *Egyptian Civilization: Daily Life*. Turin.

Douglas, M. (1970) *Purity and Danger. An Analysis of Concepts of Pollution and Taboo*. Baltimore.

Downes, D. (1974) *The Excavations at Esna, 1905-06*. London.

Drioton, E. (1926) *Rapport sur les Fouilles de Medamoud-1925*, 2. Cairo.

—, (1952) 'Les Origines Pharaoniques du Nilometre de Rodah', *BIE* 34: 291-316.

Droeper, K. and D. Wildung (1985) *Minshat Abu Omar*. Munich.

Duell, P. (1938) *Mastaba of Mereruka*, 2. Chicago.

Dumont, J. (1977) 'La Pèche dans le Fayoum Hellenistique', *CdE* 104: 125-55.

Durst, J. U. and C. Gaillard (1902) 'Studien uber die Geschichte des Aegyptischen Hausschafes', *RdT* 24: 44-76.

Edel, E. (1961-64) 'Zu den Inschriften auf den Jahreszeitenreliefs der 'Weltkammer' aus dem Sonnenheiligtum des Niuserre', *Nachrichten die Akademie der Wissenschaft in Göttingen*. Göttingen.

—, (1955) 'Inschriften des Alten Reiches. V. Die Reiseberichte des *Ḥrw-ḫwjt*', *Ägyptologische Studien* (O. Firchow, ed.). Berlin.

Efremov, J. A. (1940) 'Taphonomy: New Branch of Paleontology', *Pan American Geologist* 74.2: 81-93. Iowa.

Eggebrecht, A. (1973) *Schlachtungsbrauche im Alten Agypten und ihre Wiedergabe im Flachbild bis zum Ende des Mittleren Reiches*. Munchen.

Ebbell, B. (1937) *The Papyrus Ebers*. Copenhagen.

Emery, W. B. (1938) *Tomb of Hemaka*. Cairo.

—, (1949) *Great Tombs of the First Dynasty*, 1. Cairo.

—, (1962) *A Funerary Repast in an Egyptian Tomb of the Archaic Period*. Leiden.

Engelbach, R. (1915) *Riqqeh and Memphis VI*. London.

Épron, L. and G. Daumas (1939) *Le Tombeau de Ti*, 1. Cairo.

Epstein, H. (1971a) *The Origin of the Domestic Animals of Africa*, 1. New York.

—, (1971b) *The Origin of the Domestic Animals of Africa*, 2. New York.

—, (1971c) *Observations on Frozen Beef from Various Sources*. Bet Dagan.

Erman, A. (1927) *The Literature of the Ancient Egyptians*. London.

Evans, J. B. and C. F. Niven, Jr. (1960) *Bacteriology — The Science of Meat and Meat Products*. New York.

Evans, W. (1990) Personal Communication.
Evans-Pritchard, E. (1940) *The Nuer.* Oxford.
Excavation Notes (1924-31) from el-Amarna. Cambridge.
Fabish, S. (1989) Personal Communication.
—, (1990) Personal Communication.
Fairgrieve, S. (1992) Personal Communication.
Fairman, H. W. (1958) 'The Kingship Rituals in Egypt', *Myth, Ritual and Kingship* (S. H. Hooke, ed). Oxford.
Fakhry, A. (1973) *The Oases of Egypt*, 1. Cairo.
Farb, P. and G. Armelagos (1980) *Consuming Passions.* New York.
Fathy, H. (1969) *Gourna: A Tale of Two Villages.* Cairo.
—, (1973) *Architecture for the Poor.* Chicago.
Faulkner, R. O. (1940) 'Egyptian Seagoing Ships', *JEA* 26: 3-9.
—, (1942) 'The Battle of Megiddo', *JEA* 28: 2-15.
—, (1973-8) *Ancient Egyptian Coffin Texts*, 1-3. Warminster.
—, (1976) *Concise Dictionary of Middle Egyptian.* Oxford.
Fischer, H. G. (1960) 'The Butcher Pḥ-r-nfr', *Orientalia* 29: 168-87. Rome.
—, (1968) *Ancient Egyptian Representations of Turtles.* New York.
—, (1975) 'Two Tantalizing Biographical Fragments of Historical Interest', *JEA* 61: 33-7.
—, (1976) 'Milk in Everything Cooked', *Varia*, I: 97-99. New York.
—, (1978) 'Five Inscriptions of the Old Kingdom', *ZÄS* 105: 42-59.
Flower, B. and E. Rosenbaum. (1958) *Apicius-The Roman Cookery Book.* London.
Flower, S. (1910) *List of Animals.* Cairo.
Foaden, G. P. and F. Fletcher (1910) *Text-Book of Egyptian Agriculture*, 2. Cairo.
Forbes, R. J. (1965) *Studies in Ancient Technology*, 3. Leiden.
—, (1966) *Studies in Ancient Technology*, 5. Leiden.
Fox, B. A. and A. G. Cameron (1961) *A Chemical Approach to Food and Nutrition.* London.
Frankfort, H. (1927) 'Report on Excavations at Tell el-Amarnah: 1926-7', *JEA* 13: 209-18.
Frankfort, H. and J. D. S. Pendlebury (1933) *City of Akhenaten*, II. London.
Gaillard, C. and G. Daressy (1905) *La Faune Momifiée de l'Antique Égypte.* Cairo.
Gaillard, C. (1912) 'Les Tatonnements des Egyptiens de l'Ancien Empire a la Recherche des Animaux a Domestiquer', *Revue d'Ethnographie et de Sociologie*, 3.11-12: 329-48. Paris.
—, (1923) *Recherches Sur les Poissons Representes dans Quelques Tombeaux Egyptiens de l'Ancien Empire.* Cairo.
—, (1934) *Contribution à l'Étude de la Faune Préhistorique de l'Égypte.* Lyon.
Gamer-Wallert, I. (1970) *Fische und Fischkulte im alten Ägypten.* Wiesbaden.
Gardiner, A. H. (1915) *The Tomb of Amenemhet.* London.
—, (1941) 'Ramesside Texts Relating to the Taxation and Transport of Corn', *JEA* 27: 19-73.
—, (1947a) *Ancient Egyptian Onomastica*, I. Oxford.
—, (1947b) *Ancient Egyptian Onomastica*, II. Oxford.
—, (1948) *Ramesside Administrative Documents.* Oxford.

—, (1979) *Egyptian Grammar*. Oxford.
—, (1932) Notes on the Food Cases from the Tomb of Tutankhamun. Griffith Institute, Oxford.
Garnsey, P. (ed.) (1989) *Food, Health and Culture in Classical Antiquity*. Cambridge.
Garstang, J. (1907) *The Burial Customs of Ancient Egypt*. London.
—, (1902) *Mahasna and Bet Khallaf*. London.
Gautier, A. (1980) 'Contributions to the Archaeozoology of Egypt', *Prehistory of the Eastern Sahara*: 317-44 (F. Wendorf and R. Schild, eds.). New York.
—, (1987) 'Fishing, fowling and Hunting in Late Palaeolithic Times in the Nile Valley in Upper Egypt', *Palaeoecology of Africa and the Surrounding Islands*, 18: 429-40. (J. A. Coetzee, ed.). Rotterdam.
Gerrard, F. (1964) *Meat Technology*. London.
Ghalioungui, P. (1983) *The Physicians of Pharaonic Egypt*. Mainz.
—, (1973) *Magic and Medical Science in Ancient Egypt*. Amsterdam.
Ghazouli, E. B. (1964) 'The Palace and Magazines Attached to the Temple of Sety I at Abydos and the Facade of this Temple', *ASAE* 58: 99-186.
Ghoneim, W. (1977) *Die Okonomische Bedeutung des Rindes im Alten Agypten*. Bonn.
Gibson, C. E. (1948) *The Story of the Ship*. New York.
Giddy, L. L. (1987) *Egyptian Oases*. Warminster.
—, (1990) 'Memphis, 1989', *JEA* 76: 1-16.
Gilbert, A. S. (1988) 'Zooarchaeological Observations on the Slaughterhouse of Meketre', *JEA* 74: 69-89.
Girard, M. P. S. (1813) 'Memorie sur l'Agriculture, l'Industrie et le Commerce de l'Egypte', *Description de l'Egypte*, 2. Paris.
Godley, A. D. (1946) *Herodotus*, I. London.
Goedert, A. W. and S. K. Maddox. (1932) *The Packers' Encyclopaedia*. Chicago.
Goodman, S. (1987) 'Victual Egyptian Bird Mummies from a Presumed Late 17th or Early 18th Dynasty Tomb', *JSSEA* 17.3: 67-77.
—, (1990) Personal Communication.
Gottlicher, A. and W. Werner (1971) *Schiffsmodelle im Alten Agypten*. Wiesbaden.
Graf, W. (1949) 'Preserved Histological Structures in Egyptian Mummy Tissues and Ancient Swedish Skeletons', *Acta Anatomica* 8: 236-50. Basel.
Grdseloff, B. (1951) 'Nouvelles Données Concernant La Tente de Purification', *ASAE* 51: 129-40.
Green, A. A. (1967) 'The Protection of Dried Sea-fish in South Arabia from Infestation by *Dermestes frischii* Kug', *JSPR* 2.4: 331-50.
Green, L. (1992) Personal Communication.
Griffith, F. Ll. (1891) 'The Account Papyrus no. 18 of Boulaq', *ZÄS* 29: 102-16.
—, (1896) *Beni Hasan*, 3. London.
—, (1898) *Hieratic Papyri from Kahun and Gurob*. London.
—, (1909) *Catalogue of the Demotic Papyri in the John Rylands Library Manchester*, 3. London.
—, (1926) 'A Drinking Syphon from Tell el-Amarnah', *JEA* 12: 22-3.
—, (1927) 'The Abydos Decree of Seti I at Nauri', *JEA* 13: 193-208.

Griffith, F. Ll. and J. J. Tylor (1894) *The Tomb of Paheri*. London.
Grinsell, L. V. (1972) *Guide Catalogue to the Collections from Ancient Egypt*. Bristol.
Handoussa, T. (1988) 'Fish Offerings in the Old Kingdom', *MDAIK* 44: 105-10.
Harpur, Y. (1987) *Decoration in Egyptian Tombs of the Old Kingdom*. London.
Harris, J. R. (1961) *Lexicographical Studies In Ancient Egyptian Minerals*. Berlin.
Harris, M. (1985) *The Sacred Cow and the Abominable Pig*. New York.
—, (1980) *Culture, People, Nature*. New York.
—, (1974) *Cows, Pigs, Wars, and Witches: The Riddles of Culture*. New York.
Harris, R. S. and H. von Loesecke (eds.) (1971) *Nutritional Evaluation of Food Processing*. Westport, Conn.
Hassan S. (1932-44) *Excavations at Giza,* 1-5. Cairo.
—, (1946) *Excavations at Giza,* 6.1. Cairo.
—, (1950) *Excavations at Giza*, 6.3. Cairo.
—, (1938) 'Excavation at Saqqara', *ASAE* 38: 503-22.
—, (1955) 'The Causeway of *Wnis* at Sakkara' *ZÄS* 80: 136-9.
Hayes, W. C. (1951) 'Inscriptions from the Palace of Amenhotep III', *JNES* 10: 35-56; 82-112; 156-83.
—, (1953) *The Scepter of Egypt,* 1. New York.
—, (1959) *The Scepter of Egypt*, 2. New York.
Heath, J. (1990) Personal Communication.
—, (1992) Personal Communication.
Hecker, H. M. (1982) 'A Zooarchaeological Inquiry into Pork Consumption in Egypt from Prehistoric to New Kingdom Times', *JARCE* 19: 59-71.
Helck, W. (1961-69) *Materialien zur Wirtschaftsgeschichte des Neuen Reiches*, 1-6. Wiesbaden.
—, (1964) 'Feiertage und Arbeitstage in der Ramessidenzeit', *JESHO* 7: 136-66.
—, (1977) 'Die 'Weihinschrift' aus dem Taltempel des Sonnenheiligtums des Konigs Neuserre bei Abu Gurob', *SAK* 5: 47-77.
—, (1967) 'Eine Briefsammlung aus der Verwaltung des Amuntempels', *JARCE* 6: 135-51.
Hill, A. (19) 'On Carnivore and Weathering Damage to Bone', *Current Anthropology* 12.ii: 335-36. Chicago.
Hodges, H. (1970) *Technology in the Ancient World*. London.
Hoffman, M. (1979) *Egypt Before the Pharaohs*. New York.
Hölscher, U. (1912) *Das Grabdenkmal des Königs Chephren*. Leipzig.
—, (1930) 'Medinet Habu Studies 1928/29: The Architectural Survey', *OIC* 7.
—, (1931) 'Medinet Habu Reports: The Architectural Survey 1929/30', *OIC* 10.
—, (1932) 'Excavations at Ancient Thebes 1930/31: The Ramasseum',*OIC* 15.
—, (1934) *Excavations of Medinet Habu*, 1. Chicago.
—, (1941) *The Mortuary Temple of Ramses III*. Chicago.
Hope, C. A. (1977) *Malkata: Jar Sealings and Amphorae*. Warminster.
—, (1989) *Pottery of the Egyptian New Kingdom: 3 Studies*. Victoria, Australia.
Hornblower, G. D. (1927) 'An Humped Bull of Ivory', *JEA* 13: 222-5.
Hornell, J. (1943) 'On the Carrying Capacity of Ramesside Grain-Ships', *JEA* 29: 76-8.

—, (1947) 'Naval Activity in the Days of Solomon and Rameses III', *Antiquity* 21: 66-73.
Houlihan, P. F. and S. M. Goodman. (1986) *The Birds of Ancient Egypt*. Warminster.
Houlihan, P. F. (1995) *The Animal World of the Pharaohs*. Cairo.
Howie, J. W. (1962) 'Pathogenic Organisms in Food', *Recent Advances in Food Science* (J. Hawthorn and M. Leitch, eds.). London.
Huang, A. (1992) Personal Communication.
Huard, P. (1964) 'À Propos des Bucrânes à Corne Déformée de Faras', *Kush* 12: 63-81.
Hume, R. and M. Downes (1963) *The Cordon Bleu Cookery Book*. Bungay.
Ingram, M. (1962) 'The Bacteriology of Meat Curing Brines', *Recent Advances in Food Science* (J. Hawthorn and M. Leitch, eds.). London.
Jackson, Dr. (1992) Personal Communication.
Jackson, R. (1988) *Doctors and Disease in the Roman Empire*. London.
James, T. G. H. (1962) *The Hekanakhte Papers and Other Middle Kingdom Documents*. New York.
Janssen, J. J. (1961) *Two Ancient Egyptian Ship's Logs*. Leiden.
—, (1975a) *Ancient Egyptian Commodity Prices*. Leiden.
—, (1975b) 'Prolegomena to the Study of Egypt's Economic History During the New Kingdom', *SÄK* 3: 127-85.
—, (1979) 'The Role of the Temple in the Egyptian Economy During the New Kingdom', *State and Temple Economy in the Ancient Near East*, 2: 505-16 (E. Lipinski, ed.). Leuven.
Jeffreys, D. G. and L. L. Giddy. (1989) 'Memphis, 1988', *JEA* 75: 1-12.
Jeffreys, D. G. et al. (1986) 'Memphis, 1984', *JEA* 72: 1-14.
Jéquier, G. (1921) *Les Frises d'Objets des Sarcophages du Moyen Empire*. Cairo.
—, (1929) *Tombeaux de Particuliers Contemporains de Pepi II*. Cairo.
—, (1940) *Le Monument Funéraire de Pepi II*, 3. Cairo.
Jones, H. L. (1959) *The Geography of Strabo*, 8. London.
Jones, P. E. (1976) *The Butchers of London*. London.
Jourdan, L. (1980) *Sacrifice de Moutons dans les tombes Kerma de l'Ile de Sai*. Valbonne.
Junker, H. (1910) 'Die Schacht- und Brandopfer und ihre Symbolik im Tempelkult der Spatzeit', *ZÄS* 48: 69-77.
—, (1938) *Giza*, 3. Vienna.
—, (1943) *Giza*, 6. Vienna.
—, (1953) *Giza*, 11. Vienna.
Kamal, A. B. (1906-09) *Tables d'Offrandes*, 1, 2. Cairo.
Kanawati, N. (1980) *The Rock Tombs of El-Hawawish*, 1. Sydney.
—, (1981) *The Rock Tombs of El-Hawawish*, 2. Sydney.
Kane, J. (1967) 'Short Communication', *JSPR* 2.3: 251-5.
Kantor, H. (n.d.) *Sonderdruck aus der Propyläen Kuntstgeschichte*. Berlin.
Kees, H. (1933) *Kulturgeschichte des Alten Orients: Agypten*. Munich.
—, (1942) 'Bemerkungen zum Tieropfer der Ägypter und seiner Symbolik', *Nachrichten von der Akademie der Wissenschaften im Göttingen, Philologisch-historische Klasse*, 1942: 71-88. Göttingen.
Keimer, L. (1935) 'Le Sanglier Egyptien', *CdE* 19: 26-33.

—, (1937) 'Remarques sur le Porc et le Sanglier dans l'Egypte Ancienne', *BIE* 19: 147-56.
—, (1939) 'La boutargue dans l'Egypte Ancienne', *BIE* 21: 215-43.
—, (1940) 'Sur un Monument Egyptien du Musée du Louvre', *RdE* 4: 45-65.
—, (1948) 'Quelques Representations Rares de Poissons Egyptiens Remontant a l'Epoque Pharaonique', *BIE* 29: 263-74.
—, (1949/50) 'Notes Prises Chez les Bisarin et les Nubiens des Assouan', *BIE* 32: 49-101.
—, (1951) 'Les Oignons et les Poissons de Cham-en-Nessim',*Cahiers d'Histoire Egyptienne*, 3: 4: 1-5. Paris.
Kelterborn, P. (1990) 'In Search of a Typology for Egyptian Flint Knives', *Cahiers du Quaternaire no 17 — Le silex de sa Genès à l'Outil. Actes du V°* Colloque International sur le Silex : 599-602. Paris.
Kemp, B. J. and D. O'Connor (1974) 'An Ancient Nile Harbour, University Museum Excavations at the 'Birket Habu', *The International Journal of Nautical Archaeology and Underwater Exploration*, 3: 101-36. London.
Kemp, B. J. (1968) 'The Osiris Temple at Abydos', *MDAIK* 68:138-55.
—, (1972) 'Temple and Town in Ancient Egypt', *Man, Settlement and Urbanism*: 657-78 (P. J. Ucko et al, eds.). London.
—, (1984-89) *Amarna Reports*, 1-5. London.
—, (1991) *Ancient Egypt: Anatomy of a Civilization*. London.
—, (1992) Personal Communication.
Khattab, M. (1991) Personal Communication.
Killen, G. (1980) *Ancient Egyptian Furniture*, 1. Warminster.
—, (1992) Personal Communication.
Kitchen, K. A. (1974) *Ramesside Inscriptions: Historical and Biographical*, I: 6. Oxford.
Klebs, L (1915) *Die Reliefs und Malereien des Alten Reiches*. Heidelberg.
—, (1922) *Die Reliefs und Malereien des Mettleren Reiches*. Heidelberg.
—, (1934) *Die Reliefs und Malereien des Neuen Reiches*. Heidelberg.
Knowles, N. R. (1962) 'The Preservation of Eggs', *Recent Advances in Food Science* (J. Hawthorn and M. Leitch, eds.). London.
Koenig, Y. (1979) *Catalogue des Etiquettes de Jarres Hieratiques de Deir el Medina, #6000-6241*. Cairo.
Koraiem, S. and I. (1989) Personal Communication.
Koraiem, S. (1990) Personal Communication.
Kronenberg, A. (1961) 'The Longarim Favourite Beast', *Kush* 9: 258-77.
Krauss, R. (1984) 'Reisegeschwindigkeit' *LÄ* V: 222-3.
Kruchten, J. M. (1981) *Le Décret d'Horemheb*. Brussels.
Kuentz, C. (1981) *Bassins et Tables d'Offrandes*. Cairo.
Lacau, P. (1903) *Sarcophages Anterieurs au Nouvel Empire*. Cairo.
—, (1913) 'Suppressions et Modifications de Signes dans les Textes Funéraires', *ZÄS* 51: 1-64.
Lacovara, P. (1990) *Deir el-Ballas*. Indiana.
Landstrom, B. (1970) *Ships of the Pharaohs*. London.
Lane, E. W. (1944) *The Manners and Customs of the Modern Egyptians*. London.
Lapedes, D. N. (ed.) (1977) *McGraw-Hill Encyclopedia of Food, Agriculture and Nutrition*. New York.

Larsen, H. (1935) 'Vorbericht uber die Schwedischen Grabungen in Abu Ghalib 1932-34' *MDAIK* 6: 41-87.

Lawrie, R. A. (1976) 'Problems of Beef Quality in Relation to Transportation, Slaughter and Storage', *Beef Cattle Production in Developing Countries* (A. J. Smith, ed.). Edinburgh.

Leahy, M. A. (1978) *Excavations at Malkata and the Birket Habu 1971-74: The Inscriptions*. Warminster.

Leclant, J. (1951) 'Le Role du Lait et de l'Allaitement, d'Apres les Textes des Pyramides', *JNES*, 10: 123-27.

—, (1979) *Récherches dans la Pyramide et au Temple Haut du Pharaon Pepi I à Saqqarah*. Leiden.

Leek, F. F. (1966) 'Observations on the Dental Pathology Seen in Ancient Egyptian Skulls', *JEA* 52: 59-64.

Lefebure, M. E. (1889) *Les Hypogées Royaux de Thebes*. Paris.

Lefebvre, G. (1960) 'Lait de Vache et Autres Laits en Égypte', *RdE* 12.: 59-65.

Legge, A. J. (Unpublished) 'Animals and Economy at Malkata'.

Legge, A. J. and P. A. Rowley-Conwy (1988) *Star Carr Revisited*. London.

Leibovitch, J. (1953) 'Une Scène de Sacrifice Rituel Chez les Anciens Egyptiens', *JNES* 12: 59-60.

Leprohon, R. L. (1988) 'Cultic Activities in the Temples at Amarna', *The Akhenaten Temple Project*, 2 (D. B. Redford, ed.). Toronto.

Lichtheim, M. (1976) *Ancient Egyptian Literature*, 2. Los Angeles.

—, (1980) *Ancient Egyptian Literature*, 3. Berkeley.

Lloyd, A. (1976) *Herodotus: Book II Commentary*, 1. Leiden.

Lloyd, A. et al. (1990) *Saqqara Tombs*, 2. London.

Loret, V. (1898) 'Le Tombeau de Touthmosis III et Le Tombeau d'Amènophis II', *BIE* 9: 91-112.

Lortet, L. and C. Gaillard (1905-09) *La Faune Momifiée de l'Ancienne Egypte*. Lyon.

Lucas, A. (1962) *Ancient Egyptian Materials and Industries*. London.

Lushbough, C. H. and B. S. Schweigert. (1960) 'The Nutritive Content and the Nutritional Value of Meat and Meat Products', *The Science of Meat and Meat Products* (American Meat Institute Foundation, ed.). New York.

Lyons, Sir Henry. (1926) 'Two Notes on Land Measurement in Egypt', *JEA* 12: 242-44.

Lythgoe, A. M. and D. Dunham (1965) *The Predynastic Cemetery N7000; Naga-ed-Der*, 4. Berkeley.

—, (1902-04) *Field Notes*. Berkeley and Boston.

Mace, A. and H. Winlock (1916) *The Tomb of Senebtisi at Lisht*. New York.

Macdonald, K. C. (1992) 'The Domestic Chicken (*Gallus gallus*) in Sub-Saharan Africa: A Background to its Introduction and its Osteological Differentiation from Indigenous Fowls (Numidinae and *Francolinus* sp.), *JAS* 19: 303-17.

Mackay, E. et al. (1929) *Bahrein and Hemamieh*. London.

Malek, J. (1988) 'The 'Altar' in the Pillared Court of Teti's Pyramid-Temple at Saqqara', *Pyramid Studies and Other Essays*, (J. Baines, ed.). London.

Manniche, L. (1989) *An Ancient Egyptian Herbal*. London.

Mariette, A. (1869) *Abydos: Description des Fouilles*, 1. Paris.

—, (1880) *Abydos: Description des Fouilles*, 2. Paris.

—, (1889) *Les Mastabas de l'Ancien Empire*. Paris.
Marshall, F. (1989) 'The Role of *Bos indicus* in Sub-Saharan Africa', *Current Anthropology* 30: 235-40. Chicago.
Martin, G. T. (1979) *The Tomb of Hetepka*.. London.
—, (1987) *Corpus of Reliefs of the New Kingdom from the Memphite Necropolis and Lower Egypt*. London.
—, (1989a) *The Memphite Tomb of Horemheb*, 1. London.
—, (1989b) *The Royal Tomb at el-ʿAmarna*, 2. London.
Mason, I. L. (ed.) (1984) *Evolution of Domesticated Animals*. London.
McElroy, A. and P. K. Townsend (1985) *Medical Anthropology in Ecological Perspective*. Boulder.
McLaren, D. S. and M. M. Meguid (1988) *Nutrition and its Disorders*. London.
McNitt, J. I. (1983) *Livestock Husbandry Techniques*. London.
Meeks, D. (1972) *Le Grand Texte des Donations au Temple d'Edfou*. Cairo.
—, (1979) 'Les Donations aux Temples Dans l'Egypte du Ier Millenaire Avant J-C', *State and Temple Economy in the Ancient Near East*, 2: 605-88 (E. Lipinski, ed.). Leuven.
—, (1980) *Année Lexicographique*, 1. Paris.
Megally, M. (1974) 'À Propos du Papyrus CGC 58070', *BIFAO* 74: 161-9.
—, (1975) 'Le Papyrus CGC 58081 Suite du Papyrus CGC 58070', *BIFAO* 75: 165-81.
—, (1977) *Récherches sur l'Economie, l'Administration et la Comptabilite Egyptiennes à la XVIIIe Dynastie*. Cairo.
Miller, R. (1990) 'Hogs and Hygiene', *JEA* 76: 125-40.
—, (1990) Personal Communication.
Millet, N. et al. (1980) 'ROM I: Mummification for the Common People', *Mummies, Disease and Ancient Cultures*, (A. and E. Cockburn, eds). Cambridge.
Minault-Gout, A. (1980) 'Fouilles du Mastaba II' *BIFAO* 80: 271-86.
—, (1981) 'Fouilles du Mastaba II a Balat' *BIFAO* 81: 207-14.
Moens, M-F. and W. Wetterstrom. (1988) 'The Agricultural Economy of an Old Kingdom Town in Egypt's Western Delta: Insight from the Plant Remains', *JNES* 47: 159-73.
Montagne, P. and Dr. Gottschalk. (1961) *Larousse Gastronomique*. London.
Montet, P. (1910) 'Les Scenes de Boucherie dans les Tombes de l'Ancien Empire', *BIFAO* 7: 41-65.
—, (1914) 'Les Poissons Employés dans l'Ecriture Hieroglyphique', *BIFAO* 11: 39-48.
—, (1925) *Les Scènes de la Vie Privée Dans les Tombeaux Egyptiens de l'Ancien Empire*. Paris.
—, (1954) 'Les Bœufs Égyptiens', *Kemi* 13: 43-58.
Morgensen, M. (1921) *Le Mastaba Egyptien de la Glyptotheque Ny Carlsberg*. Copenhagen.
Morkot, R. (1990) '*Nb-M3ʿt- Rʿ* — United-With-Ptah', *JNES* 49: 323-37.
Morris, E. O. (1962) 'Effects of Environment on Micro-Organisms', *Recent Advances in Food Science* (J. Hawthorn and M. Leitch, eds.). London.
Mossel, D. A. A. et al. (1975) 'Microbial Problems in Handling and Storage of Fresh Meats', *Meat* (D. J. A. Cole, ed.). London.

Moustafa, Y. S. (1964) 'The Domesticated Animals of the Sekhem-khet Step Pyramid', *ASAE* 58: 255-65.
Muhammed, F. N. (1989) Personal Communication.
Muller, H. W. (1940) *Die Felsengraber der Fursten von Elephantine*. Hamburg.
Murray, M. (1904) *Sakkara Mastabas*, I. London.
Murray, H. and M. Nuttall. (1963) *A Handlist to Howard Carter's Catalogue of Objects in Tutankhamun's Tomb*. Oxford.
Mushi, A. M. and H. C. Chiang (1974) 'Short Communication', *JSPR* 10.1: 57-60.
Naville, E. (1894) *Ahnas el Medineh*. London.
—, (1898) *Deir el Bahari*, 3. London.
—, (1901) *Deir el Bahari*, 4. London.
—, (1907-13) *The XI Dynasty Temple at Deir el Bahari*, 1-3. London.
—, (1930) *Détails Relevés dans les Ruines de Quelques Temples Égyptiens*. Paris.
Nelson, H. (1934) *Medinet Habu-III*. Chicago.
Nelson, H. and U. Hölscher (1934) *Work in Western Thebes 1931-33*. (*OIC*, 18).
Newberry, P. E. (1893a) *Beni Hasan*, 1. London.
—, (1893b) *Beni Hasan*, 2. London.
—, (1894) *El Bersheh,* 1. London.
—, (1900) *The Life of Rekhmara*. Westminster.
—, (1928) 'The Pig and the Cult Animal of Set',*JEA* 14: 211-25.
—, (1942) 'Notes on Seagoing Ships', *JEA* 28: 64-5.
Newton, F. G. (1924) 'Excavations at el-Amarnah, 1923-24', *JEA* 10: 289-98.
Nibbi, A. (1975) 'Egyptian Anchors', *JEA* 61: 38-41.
—, (1979) 'A Fifth Dynasty Anchor from Abusir', *GM* 32: 39-44.
—, (1992) Personal Communication.
Nimr (1992) Personal Communication.
Niven, Jr., C. F. and W. R. Shesbro (1960) 'Meat Preservation: Chemicals and Antibiotics', *The Science of Meat and Meat Products* (American Meat Institute Foundation, eds.). New York.
Niven, Jr., C. F. (1960) 'Factors Influencing Quality of Cured Meats', *The Science of Meat and Meat Products* (American Meat Institute Foundation, eds.). New York.
No Author (1938) *The Packer's Encyclopedia Sausage and Meat Specialties*, 3. Chicago.
No Author (1976) *Manual of Nutrition*. London.
Notaris, H. (1990) Personal Communication.
Nowak, R. M. and J. L. Paradiso (1983) *Walker's Mammals of the World*, 1-2. Baltimore.
Ogden, J. (1992) Personal Communication.
Oldfather, C. H. (1946) *Diodorus Siculus*, 1. London.
Oldham, L. (1990) *Sociocultural Factors Influencing the Prevalence of Diarrheal Disease in Rural Upper Egypt*. Cairo.
Osborn, D. (1991) Personal Communication.
Otto, E. (1950) 'An Ancient Egyptian Hunting Ritual', *JNES* 9: 164-77.
Page, J. (1991) Personal Communication.

Paget, F. F. E. and A. A. Pirie. (1898) *The Tomb of Ptahhotep*. London.
Papa, D. (1990) Personal Communication. Oxford.
Parker, R. A. (1950) *The Calendars of Ancient Egypt*. Chicago.
Parlebas, J. (1977) 'Les Egyptiènnes Villes', *Ktema* 2: 49-57. Strasbourg.
—, (1980) 'La Condition des Paysans Egyptiens à l'Epoque Pharonique', *Ktema* 5: 131-6. Strasbourg.
Passmore, R. and M. A. Eastwood (1986) *Davidson and Passmore: Human Nutrition and Dietetics*. London.
Paton, D. (1925) *Animals of Ancient Egypt*. Princeton.
Paul, A. A. and D. A. T. Southgate (1978) *The Composition of Foods*. London.
Payne, S. (1973) 'Kill-off Patterns in Sheep and Goats: The Mandibles from Asvan Kale', *Anatolian Studies* 23: 281-303.
Pearson, A. M. and F. W. Tauber. (1984) *Processed Meats*. Westport, Conn.
Peet, T. E. and L. Woolley. (1923) *The City of Akhenaten*, I. London.
Peet, T. E. (1934) 'The Unit of Value *šʿty* in Papyrus Bulaq 11', *Melanges Maspero*, 1. Cairo.
—, (1914) *The Cemeteries of Abydos*, 2. London.
—, (1931) 'An Ancient Egyptian Ship's Log', *BIFAO* 30: 481-90.
Pellet, P. L. and S. Shadarevian. (1970) *Food Composition Tables for use in the Middle East*. Beirut.
Pendlebury, J. D. S. (1951) *The City of Akhenaten*, 3.i-ii. London.
Peterson, R. (1974) *A Field Guide to the Birds of Britain and Europe*. London.
Petrie, W. M. F. and G. Brunton (1924) *Sedment*, 1-2. London.
Petrie, W. M. F. and J. E. Quibell (1896) *Naqada and Ballas*. London.
Petrie, W. M. F., E. McKay and G. Wainwright (1912) *The Labyrinth, Gerzeh, and Mazghuneh*. London.
Petrie, W. M. F. et al. (1913) *Tarkhan I and Memphis V*. London.
Petrie, W. M. F. (1889) *Hawara, Biahmu, and Arsinoe*. London.
—, (1890) *Kahun, Gurob and Hawara*. London.
—, (1891) *Illahun, Kahun and Gurob*. London.
—, (1892) *Medum*. London.
—, (1894) *Tell el-Amarna*. London.
—, (1896) *Koptos*. London.
—, (1897) *Six Temples at Thebes*. London.
—, (1898) *Deshasheh*. London.
—, (1900) *Denderah*. London.
—, (1902-03) *Abydos*, 1-2. London.
—, (1937) *The Funerary Furniture of Egypt*. London.
Pinch, G. (1992) Personal Communication.
Platt, B. S. (1962) *Tables of Representative Values of Foods Commonly Used in Tropical Countries*. Privy Council Medical Research Council Special Report Series #302. London.
Podzorski, P. (1990) *Their Bones Shall Not Perish*. London.
—, (1993) 'The Correlation of Skeletal Remains and Burial Goods: An Example from Naga-ed-der N 7000', *Proceedings from the Conference on Bio-archaeology* (V. W. Davies and R. Walker, eds.). London.
Pomeroy, R. W. (1962) 'Anatomical Distribution of Edible Tissues in Carcasses', *Recent Advances in Food Science* (J. Hawthorn and M. Leitch, eds.). London.

Posener-Kriéger, P. and J-L. de Cenival (1968) *Hieratic Papyri in the British Museum*, 1-2. London.

Posener-Kriéger, P. (1975) 'Les Papyrus de Gebelein, Remarques Preliminaires' *RdE* 27: 211-21.

—, (1976) *Les Archives du Temple Funéraire de Neferirkare-Kakai*, 1-2. Cairo.

—, (1979) 'Les Papyrus d'Abousir et l'Economie des Temples Funeraires de l'Ancien Empire', *State and Temple Economy in the Ancient Near East*, 1: 133-52. (E. Lipinski, ed.). Leuven.

Price, F. G. H. (1897) *A Catalogue of the Egyptian Antiquities in the Possession of F. G. Hilton Price*. London.

Proctor, D. L. (1972) 'The Protection of Smoke-dried Fresh-water Fish from Insect Damage During Storage in Zambia' , *JSPR* 8.2: 146-9.

Putnis, C. (1992) Personal Communication.

Pyke, M. (1970) *Man and Food*. London.

Quibell, J. E. (1898a) *El Kab*. London.

—, (1898b) *The Ramesseum*. London.

—, (1900) *Hierakonpolis*, I. London.

—, (1908) *Tomb of Yuaa and Thuiu*. Cairo.

Quirke, S. (1986) 'The Regular Titles of the Late Middle Kingdom', *RdE* 37: 107-30.

—, (1990) *The Administration of Egypt in the Late Middle Kingdom* . Kingston-upon-Thames.

Rabbi Toledano (1992) Personal Communication.

Radcliffe, W. (1921) *Fishing From the Earliest Times*. London.

Rebeyrol, Y. (1992) 'Les Beaux Restes des Outils de Pierre', *Le Monde* Wednesday, 12 February: 18. Paris.

Redford, D. B. (1988) *The Akhenaten Temple Project*, 2. Toronto.

Reed, C. A. (1959) 'Animal Domestication in the Prehistoric Near East', *Science*, 130#3389: 1629-39. Washington, DC.

Rees, B. (1989) Personal Communication.

Reeves, C. N. (1990) *Valley of the Kings*. London.

Reisner, G. A. (1905) *The Hearst Medical Papyrus*. Leipzig.

—, (1918) 'The Tomb of Hepzefa, Nomarch of Siut', *JEA* 5: 79-98.

—, (1913) *Excavation Notes from the Giza Necropolis*.

—, (1931) *Mycerinus*. Cambridge, Mass.

—, (1942) *A History of the Giza Necropolis*, 1. Cambridge, Mass.

Reymond, E. A. E. (1974) 'Fragment of a Temple Account Roll', *JEA* 60: 189-99.

Ricke, H. and S. Schott. (1950) *Beitrage zur Agyptischen Bauforschung und Altertumskunde*. Cairo.

Robins, G. (1991) 'Book Review of *Fish and Fishing in Ancient Egypt*', *Discussions in Egyptology* 19: 103-6. Oxford.

Rodinson, S. M. (1949) 'Recherches sur les documents arabes relatifs à la cuisine', *Revue des Études Islamiques*: 95-165. Paris.

Roeder, G. (1969) *Amarna-Reliefs Aus Hermopolis*. Hildesheim.

Romer, J. (1982) *Romer's Egypt*. London.

Rose, P. (1990) Personal Communication.

Rosen, A. M. (1991) 'Microartifacts and the Study of Ancient Societies', *Biblical Archaeologist* (June) 54.2: 97-103. New Haven.

Roth, A. M. (1991 unpub.) 'The Origin of the Opening of the Mouth Ritual' paper from the 5th International Congress (1991) of Egyptology, Turin.
Roth, A. M. (1993) "Fingers, stars, and the 'Opening of the Mouth': The Nature and Function of the Ntrwj-Blades", JEA 79: 57-79.
Rothenberg, B. (1988) *The Egyptian Mining Temple at Timna*. London.
Ruffer, A. (1919) 'Food in Ancient Egypt', *Mémoires Présentés à l'Institut d'Égypte* 1: 1-88. Cairo.
Rushdy, M. E. (1912) 'The Treading of Sown Seed by Swine', *ASAE* 11:162-3.
Ryan, D. P. (1989) 'The Pacific Lutheran University Valley of the Kings Project: A Synopsis of the First (1989) Season', *NARCE* 146: 8-10.
—, (1990) 'Who is Buried in KV60?', *KMT* 1: 34-9, 63.
—, (1992) Personal Communication.
Saad, Z. (1951) *Royal Excavations at Helwan*. Cairo.
Saleh, M. (1977) *Three Old Kingdom at Thebes*. Mainz.
Saffirio, L. (1972a) 'L'Alimentazione Umana nell'antico Egitto', *Aegyptus* 52: 19-66. Milan.
—, (1972b) 'Food and Dietary Habits in Ancient Egypt', *JHE* 1/3: 297-306.
Sauneron, S. (1959) 'L'Avis des Egyptiens sur la Cuisine Soudanaise', *Kush* 7: 63-70.
—, (1977) 'Une Statue-Cube de l'Epoque Bubastite', *BIFAO* 77: 23-8.
—, (ed.) (1974) *Voyage en Egypte d'Edward Brown 1673-74*. Paris.
Säve-Söderbergh, T. (1957) *Four Eighteenth Dynasty Tombs at Thebes*. Oxford.
—, (1953) *On Egyptian Representations of Hippopotamus Hunting as a Religious Motive*. Uppsala.
Schäfer, H. (1908) *Priestegraber und Andere Grabfunde vom Ende des Alten Reiches bis zur Griechischen Zeit vom Totentempel des Ne-User-Re*. Leipzig.
—, (1918) 'Vorläufiger Bericht über die Ausgrabungen bei Abu Sir', *ZÄS* 37: 1-9.
—, (1899) 'Altes und Neues zur Kunst und Religion von Tell el-Amarna', *ZÄS* 55: 1-43.
Schiaparelli, E. (1924) *Relazione sui Lavori della Missione Archeologica Italiana in Egitto*, 1: *Esplorazione della 'Valle delle Regine'*. Turin.
—, (1927) *Relazione sui Lavori della Missione Archeologica Italiana in Egitto*, 2: *La Tomba Intatta Dell'Architetto Cha*. Turin.
Schrumpf-Pierron, P. (1931) 'La Teneur en Mineraux de la Nourriture du Fellah', *BIE* 13: 153-76.
Schweinfurth, G. and L. Lewin (1897) 'Der Salzfund von Qurna', *ZÄS* 35: 142-3.
Seton-Williams, V. and P. Stocks (1988) *Blue Guide: Egypt*. London.
Shennum, D. (1977) *English-Egyptian Index of Faulkner's Concise Dictionary of Middle Egyptian*. Malibu.
Shewan, J. M. (1962) 'The Bacteriology of Fresh and Spoiling Fish and Some Related Chemical Changes', *Recent Advances in Food Science* (J. Hawthorn and J. Leitch, eds.). London.
Shinnie, P. L. (1978) *Debeira West: A Mediaeval Nubian Town*. Warminster.
Silver, I. A. (1969) 'The Ageing of Domestic Animals', *Science in Archaeology* (D. Brothwell and E. S. Higgs, eds.). London.
Simmonds, P. L. (1859) *The Curiosities of Food*. London.
Simoons, F. J. (1961) *Eat Not This Flesh: Food Avoidances in the Old World*. Madison.

Sisson, S. (1940) *The Anatomy of the Domestic Animals*. Philadelphia.
Smith, D. B. and A. H. Walters (1967) *Introductory Food Science*. London.
Smith, E. B. (1938) *Egyptian Architecture as Cultural Expression*. London.
Smith, G. E. and W. R. Dawson (1924) *Egyptian Mummies*. London.
Smith, H. S. (1969) 'Animal Domestication and Animal Cult in Dynastic Egypt', *The Domestication and Exploitation of Plants and Animals*, (P.J. Ucko and G. W. Dimbleby, eds). London.
—, (1976) *The Fortress of Buhen*. London.
Smith, R. W. and D. B. Redford (1976) *The Akhenaten Temple Project*, 1. Warminster.
Smither, P. C. (1941) 'Tax Assessor's Journal of the Middle Kingdom', *JEA* 27: 74-6.
Snead, P. G. (1972) *Beef Production in Zambia*. Lusaka.
Spalinger, A. (1985) 'Notes on the Day Summary Accounts of P. Bulaq 18 and the Intradepartmental Transfers', *SAK* 12: 179-241.
—, (1985) 'A Redistributive Pattern at Assiut', *JAOS* 105: 7-20.
Spencer, A. J. (1991) Personal Communication.
Spencer, P. (1984) *The Egyptian Temple: a Lexicographical Study*. London.
Speth, J. D. (1983) *Bison Kills and Bone Counts*. Chicago.
Spiegelberg, W. (1900) 'Die Aegyptischen Worte für 'Schaf'', *RdT* 22: 212-14.
Starkey, D. (ed.) (1987) *The English Court: from the Wars of the Roses to the Civil War*. London.
Stewart, H. (1977) *Indian Fishing: Early Methods on the Northwest Coast*. Vancouver.
Stewart, H. M. (1979) *Egyptian Stelae, Reliefs and Paintings from the Petrie Collection*, 2. Warminster.
Strang, J. (1991) *Goose Fat and Garlic*. London.
Strudwick, N. (1985) *The Administration of Egypt in the Old Kingdom*. London.
Subira, M. E. and A. Malgosa (1992) 'Multi-element Analysis for Dietary Reconstructiion at a Balearic Iron Age Site', *International Journal of Osteoarchaeology* 2: 199-204. London.
Sullivan III, A. P. (1983) 'Storage and Nonedible Resource Processing, and the Interpretation of Sherd and Lithic Scatters in the Sonoran Desert Lowlands', *Journal of Field Archaeology* 10/3: 309-24. Boston, Mass.
Susman, L. (1992) Personal Communication.
Tannahill, R. (1973) *Food in History*. London.
Taylor, T. (trans.) (1965) *Porphyry: On Abstinence from Animal Food*. London.
Thilenius, G. (1900) 'Das Aegyptische Hausschaf', *RdT*, 22: 199-212.
Thorne, S. (1986) *The History of Food Preservation*. Cumbria.
Tooley, A. M. J. (unpublished) *Middle Kingdom Burial Customs: A Study of Wooden Models and Related Materials*. 1989 Ph.D. dissertation from Liverpool University.
Trigger, B. G., et al. (1983) *Ancient Egypt: A Social History*. Cambridge.
Turton, J. D. (1969) 'The Effect of Castration on Meat Production from Cattle, Sheep and Pigs', *Meat Production from Entire Male Animals* (D. N. Rhodes, ed.). London.
Tylor, J. J. (1896) *Wall Drawings and Monuments of El Kab*. London.
Tylor, J. and S. Clarke (1896) *The Tomb of Sebeknekht*. London.

Valbelle, D. (1985) *Les Ouvriers de la Tombe*. Cairo.
Valloggia, M. (1978) 'Fouilles à Balat', *BIFAO* 78: 65-80.
—, (1979) 'Rapport Preliminaire sur la Deuxieme Campagne de Fouilles du Mastaba V à Balat', *BIFAO* 79: 51-62.
—, (1980) 'Fouilles du Mastaba V', *BIFAO* 80: 97-128.
—, (1981) 'Rapport Preliminaire sur la Quatrieme Campagne de Fouilles du Mastaba V à Balat', *BIFAO* 81: 227-36.
—, (1986) *Balat I: Le Mastaba de Medou-Nefer*. Cairo.
—, (1991) Personal Communication.
Valtz, E. (1991) Personal Communication.
van den Brink, E. C. M. et al. (1989) 'A Transitional Late Predynastic-Early Predynastic Settlement Site in the Northeastern Nile Delta, Egypt', *MDAIK* 45: 55-108.
Vandier, J. (1936) *La Famine dans l'Égypte Ancienne*. Cairo.
—, (1964a) *Manuel d'Archeologie Egyptienne*, 4.i-ii. Paris.
—, (1964b) 'Quelques Remarques sur la Preparation de la Boutargue', *Kemi* 17: 26-34.
—, (1969) *Manuel d'Archeologie Egyptienne*, 5.i-ii. Paris.
—, (1978) *Manuel d'Archeologie Egyptienne*, 6. Paris.
Vassilika, E. (1989) *Ptolemaic Philae*. Leuven.
Vercoutter, J. (1970) *Mirgissa*, 1. Paris.
Verhoeven, U. (1984) *Grillen, Kochen, Backen im Alltag und im Ritual Altagyptens*. Bruxelles.
Verner, M. (1986) 'A Slaughterhouse from the Old Kingdom', *MDAIK* 42: 181-90.
—, (1991) Personal Communication.
von den Dreisch, A. (1986) 'Tierknochenfunde aus Qasr el-Sagha/Fayum', *MDAIK* 42: 1-8.
von Bissing, F. W. and H. Kees. (1927) 'Untersuchungen ze den Reliefs aus dem Re-Heiligtum des Rathures', *Abhandlungen der Bayerischen Akademine der Wissenschaften*, 32. Munich.
von Bissing, F. W. (1905a) *Das Re-Heiligtum des Königs Ne-Woser-Re*. Berlin.
—, (1905b) *Die Mastaba des Gem-ni-kai*. Berlin.
Watson, J. P. N. (1972) 'Fragmentation Analysis of Animal Bone Samples from Archaeological Sites', *Archaeometry* 14.ii: 221-8. Oxford.
Watson, P. J. (1979) *Archaeological Ethnography in Western Iran*. Tucson.
Webster, H. (1942) *Taboo: A Sociological Study*. Stanford.
Weltfish, G. (1965) *The Lost Universe*. New York.
Weigall, A. (1915) 'An Ancient Egyptian Funeral Ceremony', *JEA* 2: 10-2.
Weill, R. (1927) 'L'unite de valeur *shat* et le Papyrus de Boulaq No. 11', *RdE* 1: 45-87.
Weinstein, J. M. (1973) *Foundation Deposits in Ancient Egypt*. Ph.D. available from University of Michigan Imprints, Ann Arbor.
Wendorf, F. , A. E. Close and R. Schild. (1987) 'Early domestic Cattle in the Eastern Sahara', *Palaeoecology of Africa and the Surrounding Islands* , 18: 441-8. (J. A. Coetzee, ed.). Rotterdam.
Wendorf, F. et al. (1991) 'The Combined Prehistoric Expedition: Results of the 1990 and 1991 Seasons', *NARCE* 154: 1-8.

Wenke, R. (1985) 'Excavations at Kom el-Hisn', *NARCE* 129: 1-11.
—, (1990) 'Excavations at Kom el-Hisn', *NARCE* 149: 1-6.
Wenke, R. J. et al. (1988) 'Kom el-Hisn: Excavations of an Old Kingdom Settlement in th Egyptian Delta', *JARCE* 25: 5-34.
White, C. and H. P. Schwarcz (1989) 'Ancient Maya Diet: as Inferred from Isotopic and Elemental Analysis of Human Bone', *JAS* 16: 451-74.
Whitfield, P. (ed.) (1984) *Longman's Illustrated Animal Encyclopedia*. London.
Wild, H. (1953) *Le Tombeau de Ti*, 2.1. Cairo.
—, (1956) *Le Tombeau de Ti*, 2.2. Cairo.
—, (1966) 'Brasserie et Panification au Tombeau de Ti', *BIFAO* 64: 95-120.
Wilkinson, C. K. and M. Hill (1983) *Egyptian Wall Paintings*. New York.
Wilkinson, J. G. (1837) *The Manners and Customs of the Ancient Egyptians*, 1. London.
—, (1854) *The Manners and Customs of the Ancient Egyptians*, 2. London.
Wilson, B., C. Grigson and S. Payne (1982) *Ageing and Sexing Animal Bones from Archaological Sites*. British Archaeological Reports series, 109. Oxford.
Wilson, J. A. (1944) 'Funeral Sevices of the Egyptian Old Kingdom', *JNES* 3: 201-18.
Wilson, W. H. (1917) *Report of the Prison Diet Committee*. Cairo.
Winlock, H. E. (1932) *The Tomb of Queen Meryet-Amun at Thebes*. New York.
—, (1941) *Materials Used at the Embalming of King Tut-Ankh-Amun*.. New York.
—, (1942) *Excavations at Deir el Bahri 1911-31*. New York.
—, (1955) *Models of Daily Life in Ancient Egypt*. Cambridge, Mass.
—, (1912) 'The Work of the Egyptian Expedition', *Bulletin of the Metropolitan Museum of Art*, VII.10: 184-90. New York.
Wissa Wassef, C. (1988) *Pratiques Rituelles et Alimentaires des Coptes*. Cairo
Wreszinski, W. (1923-36) *Atlas zur Altagyptischen Kulturgeschichte*, 1-3. Leipzig.
Wysocki, Z. (1984) 'The Results of Research, Architectonic Studies and Protective Work Over the Northern Portico of the Middle Courtyard in the Hatshepsut Temple at Deir el Bahari' *MDAIK* 40: 329-42.
Zeuner, F. (1963) *A History of Domesticated Animals*. New York.

ORIENTALIA LOVANIENSIA ANALECTA

1. E. LIPIŃSKI, Studies in Aramaic Inscriptions and Onomastics I.
2. J. QUAEGEBEUR, Le dieu égyptien Shaï dans la religion et l'onomastique.
3. P.H.L. EGGERMONT, Alexander's Campaigns in Sind and Baluchistan and the Siege of the Brahmin Town of Harmatelia.
4. W.M. CALLEWAERT, The Sarvāṅgī of the Dādūpanthī Rajab.
5. E. LIPIŃSKI (ed.), State and Temple Economy in the Ancient Near East I.
6. E. LIPIŃSKI (ed.), State and Temple Economy in the Ancient Near East II.
7. M.-C. DE GRAEVE, The Ships of the Ancient Near East (c. 2000-500 B.C.).
8. W.M. CALLEWAERT (ed.), Early Hindī Devotional Literature in Current Research.
9. F.L. DAMEN, Crisis and Religious Renewal in the Brahmo Samaj Movement (1860-1884).
10. R.Y. EBIED-A. VAN ROEY-L.R. WICKHAM, Peter of Callinicum, Anti-Tritheist Dossier.
11. A. RAMMANT-PEETERS, Les pyramidions égyptiens du Nouvel Empire.
12. S. SCHEERS (ed.), Studia Paulo Naster Oblata I. Numismatica Antiqua.
13. J. QUAEGEBEUR (ed.), Studia Paulo Naster Oblata II. Orientalia Antiqua.
14. E. PLATTI, Yaḥyā ibn ʿAdī, théologien chrétien et philosophe arabe.
15. E. GUBEL-E. LIPIŃSKI-B. SERVAIS-SOYEZ (ed.), Studia Phoenicia I-II.
16. W. SKALMOWSKI-A. VAN TONGERLOO (ed.), Middle Iranian Studies.
17. M. VAN MOL, Handboek Modern Arabisch.
18. C. LAGA-J.A. MUNITIZ-L. VAN ROMPAY (ed.), After Chalcedon. Studies in Theology and Church History.
19. E. LIPIŃSKI (ed.), The Land of Israel: Cross-Roads of Civilizations.
20. S. WACHSMANN, Aegeans in the Theban Tombs.
21. K. VAN LERBERGHE, Old Babylonian Legal and Administrative Texts from Philadelphia.
22. E. LIPIŃSKI (ed.), Phoenicia and the East Mediterranean in the First Millennium B.C.
23. M. HELTZER-E. LIPIŃSKI (ed.), Society and Economy in the Eastern Mediterranean (1500-1000 B.C.).
24. M. VAN DE MIEROOP, Crafts in the Early Isin Period.
25. G. POLLET (ed.), India and the Ancient World.
26. E. LIPIŃSKI (ed.), Carthago.
27. E. VERREET, Modi Ugaritici.
28. R. ZADOK, The Pre-Hellenistic Israelite Anthroponomy and Prosopography.
29. W. CALLEWAERT-M. LATH, The Hindī Songs of Nāmdev.
30. S. HALEVY, Coptic Grammatical Chrestomathy.
31. N. BAUM, Arbres et arbustres de l'Égypte ancienne.
32. J.-M. KRUCHTEN, Les Annales des prêtres de Karnak.
33. H. DEVIJVER-E. LIPIŃSKI (ed.), Punic Wars.
34. E. VASSILIKA, Ptolemaic Philae.
35. A. GHAITH, La Pensée Religieuse chez Ǧubrân Ḫalil Ǧubrân et Miḫâʾîl Nuʿayma.
36. N. BEAUX, Le Cabinet de curiosités de Thoutmosis III.

37. G. POLLET-P. EGGERMONT-G. VAN DAMME, Archaeological Sites of Ancient India.
38. S.-A. NAGUIB, Le Clergé féminin d'Amon thébain à la 21e dynastie.
39. U. VERHOEVEN-E. GRAEFE (ed.), Religion und Philosophie im Alten Ägypten.
40. A.R. GEORGE, Babylonian Topographical Texts.
41. A. SCHOORS, The Preacher Sought to Find Pleasing Words.
42. G. REININK-H.E.J. VAN STIPHOUT (ed.), Dispute Poems and Dialogues in the Ancient and Mediaeval Near East.
43. C. TRAUNECKER, Coptos. Hommes et dieux sur le parvis de Geb.
44. E. LIPIŃSKI (ed.), Phoenicia and the Bible.
45. L. ISEBAERT (ed.), Studia Etymologica Indoeuropaea Memoriae A.J. Van Windekens dicata.
46. F. BRIQUEL-CHATONNET, Les relations entre les cités de la côte phénicienne et les royaumes d'Israël et de Juda.
47. W.J. VAN BEKKUM, A Hebrew Alexander Romance according to MS London, Jews' College no. 145.
48. W. SKALMOWSKI-A. VAN TONGERLOO (ed.), Medioiranica.
49. L. LAUWERS, Igor'-Severjanin, His Life and Work — The Formal Aspects of His Poetry.
50. R.L. VOS, The Apis Embalming Ritual. P. Vindob. 3873.
51. Fr. LABRIQUE, Stylistique et Théologie à Edfou. Le rituel de l'offrande de la campagne: étude de la composition.
52. F. DE JONG (ed.), Miscellanea Arabica et Islamica.
53. G. BREYER, Etruskisches Sprachgut im Lateinischen unter Ausschluß des spezifisch onomastischen Bereiches.
54. P.H.L. EGGERMONT, Alexander's Campaign in Southern Punjab.
55. J. QUAEGEBEUR (ed.), Ritual and Sacrifice in the Ancient Near East.
56. A. VAN ROEY-P. ALLEN, Monophysite Texts of the Sixth Century.
57. E. LIPIŃSKI, Studies in Aramaic Inscriptions and Onomastics II.
58. F.R. HERBIN, Le livre de parcourir l'éternité.
59. K. GEUS, Prosopographie der literarisch bezeugten Karthager.
60. A. SCHOORS-P. VAN DEUN (ed.), Philohistor. Miscellanea in honorem Caroli Laga septuagenarii.
61. M. KRAUSE-S. GIVERSEN-P. NAGEL (ed.), Coptology. Past, Present and Future. Studies in Honour of R. Kasser.
62. C. LEITZ, Altägyptische Sternuhren.
63. J.J. CLÈRE, Les Chauves d'Hathor.
64. E. LIPIŃSKI, Dieux et déesses de l'univers phénicien et punique.
65. K. VAN LERBERGHE-A. SCHOORS (eds.), Immigration and Emigration within the Ancient Near East. Festschrift E. Lipiński.
66. G. POLLET (ed.), Indian Epic Values. *Rāmāyaṇa* and its impact.
67. D. DE SMET, La quiétude de l'Intellect. Néoplatonisme et gnose ismaélienne dans l'œuvre de Ḥamîd ad-Dîn al-Kirmânî (Xe-XIe s.).
68. M.L. FOLMER, The Aramaic Language in the Achaemenid Period. A Study in Linguistic Variation.
69. S. IKRAM, Choice Cuts: Meat Production in Ancient Egypt.